T0257485

Protein-Protein Interactions

Protein-Protein Interactions

Edited by **Anton Torres**

New York

Published by Callisto Reference,
106 Park Avenue, Suite 200,
New York, NY 10016, USA
www.callistoreference.com

Protein-Protein Interactions
Edited by Anton Torres

International Standard Book Number: 978-1-63239-525-2 (Hardback)

Printed in the United States of America.

Contents

Preface

Proteins have been the focus of researches for decades. Proteins are vital in almost all biological procedures. The functions of proteins are synchronized through complex regulatory networks of transient protein-protein interactions (PPIs). To study PPIs, a broad range of methods have been developed over the past few decades. Several in vitro and in vivo attempts have been implemented to discover the mechanism of these pervasive interactions. Despite noteworthy developments in these investigational approaches, many issues exist such as false-positives/false-negatives, difficulty in obtaining crystal structures of proteins, etc. In order to overcome these challenges, various technical methods have been created, that are becoming more broadly used to examine PPIs. This book primarily demonstrates various computational approaches utilized to study these interactions. The objective of this book is to bring together several researches accomplished by experts around the globe.

The information contained in this book is the result of intensive hard work done by researchers in this field. All due efforts have been made to make this book serve as a complete guiding source for students and researchers. The topics in this book have been comprehensively explained to help readers understand the growing trends in the field.

I would like to thank the entire group of writers who made sincere efforts in this book and my family who supported me in my efforts of working on this book. I take this opportunity to thank all those who have been a guiding force throughout my life.

Editor

Computational Approaches

Advances in Human-Protein Interaction - Interactive and Immersive Molecular Simulations

Nicolas Férey, et al.*

CNRS - Laboratoire d'Informatique pour la Mécanique et les Sciences de l'Ingénieur - Université Paris XI Bâtiment 508, 512 et 502 bis, 91403 Orsay Cedex France

1. Introduction

Molecular simulations allow researchers to obtain complementary data with respect to experimental studies and to overcome some of their limitations. Current experimental techniques do not allow to observe the full dynamics of a protein at atomic detail. In return, experiments provide the structures, i.e. the spatial atomic positions, for numerous biomolecular systems, which are often used as starting point for simulation studies. In order to predict, to explain and to understand experimental results, researchers have developed a variety of biomolecular representations and algorithms. They allow to simulate the dynamic behavior of macromolecules at different scales, ranging from detailed models using quantum mechanics or classical molecular mechanics to more approximate representations. These simulations are often controlled *a priori* by complex and empirical settings. Most researchers visualise the result of their simulation once the computation is finished. Such post-simulation analysis often makes use of specific molecular user interfaces, by reading and visualising the molecular 3D configuration at each step of the simulation. This approach makes it difficult to interact with a simulation in progress. When a problem occurs, or when the researcher does not achieve to observe the predicted behavior, the simulation must be restarted with other settings or constraints. This can result in the waste of an important number of compute cycles, as some simulations last for a long time: several days to weeks may be required to reproduce a short timespan, a few nanoseconds, of molecular reality. Moreover, several biomolecular processes, like folding or large conformational changes of proteins, occur on even longer timescales that are inaccessible to current simulation techniques. It can thus be necessary to impose empirical constraints in order to accelerate a simulation and to reproduce

*Alex Tek, Benoist Laurent, Marc Piuzzi, Zhihan Lu and Marc Baaden (CNRS - Laboratoire de Biochimie Théorique, Institut de Biologie Physico-Chimique, 13, rue Pierre et Marie Curie, 75005 Paris, France)
Olivier Delalande (CNRS - Interactions Cellulaires et Moléculaires - Université de Rennes 1, Avenue du Professeur Léon Bernard, 35065 Rennes cedex, France)
Matthieu Chavent (Structural Bioinformatics and Computational Biochemistry Unit, Dept. of Biochemistry, University of Oxford, United Kingdom)
Christine Martin, Lorenzo Piccinali, Brian Katz and Patrick Bourdot (CNRS - Laboratoire d'Informatique pour la Mécanique et les Sciences de l'Ingénieur - Université Paris XI Bâtiment 508, 512 et 502 bis, 91403 Orsay Cedex, France)
Ludovic Autin(Molecular Graphics Laboratory Department of Molecular Biology, MB-5 - The Scripps Research Institute, 10550 North Torrey Pines Road, La Jolla, CA 92037-1000, USA)

an experimental result in MD. These constraints have to be defined *a priori*, rendering it difficult to explore all possibilities in order to examine various biological hypotheses.

A new approach allowing to address these problems has emerged recently: Interactive Molecular Simulation (IMS). IMS consists in visualising and interacting with a simulation in progress, and provides the user with control over simulation settings in interactive time. With the recent advances in human computer interaction and the impressive increase of available computing power, the IMS approach allows a user to interact in 3D space in real time with a molecular simulation in progress. This approach provides quality control features by visualizing results of a simulation in progress and supplies interactive features, such as feeling forces involved in the simulation as well as triggering specific events by applying custom forces during the simulation in progress. These advances led to a new generation of scientific tools to better understand life science phenomena, which place the human expertise at the centre of the analysis process, complementarily to automatic computational methods.

The IMS approach emerged from the breakthrough initiated by the *Sculpt* precursor program proposed by Surles et al. (1994). Since then, the interactive molecular simulations field has been developing continuously. Initial interactive experiments using molecular mechanics techniques gave quickly rise to "guided" dynamics simulations [Wu & Wang (2002)] or *Steered Molecular Dynamics* (SMD) [Isralewitz et al. (2001)] [Leech et al. (1996)]. The interest for these methods increased with the enhancement of simulation accuracy and thanks to the exciting new possibilities for dynamic structural exploration of very large and complex biological systems. In the Interactive Molecular Dynamics (IMD) approach, steering forces are applied interactively with a chosen amplitude, direction and application point. This enables the user to explore the simulation system while receiving instant feedback information from real-time visualisation or haptic devices [Leech et al. (1997)]. Schulten's group has carried out several applications of IMS simulations to macromolecular structures [Grayson et al. (n.d.)] [Stone et al. (2001)]. This effort lead to the design of two efficient software tools facilitating the process of setting up an IMS : *NAMD* and *VMD* [Phillips et al. (2005)] [Nelson et al. (1995))]. The underlying exchange protocol is also supported by ProtoMol [Matthey et al. (2004)], LAMMPS [Plimpton (1995)], HOOMD-blue [Anderson et al. (2008)] and any software using the MDDriver library [Delalande et al. (2009)]. Similar projects proposing an interactive display for molecular simulations exist, such as the *Java3D* interface proposed in Knoll & Mirzaei (2003) and Vormoor (2001), or the *Protein Interactive Theater* [Prins et al. (1999)].

With fast generalization of new computer hardware devices and increasing accessibility to powerful computational infrastructures, IMS showes a fast and promising evolution, even for very large molecular systems (over 100.000 atoms). Such applications are now in the reach of state-of-art desktop computing. This evolution was possible given the strong increase in raw computing power leading to faster and bigger processing units (multi-processors, multi-core architectures). Currently ongoing technological developments such as GPU computing and the spread of parallelized entertainment devices (PS3, Cell) with specific graphic and processing capabilities open exciting new opportunities for interactive calculations. These approaches could provide even more processing power for highly parallelizable computational problems, for instance by differentiating the parallelisation of molecular calculations and graphical display functionalities. Given these developments, the range of accessible computational methods and representations is bound to grow. It may soon be possible to extend the IMS approach to *ab initio* or QM/MM calculations. Indeed, the precision achieved in the description of a system can be improved by switching to a more accurate physical model and/or by improving the representation of the molecular context simulated. Thus, multi-scale simulations [Baaden & Lavery (2007)] would indeed benefit from an interactive approach leading to important advantages with respect to the study of complex biological systems. However, the raw increase in computer speed alone is not sufficient to grant a successful future evolution of the IMS approach. In addition, it is necessary to develop adapted software solutions, which are generally more efficient [Grayson et al. (n.d.)], as it is commonly admitted in the numeric simulation field. Finally, the most recent and famous work illustrating the revolution of this approach is the "Fold It" serious game, which allows a user to interactively propose a protein folding solution [Cooper et al. (2010)].

We will describe in this chapter the recent advances relating to these IMS approaches previously described. As IMS implies to efficiently combine simulation and interaction features, we will explain how we designed specific simulation, visualisation, and interaction techniques to solve the real time constraint, to study complex biomolecular systems, and to address a larger simulation timescale. Then we will discuss software architectures to efficiently put the different building blocks together. Finally, we will explain how we apply IMS to different fields of research including various topics such as protein-protein docking in a virtual reality and multimodal context, an ion substitution study using an haptic device, and a study about the opening and closure of the Guanylate Kinase enzyme.

2. Multiscale and multiphysics protein simulation models

In structural biology, recent advances in experimental techniques allow us to solve larger and larger protein 3D structures. However, even if structure is known to be strongly linked to biological function, static states often lack in providing dynamical informations that are crucial for the understanding of the subtle mechanisms occuring at the molecular level. Thus, molecular simulations are nowadays used to complete experimental biostructural studies, especially to better understand the dynamic behaviour and the fundamental mechanisms involved in a protein complex. In spite of the increasing computational resources, classical simulation tools are not well adapted to quickly obtain insight into the global biomechanical properties, because of the limited timescale covered by all-atom or coarse-grained simulations. For these reasons, it is necessary to develop new modeling approaches at a larger scale, complementary to all-atom and coarse-grained models, especially designed to interactively study protein complex formation and biomechanical properties of large biomolecular structures. We present in this part unconventional approaches that could address these requirements. The first one, based on a rigid body model of a protein, was especially designed to study protein-protein interactions for an interactive rigid docking application. The second one, based on a spring netwok model, takes into account protein flexibility in order to study biomechanical behavior of large protein structures in interactive time.

2.1 A rigid body simulation model to interactively study protein-protein interactions

At a larger scale, it is sometimes not necessary to model and simulate the flexibility of a protein, but sufficient to consider the protein as a rigid body. Using a simple but accurate model at the macroscopic scale allows us to overcome the main constraint to provide an interactive time biophysical simulation as required for IMS: taking into account the user interaction during a simulation in progress. To present our rigid body simulation model dedicated to IMS, especially interactive rigid docking, we have to focus on the main phenomena that are involved in the protein interactions.

2.1.1 Geometry and surface

Proteins can be viewed as both the building blocks and the workforce of cells. They are synthesized based on portions of DNA (Deoxyribonucleic Acid) called coding sequences or genes. Genes are transcribed in the form of mRNA (Messenger RiboNucleic Acid), which is then translated by ribosomes in the form of a protein, following a specific coding scheme (figure 1A). Each triplet of mRNA bases corresponds to one AA (Amino Acid) or residue, of which there are twenty basic types. The various physicochemical properties of AAs give rise to interactions at the atomic level, inducing protein folding which contributes in turn to protein stability (figure 1B). These properties also play a crucial part in protein-protein interactions.

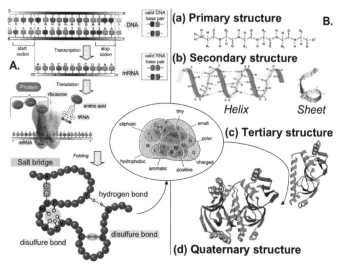

Fig. 1. (**A**) - Overall view of protein synthesis: transcription of DNA to messenger RNA (mRNA) and translation of mRNA to amino acid sequences chosen from 20 possible varieties, here shown according to their physicochemical properties (using a Venn diagram). (**B**) - Based on the chemical nature of component amino acids, resulting interactions cause the protein to fold to a favorable arrangement in space. This 3D shape can be described according to four levels: the (a) primary, (b) secondary, (c) tertiary and (d) quaternary structure.

Proteins, therefore, can be seen as long chains composed of successive amino acids folded in space, which are the product of the expression of an organism's genetic makeup. But in order to execute their functions within cells, proteins must undergo folding and take a specific 3D form. This form may be characterized following four levels of structure (see figure 1B). The order in which residues are linearly arranged, *i.e.* their sequence, constitutes the protein's primary structure. (see figure 1B-a). Some of the structure's segments organize themselves into sequences of specific substructures called secondary structures (see figure 1B-b). These structures, stabilized by hydrogen bonds, can be divided into two groups: regular secondary structures, called alpha helices and beta sheets, which are linked together by irregular structures called loops. The arrangement of these secondary structures thus constitutes the 3D, or tertiary structure of the protein (see figure 1B-c), which determines protein function within the cell.

Once folded, proteins carry out various functions within the cell, such as transporting molecules to and from various components of the organism (*e.g.* hemoglobin, chaperone proteins), inter- and intracellular signaling and communications (*e.g.* hormones, neurotransmitters, ions), immune defense functions (immunoglobulins, adhesion molecules) or cellular metabolism (chlorophyll, apoptosis proteins, transcription factors, ATP synthesis). These cellular functions are closely linked to the protein's tertiary structure, but also to its interactions with other proteins.

In short, better understanding of protein-protein interactions is a major stake for biomedical research. Indeed, designing new drugs increasingly involves targeting specific protein-protein interactions [Villoutreix et al. (2008)], or alternatively, involves synthesizing recombinant proteins meant to emulate interactions with the original native protein [Pipe (2008)]. It becomes more and more necessary, therefore, to identify the 3D structure of protein complexes. Two main experimental methods are currently used to determine the 3D structure of a protein complex. These are X-ray crystallography and Nuclear Magnetic Resonance (NMR) spectroscopy. All known publicly available protein structures are currently housed on the website of the Protein Data Bank (PDB) [Berman et al. (2000)]. This database contains now several hundred thousand protein structures for many organisms. However, this number remains small in comparison to estimates of the number of existing proteins in the natural world. This is because experimental determination of protein structure is often difficult, and in some cases impossible. Indeed, solving a problem of this kind involves mass production and purification of the protein, and in the case of crystallography, production of diffractive crystals. In determining the structure of a protein complex, difficulties in production and purification are all the more critical, because partner proteins must be produced at the same time for complexes to form. Furthermore, the time necessary for crystallization may be incompatible with the lifespan of some complexes. For all these reasons, many scientists have attempted to predict the structure of such complexes using computational tools through methods and algorithms for molecular docking.

Current techniques for the experimental study of the 3D structure of protein complexes (crystallography, NMR, electron cryomicroscopy, SAXS, etc.) have several limitations (in terms of size and type of proteins) and are costly in terms of time and money. For that reason, computer-based (*in silico*) docking methods have been developed in the past, to deduce the functional 3D structure of a complex based on single molecules, which turns out to be considerably easier and cheaper than experimental *in vitro* methods. Current approaches are strictly computational and results are evaluated using visualization tools. These approaches can be divided into 4-5 successive stages (figure 2): (1) choice of the representation mode for proteins (atomic view, pseudo-atoms, grid, etc.); (2) conformational exploration (taking into account position, orientation, and shape of the ligand); (3) minimization of the function used to evaluate binding energy (i.e. *score*) for conformations derived from the exploration; (4) grouping by similarity and classification through evaluation or fine-tuning of the *scores*, augmented with a manual stage of visualization when the score alone doesn't allow native conformations (i.e. the ones present in nature) to be discriminated from other generated conformations; (5) an optional stage for fine-tuning selected complexes, through energy minimization or molecular dynamics.

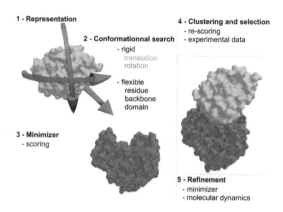

Fig. 2. The 5 stages of the docking task

A large number of fully automatic computational docking algorithms depend on an comprehensive approach of conformational exploration, the main problem being combinatorial explosion of the number of possible solutions. These approaches can be sorted into three categories: those based on systematic sampling, on molecular dynamics techniques and on classification interaction modes between proteins. An ideal function would yield, for a given mode of interaction, the binding energy of two proteins involved in a complex (see section 2.1.3). Such functions aim to reproduce experimental values of free binding energy, and through minimization, to reach the overall minimum energy in the set of all possible protein-protein complexes.

Consequently, in real life cases, automatic docking algorithms, such as *ClusPro* [Comeau et al. (2004)] or *Hex* [Ritchie (2003)], must manage two difficulties in order to reach a relevant result. The first is to process a space of potential solutions which increases in size along with the number of degrees of freedom in describing protein position and conformation, thus running the risk of not beeing processed in an acceptable amount of time. The second problem is that search algorithms produce local minima, and cannot easily find the global minimum that is associated to the native form of the complex [Wang et al. (2003)].

To finalize a docking simulation, experts rely upon a manual stage of visualization to analyse the generated complexes. This task consists in a detailed analysis of residues and atoms involved in the interface each complex, through the identification of hydrogen bonds, salt bridges, and especially the presence of hotspots, *i.e.* amino acids at the interface, known from experimental studies to be an essential part of this interface. However, it can be difficult to manipulate two 3D structures at the same time to observe the interface with traditional interaction tools, since one protein usually hides the other. Therefore docking assisted by user interaction is a useful alternative to improve the work of experts in this field. Such techniques might allow a more intuitive interaction with 3D protein structures.

Finally, two approaches are used to "thin the herd" of selected complexes. One consists in minimizing the rigid bodies and lateral chains of amino acids present at the interface. This approach is implemented in several applications such as *ICM-DISCO* [Fernandez-Recio et al. (2003)], *MMTK* [Hinsen (2000)], *FireDock* [Andrusier et al. (2007)], *PELE* [Borrelli et al. (2005)], *ATTRACT* [Zacharias (2005)], etc. The other approach involves studying the dynamic behavior of the selected complex. The software program *Gromacs* [Hess et al. (2008)], for example, allows evaluation of atomic positions over time based on their physicochemical properties. This approach allows first to evaluate the complex stability, as well as possible conformational changes induced by the interaction, *e.g.* loop deformation. We should add, however, that this approach remains very costly in terms of processing time, compared to minimizers which allow users to process a given configuration very quickly.

As the automatic docking software programs previously presented did not respond to the interactive time constraint, we developed a new simulation tool dedicated to interactive protein-protein rigid docking. Our protein docking method is essentially based on two sets of criteria: geometric/topological criteria, and biophysical criteria.

2.1.2 Interactive time evaluation of geometry and surface complementarity

One of the earliest criteria identified in protein-protein interaction is surface topology of the proteins involved. In most known structures of 3D complexes, partners exhibit good surface complementarity. Studies have also shown that the surface of the protein-protein interface generally covers between 1000 and 2500 square Angstroms. This criteria allowed the development of first-generation docking software, based solely on shape recognition [Connolly (1983)] (*i.e.* complementarity of molecular surfaces). This approach is well adapted to "hard" rigid protein docking. We used these geometric/topological criteria in our multimodal immersive environment in two ways:

Surface collision. For each protein, a surface mesh is computed using the *MSMS* software before interactive docking occurs [Sanner et al. (1996)]. The resolution of this mesh can be adjusted using parameters. Collision detection during interaction then uses the *RAPID* library [Gottschalk et al. (1996)], which allows real-time computation of a list of colliding triangles among the two protein surface meshes during docking. This set of triangles can be used to generate feedback based on triangle normals and on the intersection volume of the two protein surfaces.

Atomic surface complementarity. Atomic surface complementarity is estimated essentially as a calculation of the variance of the inter-atomic distances on the two protein surfaces. We use this overall atomic surface complementarity score in audio or visual feedback.

2.1.3 Interactive time computation of physicochemical properties and energies

However, geometric criteria turned out to be insufficient to predict the structure of a complex. Thus, we had to rely on methods including energy criteria. Protein-protein complexes seem to follow the rule of thumb that the active configuration is the one whose level of free energy is lowest [Wang et al. (2003)]. In order to evaluate free energies between two proteins, we rely on molecular mechanics methods. For this purpose, atoms are viewed as spheres, and interactions between atoms can be computed using the van der Waals and electrostatic potentials. The free energy for protein-protein interaction can then be approximated by the sum of these potentials, which is known as the *score*. In the context of real-time immersive docking, the choice of equations and methods to evaluate the energy of a complex and hence its score is a crucial issue [Wang et al. (2003)].

Van der Waals interactions. Van der Waals interactions are an empirical approximation of atomic interactions. The van der Waals force, obtained by constructing a gradient of the potential field, is defined by the Lennard-Jones potential equation (equation 1). In this equation, r_{ij} is the distance between two atoms i and j, σ the interatomic distance for which the potential becomes zero, and ϵ the depth of the potential well. ϵ and σ are determined empirically and depend on what pair of atoms is considered. The van der Waals potential includes an attractive component when atoms are bound, and a repulsive component when atoms are too close to each other. It prevents two proteins from penetrating into each other during interactive docking, through calculation of interatomic forces at the protein-protein interface.

$$U_{vw}(r) = 4\epsilon[(\frac{\sigma}{r_{ij}})^{12} - (\frac{\sigma}{r_{ij}})^6] \tag{1}$$

These forces apply only to very short distances and mostly concern surface atoms. As computing distances between all pairs of atoms has a quadratic complexity, we apply specific filtering rules to keep only surface atoms and opposite atoms from each protein (see figure 3). The resultant translational and rotational components of van der Waals's forces on each atom are calculated and applied to the barycenters of the proteins.

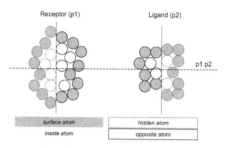

Fig. 3. Dynamic and static atom filtering for optimized computing of van der Waals interactions

Electrostatic interactions Unlike van der Waals interactions, electrostatic interactions even operate when "long" distances (about 10 Angstrom) separate groups of electrically charged atoms. Indeed some amino acids or atoms may present a positive or negative electric charge, which gives rise to electrostatic phenomena allowing formation of a protein-protein complex. Two approaches have been implemented to compute electrostatic phenomena.

We consider the interaction between two point charges in vacuum, and we use Coulomb's law (equation 2) with r_{ij} being the distance between the barycenters of charges q_i and q_j of the atoms considered, and ϵ_0 is the constant of the permittivity of vacuum. This potential can be translated to a force (F_{el}) usable for haptic interaction for example. This first approach involves calculating the forces to apply to each electrically charged particle considering only pairs of charged particles. This computation has quadratic complexity, because all distances between atoms must be computed. But it remains relevant in the case of medium-sized proteins, since the number of charged particles in a protein is limited in several models.

$$U_{el}(r_{ij}) = \frac{1}{4\pi\epsilon_0} \frac{q_i q_j}{r_{ij}} \tag{2}$$

In the second approach (see figure 4, designed for a more efficient optimised calculation, the overall field of the electrostatic potential of the target protein (receptor) is computed beforehand using the *APBS* software [Baker et al. (2001)]. It allows to generate a 3D electrostatic potential grid, which can be used as a 3D texture. The gradient of the electrostatic potential allows computation of force field vectors for each point of the grid. Atoms from the ligand protein are then "immersed" in this 3D force field surrounding the receptor. This method allows us to compute electrostatic forces for each atom in linear time, depending only on the number of charged atoms in the ligand. In both cases, we are able to obtain overall electrostatic energy and electrostatic forces on each atom.

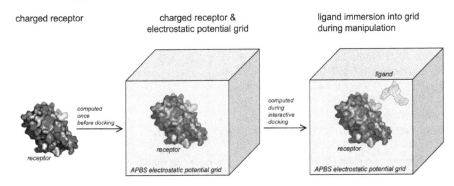

charged receptor

charged receptor &
electrostatic potential grid

ligand immersion into grid
during manipulation

Fig. 4. Ligand immersion in the electrostatic potential grid of the receptor

2.1.4 Other criteria

In order to reach a finer description of protein-protein interactions, other criteria, based on energy, can be taken into account. To geometric/topological and biophysical criteria, one can add other criteria of utmost importance to protein-protein interactions, such as hydrogen bonds or the hydrophobic effects.

Hydrogen bonds. Hydrogen bonds (*e.g.* figure 1 in the bottom left corner) may strongly contribute to the favorable interactions of the complex binding energy. On average, there are 5-6 hydrogen bonds per protein-protein interface. In our application, when several atoms (nitrogen and oxygen) on the surface of each protein are close enough, closer than a distance of 3 Angstroms, and when their chemical environment is adequate, hydrogen bonds are created between these atoms. We use the same methods as described above for van der Waals interactions to filter surface atoms in order to decrease the complexity of calculating distances between atoms.

Hotspots at the interface. The number of "hotspots" at the complex interface refers to the list of amino acids present within the current interface region and previously identified using experimental methods as being important actors to stabilize the protein-protein complex.

2.1.5 Conclusion

This simulation model, based on rigid body docking, including optimisation to efficiently compute geometrical as well as biophysical properties, allows us to present these properties in real time during the interactive building of a protein complex performed by the user. Designing real time simulations is a first step of the IMS approach, providing a user with interactive control on the simulated object in real time.

2.2 A multiphysics and multiscale approach based on elastic networks

Simulation models dedicated to IMS must also deal with the intrinsic flexibility of proteins, and especially take into account local moves as well as large conformational changes. The representation used for our interactive simulation approach is quite simple, yet innovative, and has proven efficient on large biomolecular structures. Our method is based on a spring network simulation, inspired by the success of the Normal Mode Analysis method (NMA), known to accurately reproduce the elastic behavior [Cui & Bahar (2006)]. Moreover NMA is not sensitive to the scale of representation used for modeling. Hence this method can be applied to all-atom, coarsed-grain and residue/CA representations. We augmented this spring network model with non-bonded interactions and we propose to surround the charged spring network by an electrostatic field, allowing us to study conformational changes guided by electrostatic constraints. We called our implementation of this method *BioSpring* and describe it in this section. *BioSpring* allows us to simulate large structures in real time, fulfilling the most important constraint to provide IMS features.

2.2.1 BioSpring : an enhanced interactive spring network model

The first step of our approach is to build the spring network according to the 3D structure of the biomolecular system [Berman et al. (2000)]. At this stage, the user needs to choose a scale, targeting for example an all atom (AA), a coarse-grained (CG), or an alpha-carbon representation (CA). In this context, individual atoms can be considered as separate particles (AA model), or can be grouped into a single pseudo-particle, according to rules defined by the user (e.g. at the CG or CA level). In this way, we can adapt our approach to most commonly-used modeling approaches in theorical biochemistry.

The second step, is to connect the particles by springs obtained in the previous step. For this purpose, we define a distance cut-off, and we add a spring between two particles if the distance between them is less than the cut-off distance. This cut-off will depend on the scale and the representation mode (all-atom, coarse-grained, alpha-carbon, ...). For example, a cut-off between 7 and 15 angstrom is classically used for the CA representation [Cui & Bahar (2006)]. This process can be computationnally very time-consuming, especially on large structures. For each particle, we need to test if any other particles are closer than the cut-off distance. This approach has a quadratic complexity according to the number of particles. In order to deal with large structures and to decrease the complexity of the previous approach, we use a classical technique based on a regular 3D grid to partition a three dimensional space into cubes, also named "voxels". The grid covers the entire space occupied by the particles. The size of each voxel is the cut-off distance. According to its coordinates in space, each particle is projected into its voxel on the grid. In order to determine for a given particle p, all particles near p within the cut-off distance, we have to test the particles in the same and in the direct neighbouring voxels. This method has linear complexity according to the number of particles, allowing us to address very large structures. We will see in the next part of this paper how we can use the same grid to efficiently compute non-bonded interactions between particles.

After building the initial spring network, interactive manipulation of this molecular structure is provided using a classical newtonian particle-based simulation, taking into account spring forces (see equation 3) between particles and external forces $\vec{F}_{control}(p)$ (see 5 equation) provided by the user on a particles p through a specific graphical user interface. In the equation 3, $k_{stiffness}$ is a global stiffness for all the springs and the force between two particles p and p' linked by a spring depends on the distance between these particles. If this distance between p and p' is equal to the equilibrium spring length $e_{pp'}$, which is the distance between these particles in the initial structure used to compute the network, this force is null by definition. In the other cases, when the distance $d_{pp'}$ between p and p' changes because of external forces, the generated spring forces tend to bring the structure back to its equilibrium conformation at $e_{pp'}$. Damping forces are used to stabilize the system (see equation 4). This is necessary because the user injects energy into the system by adding external arbitrary forces. It should be noted that some experimental and theorical studies provide estimates for $k_{stiffness}$, which allows us to work with magnitudes of forces in the simulation that are relevant from a biophysical point of view.

$$\vec{F}_{spring}(p) = \sum_{p' \in Springs(p)} k_{stiffness} \vec{u}_{pp'} (d_{pp'} - e_{pp'}) \tag{3}$$

$$\vec{F}_{damping}(p) = -k_{damping} \vec{V}(p) \tag{4}$$

$$\vec{F}(p) = \vec{F}_{spring}(p) + \vec{F}_{damping}(p) + \vec{F}_{control}(p) \tag{5}$$

At each time step of the simulation, to compute new positions and velocities according to the spring and external forces applied on the particles, we use a velocity verlet integrator described in equation 6.

$$\vec{P}(t + \Delta t) = \vec{P}(t) + \vec{P}(t)\Delta t + \tfrac{1}{2}\vec{A}(t)\Delta t^2$$

$$\vec{V}(t + \Delta t) = \vec{V}(t) + \frac{\vec{A}(t) + \vec{A}(t + \Delta t)}{2}\Delta t \tag{6}$$

Finally, the graphical user interface must provide interactive simulation features, allowing a user to visualise the spring network simulation in progress and to interactively apply external forces $\vec{F}_{control}(p)$ on particles. Combining these interactive simulation features and our interactive spring network simulation approach, a user can manipulate some parts of a large biomolecular system, and interactively observe the effects of this manipulation highlighting biomechanical properties such as rigid vs. flexible areas or allosteric effects.

However, even if the spring network model embeds an approximation of bonded and non-bonded interactions at the local scale, this model is not able to deal with long range and steric interactions during a simulation, because the spring 'particles' (atom, coarse grain, or residue-level) are considered as points. For example, domain interpenetration is allowed in the default spring network model. This is not a problem when the goal is to highlight local flexibility or rigid areas, but it is a critical issue for our objective of interactive modeling of large biomolecular systems. Similarly, it is also necessary to take into account electrostatic interactions during interactive modeling. For these reasons, in addition to spring forces, we introduce classical non-bonded forces to take into account steric and electrostatic interactions between particles in our model.

In order to meet the specific needs of the user, BioSpring provides a variety of terms to represent steric interactions. The most simple term is the linear steric model (see equation 7) which can be used to avoid atom or pseudo-atom collisions and take into account the 3D shape of the biomolecular model. This avoids domain interpenetration during the interactive manipulation, without taking into account complex realistic steric energy considerations and in particular attractive terms. More classical models such as Lennard-Jones (see equation 8) are also avalaible, in order to take into account both attractive and repulsive interactions between atoms or pseudo-atoms, and to compute a more relevant steric energy during an interactive simulation. Atom radius r, epsilon ϵ and sigma σ parameters can be set up using configuration files, allowing us to use many of the currently available forcefields. However, this method has a complexity in $O(n^2)$ which is quadratic according to the number n of particles, because for each particle we need to compute the distance with respect to all other particles in the simulation. To address larger biomolecular systems, we necessarily have to decrease the complexity. We can remark that beyond a certain distance, several pairwise interactions become null or negligible. This is especially the case for linear or Lennard-Jones steric interactions. In this case, according to this distance cutoff, we can use the same optimisation techniques as in section 2.2.1, projecting particles into a 3D grid to accelerate the distance computation between particles at each time step of the simulation, by reducing quadratic complexity to linear complexity according to the number of particles in the simulation. The complexity is in $D.O(n)$ which is linear according to the number n of particles and the mean number D of particles in a voxel, which can be considered as a constant because it is related to the mean density of particles at a molecular scale.

We also use another way to optimize our simulation method as in many cases some part of the biomolecular complex can be considered as a rigid component. In this case, we can consider that these particles are static, *i.e.* have a constant position in space, because they belong to a rigid component. Hence it is useless to compute their interactions and to apply a positional integration on these static particles. We only have to take into account interactions between the dynamic particles belonging to the flexible part, and the interactions originating from the static particles and acting on the dynamic ones. This is a simple way to decrease complexity.

The following equations 7 to 9 describe the last two optimisations. *Dynamic* is the dynamic particle set, which contains all the particles belonging to the flexible part. The complexity is in $D(|Dynamic|)$ which is linear according to the number $|Dynamic|$ of particles in the flexible part.

$$s_{pp'} = (r_p + r_{p'}) - d_{pp'}$$

$$\vec{F}_{linearsteric}(p \in Dynamic) = \begin{cases} \vec{0} & \text{if } s_{pp'} \leq 0 \\ \sum_{p' \in Neighbors(p)} -k_{steric}\vec{u}_{pp'}s_{pp'} & \text{else} \end{cases} \tag{7}$$

$$\vec{F}_{lennardjonessteric}(p \in Dynamic) = \sum_{p' \in Neighbors(P)} \vec{u}_{pp'}4\epsilon_{pp'} \left[\left(\frac{\sigma_{pp'}}{9d_{pp'}} \right)^9 + \left(\frac{\sigma_{pp'}}{7d_{pp'}} \right)^7 \right] \tag{8}$$

$$\vec{F}_{coulomb}(p \in Dynamic) = \sum_{p' \in Neighbors(p)} -\vec{u}_{pp'}\frac{q_p q_{p'}}{4\pi\epsilon_0 d_{pp'}^2} \tag{9}$$

We can highlight another important fact: the previous approach is well-adapted to efficiently compute steric interactions by defining a distance cut-off. For long range interactions such as electrostatic ones, we must be extremely careful with this cut-off. It is preferable to avoid the use of cut-offs to stay biophysically relevant, but in this case, we fall down to quadratic complexity. We thus propose an efficient alternative to take into account long range electrostatic interactions, considering that some parts of a biomolecular complex are rigid. A charge distribution can be translated into an electrostatic potential map, using the APBS tools [Baker et al. (2001)] for example. Charged particles belonging to the rigid components of our complex can be considered as a charge distribution, and are used as an input for APBS. The results of APBS can be interpreted as a 3D grid, and each voxel $V_{i,j,k}$ of this grid contains an electrostatic potential $E_{i-1,j,k}$. For a dynamic particle belonging to $V_{i,j,k}$, using this potential, we can compute an electrostatic force $\vec{F}_{map}$ using the charge of the particle V_p, by spatial derivation of this electrostatic potential (see equation 10). The potential forces $\vec{F}_{map}$ act on the flexible part and originate from the electrostatic potential map. They are defined by computing of the electrostatic potential gradient using the finite central difference method. In equation 10, we consider particle p belonging to the voxel $V_{i,j,k}$ of the electrostatic potential grid, and $E_{i,j,k}$ the value of the potential in this voxel. We define the gradient as the mean of the difference between the $E_{i,j,k}$ potential and the potentials of the six adjacent voxels, two for each axis. This method of computing the gradient reduces the bias related to the discretization of the grid. As regular grids are usually provided by tools such as APBS, Δ_x, Δ_y and Δ_z is the size of the voxel.

$$\vec{F}_{map}(p \in V_{i,j,k}) \simeq \begin{bmatrix} \frac{E_{i+1,j,k}-E_{i-1,j,k}}{2\Delta_x} \\ \frac{E_{i,j+1,k}-E_{i,j-1,k}}{2\Delta_y} \\ \frac{E_{i,j,k+1}-E_{i,j,k-1}}{2\Delta_z} \end{bmatrix} \tag{10}$$

To summarize, this last optimization technique is particularly well-adapted to study the behaviour of flexible biomolecules interacting with a large rigid biomolecular complex. Flexible parts are immersed into a grid and guided by a potential field induced by the rigid component, computed before the simulation. We have combined Eulerian (particle-based) and Lagrangian (grid-based) representations for molecular simulations, inspired by Joe Stam's works on Computational Fluid Dynamics [Stam (1999)]. This approach is also called semi-lagrangian or semi-eulerian method.

During the simulation, forces are computed and applied on the dynamic particle set ($P_{dynamic}$). We explicitly consider potential, van der Waals, Coulomb and external forces. Finally, these new forces are summed with an external force $\vec{F}_{control}(p)$ provided by the user through the graphical interface during the simulation.

$$\vec{F}(p) = \vec{F}_{spring}(p) + \vec{F}_{damping}(p) + \vec{F}_{map}(p) + \vec{F}_{steric}(p) + \vec{F}_{coulomb}(p) + \vec{F}_{control}(p) \tag{11}$$

2.2.2 Conclusion

BioSpring allows a user to quickly study the biomechanical properties, by interactively highlighting rigidity, flexibility, and allosteric effects, in order to provide new hypotheses about a biomolecular system. Moreover, our approach is also designed to help the user in the complex task of modeling large biomolecular complexes before using more classical (and more time-consuming) simulation tools.

3. Multimodal interaction models

In order to interact with biomolecular complexes during a particle-based interactive simulation such as *Gromacs* [Hess et al. (2008)], *NAMD* [Phillips et al. (2005)] or *BioSpring* (see section 2.2.1) in progress, it is common to use a mouse for adding force constraints on particles, providing two degrees of freedom (2DoF), e.g. the x- and y-axes, for the interaction. Using a 3DoF device such as a 3D mouse or a 3D haptic device is even better adapted to this task, in particular for selecting and moving particles in 3D space. Such a device with three instead of two degrees of freedom is more intuitive and efficient for interacting with a complex three-dimensional object, especially when stereoscopic features are used to improve the spatial perception. Furthermore, the immediate force feedback using a haptic device when a particle is actually picked significantly improves the user experience and greatly helps to immerse the user in the molecular scene. If visual feedback is essential especially during the selection and picking task of a particle, the user often asks for additional explanations before getting started. With force feedback, this barrier is lifted, as the interactive simulation becomes more intuitive and is comparable to intuitive dextrous manipulations such as those carried out in daily life. Hardware requirements are modest. In our experience, this approach is viable using a small and affordable haptic device, providing 3D positions and handling 3D directional force feedback. Such an entry-level solution designed for a desktop use is targeted at a large user community and is very easy to set up.

3.1 Pick and pull particle interaction models

The haptic device is used in order to control the direction of the forces applied to selected particles and to adjust the amplitude of these forces. This interaction method contains two stages. The first stage comprises the selection of a single probe particle or a set of particles that we will name $P_{selection}$, using a 3D tool attached to a haptic device and its buttons. In a second stage, the model described by equation 12 is used in order to compute the forces $F_{control}$ applied to the selected particles and sent to the BioSpring simulation as control force (see section 2.2.1). $F_{control}$ is proportional to the distance between the geometric centre of the particle set and the tracker position $P(tool)$. For computing force feedback, the main idea of this approach is to link the selected atoms and the 3D haptic tool with a spring. Instead of providing direct haptic rendering of forces computed in the simulation, the force feedback $F_{feedback}$ only depends on the spring length according to equation 13, which in turn is influenced by the way the simulation reacts to the applied force.

$$\vec{F}_{control}(p \in Selection) = -k_{control}[\vec{P}(tool) - \frac{1}{|Selection|} \sum_{p' \in Selection} \vec{P}(p')] \qquad (12)$$

$$\vec{F}_{feedback}(tool) = -k_{feedback}[\vec{P}(tool) - \frac{1}{|Selection|} \sum_{p' \in Selection} \vec{P}(p')] \qquad (13)$$

The resulting forces are rendered by haptic feedback if a haptic device is used, and by visual feedback such as the blue arrows shown in the Molecular User Interface (MUI), top left part of Figure 5. These forces are simultaneously sent to the interactive simulation. It will take these forces applied by the user on the selected atoms set into account as a control force as described in section 2.2.1.

We emphasize that the haptic loop computation frequency must be between at least 300 to 1000 Hz in order to provide a haptic rendering of good quality. A strong point of the approach described above is that a low physical simulation framerate does not cause instabilities and does not affect the quality of the haptic feedback. With this decoupled spring model, force feedback can be computed at a very high frequency required by the haptic device.

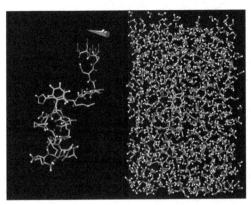

Fig. 5. Dynamic haptic control of a simple polypeptide with (right) or without (left) solvent - "ball and stick" representation

3.2 Interaction models for manipulating proteins as rigid body

The interaction method is more complicated when we want to provide controls and feedback during a rigid body based simulation, comparing to pick and pull a particle set as described in the last section. In order to manipulate both individual proteins and attempt to interactively study interactions between two proteins, the user may rely on various devices and interaction paradigms. A first paradigm associates the position and orientation of the protein with a 6DoF (6 degrees of freedom, 3 for translation, 3 for rotation) devices, such as a 3D mouse or a haptic device. Commonly used in the Virtual Reality domain, haptic devices are specifically used for manipulation and assembly tasks. Collision feedback rendered by 6 Degrees of Freedom (6DoF) haptic devices helps users to assemble 3D objects (see figure 6).

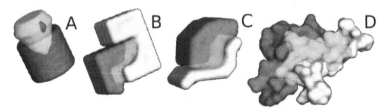

Fig. 6. Different kinds of assembly

3.2.1 Related works dealing with device workspace limitations

All devices have a limited workspace, a limited precision, and limited rotational movements. In order to overcome these limitations, a basic manipulation control is the clutching/unclutching interaction technique, which is however time consuming and does not allow a user to focus on his task. When the user is physically stopped in his movement, by reaching either a boundary of the device or an uncomfortable wrist position, he can press the clutching button to find a better position without moving the virtual object. When releasing, the object is re-attached with the same position and orientation. In this technique, the position and orientation of the 3D virtual object is an isomophic mapping of the position and orientation of the device.

Other solutions avoiding clutching/declutching, such as the *Bubble* technique, [Dominjon et al. (2005)] propose, to perceive (via haptic and visual feedback) the hardware limitations of the device, and provide a rate control based on an isomorphic mapping when the device is far from its workspace boundaries, and on a non-isomorphic mapping near the boundaries, also proposed LaViola & Katzourin (2007).

3.2.2 Related works that deal with high precision assembly

Morover protein-protein docking tasks require high precision. Haptic-guidance based approaches are often used in order to help the user reach a precise and predefined assembly goal. However, there are a few haptic interaction techniques designed to facilitate microassembly tasks for which haptic guidance is unsuitable, such as protein docking. The objective is to find an optimal but precise 3D configuration by interactive exploration.

3.2.3 A haptic interaction paradigm for rigid body based biophysical simulation

We propose an innovative technique to both overcome the physical limitations of the device and to reach the high accuracy required by micromanipulation tasks without a predefined goal, as it is the case during interactive docking simulations. This approach is based on a non-isomorphic mapping around a neutral referential retrieved by an elastic haptic feedback, in addition to external haptic feedback computed by the biophysical rigid body simulation.

In contrast to the method based on the haptic workspace boundaries, our approach is based on a *neutral referential*. Our solution is an implementation of a rate control technique with a 6DoF feedback device, based on a neutral referential, inspired by Bourdot & Touraine (2002).

Our contribution is to use the elastic force feedback to help the user return to the position/orientation of the neutral referential. First, we define a neutral referential with an origin corresponding to the most convenient position/orientation for the user holding the device. For each movement, we calculate a feedback force and torque to bring the user back to this neutral orientation/position (see figure 7 A). In order to compensate the inherent imprecisions of the device, we define a "dead zone" near the neutral referential, in which no movement occurs. The device is then physically restrained inside a comfortable workspace, while the virtual objects have an infinite motion space.

The rate control is based on the difference between the position/orientation of neutral referential initially chosen by the user and the position/orientation of the device during manipulation. The interpolation of movements is obtained by a downscale factor for translation and by a quaternion interpolation for rotation. The level of interpolation varies according to the distance of the two objects to be assembled. Concerning translational motion, the interpolation is done by rescaling the distance vector representing the position of the device from the origin of the neutral referential. In the following equation, $\vec{S}$ is the interpolated translation, $\vec{p}$ the current device's vector position and i the scaling factor. Concerning rotation, at any time of the manipulation, the rotational motion of the device controls the angular velocity of the object. The orientations of the device q_d and the object q_o are represented by quaternions. The rotational motion q_s of the object is then given by the multiplication of the two quaternions (see figure 7 B). The *SLERP* interpolation [Shoemake (1985)] is traditionally used to calculate intermediate frames between two quaternions (start and end orientations) in order to produce smooth rotation. Here, we use the *SLERP* interpolation to calculate a range of quaternion orientations, between the current object's orientation and the one it would adopt after the motion. Then, an orientation at the time t can be picked according to the desired level of attenuation. In the following equation, q_s is the quaternion representing the rotational motion applied to the object without interpolation, q_d is the quaternion of the device and q_o the quaternion of the object. q_a is the softened speed using the *SLERP* function applied between q_o and q_s at the time t of the interpolation. Here, t and i are functions of the distance between the manipulated object and the area of assembly. The attenuation increases when this distance decreases.

$$\vec{S} = \vec{p} \cdot i \tag{14}$$

$$q_s = q_d \cdot q_o \tag{15}$$

$$q_a = \text{slerp}(q_o, q_s, t) \tag{16}$$

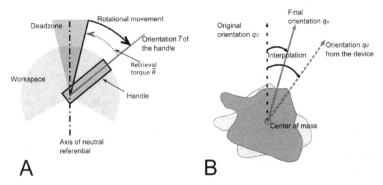

Fig. 7. **Results of a device rotation.** A, the device is rotated from the axis of the neutral referential to the orientation $\vec{r}$, taking into account the deadzone. The retrieval torque $\vec{R}$ is thus equal to $-\vec{r}$. B, the motion of the device is mapped on the object, q_o representing the neutral referential. The final orientation q_a is obtained by applying the *SLERP* interpolation

3.2.4 External feedback

The interaction between the manipulated protein and the other one during the docking task produces biophysical interactions provided by the rigid body based biophysical simulation. These forces are not directly applied to the 3D object in the scene, but are rendered by a haptic force-feedback summed with the elastic feedback used to retrieve the neutral referential.

3.2.5 Conclusion

One of the challenges lies in the protein interaction paradigm simulated by rigid body biophysical simulation. We developed a new method to provide a fine control of the protein with a 6DoF force-feedback device, providing simultaneously biophysical feedback coming from rigid body based simulation. According to the results of an ergonomic study, our technique provides at least the same precision (RMSD) and performance (task time) as direct manipulation with clutching/declutching and successfully overcomes the physical limitations of the device. Moreover subjective results show that users feel more comfortable with our method which avoids the clutching mechanism. We suspect that these results come from the fact that the user is more focused on the assembly task, instead of spending time in clutching/declutching. Further evaluation must be lead in this way. Participants found our technique less disturbing than clutching, appreciating the fact that there is no button to press to manipulate the object. Furthermore, their arm was never in an uncomfortable posture. They furthermore liked the adaptive interpolation. The slowness of the interaction when the two objects are very close was judged pertinent in order to accurately assemble the objects. Another interesting observation is that the most negative comments were not about the manipulation technique itself, but concerned difficulties with the 3D visual perception of a complex protein surface (see figure 6 D). Our approach could thus be an alternative to classical ones and provide at least the same efficiency. We are working on improving the precision of our approach by dynamically tuning the scaling factor used to control rotational and translational velocity. This could be done using the minimal distance between the two objects during the assembly. Finally, we highlight the fact that our approach addresses most problems of the physical limitations of haptic devices (workspace size, precision, mechanical constraints), avoids the use of a clutching/declutching mechanism, is well-adpated to both manipulation and navigation, and could be applied to other 6DoF devices, and does not require complementary visual feedback.

4. Multimodal rendering models

Given the large quantity of biophysical or geometrical information provided by IMS and conveyed to the user in real time, it seems relevant to supplement visual feedback with audio and haptic feedback. Haptic rendering is known to improve the quality of operator interactivity in an immersive environment, as well as his perception of the objects handled or data analyzed [Seeger & Chen (1997)]. Likewise, audio renderings may improve communication of complex information [Barass & Zehner (2000)]. Furthermore, substitutions and redundancy between these channels of communication may have beneficial results on user performance, as long as the choice of modalities is relevant to the task at hand. Richard et al. (2006) and Kitagawa et al. (2005) showed that specific audio and visual renderings can effectively convey information that is presented using haptic modalities. In this part, we provide some examples of haptic audio rendering especially dedicated to study protein interactions using rigid body based biophysical simulation.

4.1 Visual rendering

To represent protein structures, the community of biologists uses standard representations that any specialist can understand 8. They range from a per-atom representation 8(A, B, C) to molecular surfaces 8(H). Some high-level metaphors with ribbons and arrows 8(E,F,G) can describe the secondary structure in a schematic way. Color schemes for atoms respect different standards to simplify the distinction between the different elements of the molecule.

IMS can act at different scales, from whole proteins to precise atomistic interactions, sometimes in the same simulation run. The visuals must then follow the user needs. Three main features have to be fulfilled : interactive frame rate, display of potentially huge molecules and coherent visual information.

For rigid-body docking, pre-computed triangulated surfaces of the proteins and secondary structure representations can be used. But if the atoms are allowed to move inside the structure, computing their surface in real-time is too time consuming in most of the cases.

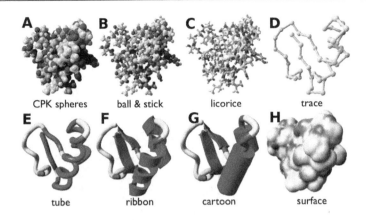

Fig. 8. Standard representation of protein structures. A-D : atomistic representation with spheres and bonds. Colors depend on the atom element. D : Backbone of the protein. E-G : secondary structure visualisation with high-level metaphor objects. Beta-sheets are in red, helices in blue and turns in green

So when it comes to more precise interaction and soft docking, spheres and bonds are more tractable.

Representation of proteins by spheres and bonds using common graphical primitives is easy to implement but generally not appropriate to reach an interactive frame rate. Each primitive is composed of many triangles, then displaying spheres or cylinders consumes a lot of computation time.

Other methods use Graphics Processing Unit (GPU) programming capabilities to draw the spheres and bonds directly on the GPU with no other information than the size and position of the particles.

Different textures and effects can be applied to emphasize interesting locations, collisions or other physical properties.

4.1.1 GPU shaders and HyperBalls

The computer visualization field evolves very quickly due to continuously renewed graphics hardware capabilities. So, the latest contributions from this domain of research has clearly helped scientists to display more and more complex systems. The latest graphics techniques can provide an improved visual perception which could drastically impact the way to visualize molecular structures [Chavent, Lévy, Krone, Bidmon, Nominé, Ertl & Baaden (2011)]. For example, using GPU shaders, i.e. code used to directly program the GPU, it is possible to accelerate and enhance the quality of well known molecular representations such as Molecular Surfaces (figure 9 A), Ball & Stick (figure 9 B), Van der Waals (figure 9 D and E) or protein Secondary Structure (figure 9 C). It is also possible to add lighting effects in real time in order to improve the perception of molecular shape or highlight molecular contours (figure 9 D and E). Furthermore, one can add effects such as blur to depict protein flexibility (figure 9 B). All these graphics techniques, available in real time, will be a great help for the users to interact in a wiser manner with their molecular structures.

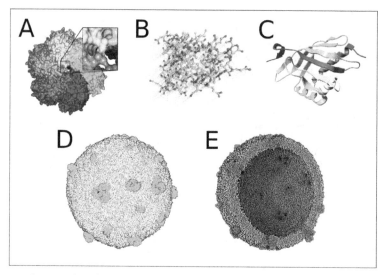

Fig. 9. New and revived molecular metaphors. (A-D) several molecular surface representations: (E) illustrate lighting effect to enhance molecular structure perception.

We have recently developed our implementation of molecular representations on the GPU [Chavent, Vanel, Tek, Lévy, Robert, Raffin & Baaden (2011)]. In this work, we introduced a visual molecular model, the HyperBalls representation, that offers a continuous representation smoothly connecting between classical representations such as licorice or ball-and-stick (figure 10). This representation takes benefit of a GPU ray-casting implementation to visualize molecular systems efficiently. The proposed implementation of the HyperBalls method is efficient for both static and dynamic visualizations of a large number of molecules and is particularly well adapted to visualize huge molecular systems. At present, without further optimization, we can smoothly and interactively render systems with more than 560,000 atoms, reaching some limits for systems comprising a few million spheres. We can expect that our implementation will benefit from the future GPU architectures, where performance increases drastically from a generation to another. This HyperBalls implementation is clearly well suited for an interactive and immersive approach due to the quality rendering and the display efficiency. Furthermore, it is possible to see in real time atomic bond evolution that can be beneficial for interactive docking (see figure 10).

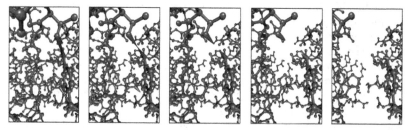

Fig. 10. HyperBalls representation to depict hydrogen bond disruption at a protein interface

4.1.2 Point-sprites

A simple way to represent molecular structure is to depict it as a collection of spheres. To represent spheres, it is possible to use only one square per sphere, always oriented perpendicular to the screen plane. Then an image (also called sprite) of a sphere is pasted on this square (figure 11). This method is usually used to depict visual effects such as flames, smoke or dust where one needs to display a big amount of animated particles. This method is really efficient and commonly implemented in 3D graphics libraries. The main drawback is that sprites are superimposed on each other, so there is no intersection between the individual spheres and it implies to sort the particles along the depth axis.

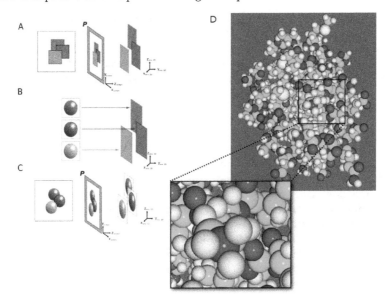

Fig. 11. Point-Sprites method to represent atoms of a protein

4.1.3 Benchmarks

These methods, as well as HyperBalls, have been implemented in an Unity3D (http://unity3d.com/) application and evaluated in terms of frame rate. The HyperBalls GPU shaders were adapted to fit the constraints of Unity3D but the performance was not as good as the initial implementation. The benchmarks show that the point-sprites method is far more efficient than the others. However, the frame rate is not constant when the camera is moving. In fact, the particles must be sorted to be displayed correctly which takes some time when there is a huge number of particles. Domain decomposition can be used to reduce this effect but then some visual glitches at the frontiers of the domains can occur.

The visual result is quite different depending on the methods. Point-sprites can be confusing as the bonds are missing and the spheres are superimposed (figure 11 D). But from a far point of view, the general form of big proteins is kept and using a good color scheme helps to distinguish the interesting areas of the molecules. The traditional primitives can be used along with visual effects such as ambient occlusion, shading or texturing that are often pre-implemented for triangulated objects.

So what we suggest is to combine those methods in an interactive way, according to the size of the system and the user actions. Triangulated primitives are easy to implement for quick development and nice visual effects. For big systems and when the proteins are far away from the camera, particles are particularly suited. When the user zooms into a specific area, a more precise representation such as HyperBalls is adapted, especially to depict interactions.

4.1.4 Visual effects

To allow an accurate interaction with the particles, the position of each one in space must be easily discriminated by the user. The best option is to use stereoscopy but 3D display devices are not common yet. However it is possible to add some visual effects on the objects to overcome this problem. Shading, depth-cueing and ambient-occlusion are commonly used to add realistic lighting and depth perception to a 2D image (9 E). Also, texturing and contouring can help to highlight particular areas and particles and blurring effects can be used to emphasize some movements (9 B).

4.2 Haptic rendering

Currently, there are very few IMS frameworks that include large-scale haptic feedback (force/tactile feedback). This is mainly due to the complexity of computing operations of molecular simulation, which makes it difficult to comply with constraints in terms of refresh rates for real time haptic feedback (from 200 Hz to 1 kHz). Another difficulty is to render various kinds of physico-chemical interactions such as steric or electrostatic interaction. In order to obtain a consistent haptic feedback, only one type of rendering is provided to the user at a time. However one should note that at the perceptual level, steric interaction rendered using haptic feedback are similar to surface collision renderings since it prevents molecular interpenetration.

Most of haptic feedback presented in this section are computed using the rigid body simulation model described in section 2.1. In all rendering, the haptic-device controlled protein, which we will call ligand, can be considered as a big probe against the other protein, which we will call receptor.

4.2.1 Steric and electrostatic interactions

This rendering is used to provide haptic feedback of non-bonded interaction. Haptic rendering of physicochemical interactions consists in feeding the haptic device with the resultant forces computed as described in section 2.1. Forces can be computed and rendered independently or summed up to obtain a total resultant force. Exploration of the receptor by the ligand thus aims at finding stable areas. When the two proteins are in an unstable conformation it renders an unsteady feedback, thus leading the user to drag the ligand towards the surface of the receptor to find a better position and orientation. However the complexity of the force fields induce very irregular directional forces affecting the precision of the manipulation. It appears especially with steric interactions because of the non-linearity in the *Lennard-Jones* potential used to model these forces.

4.2.2 Surface collision

Two approaches were explored to render collisions between both proteins considering their surfaces. The first consists in computing a repulsive force. The direction of this force is the opposite of the direction provided and the module is proportional to the number of colliding triangles determined by the *RAPID* computation as explained in section 2.1. This force can also be weighed by a distance or a volume of interpenetration. Therefore the feedback is more relevant, but the complexity of the computation induces lower refresh rates which could lead to lags in feedback. Rather than repulse the two molecules from each other, the second approach, also based on distance computation, aims to prevent collisions locally by modeling contacts points as springs. The method is introduced by Johnson & Willemsen (2003) and allows fast computation of local minimum distances based on the geometry of the model as well as resulting force and torque. Interestingly the spring model described can be easily adapted to model atomic clashes, such as steric ones in our case. Instead of using the complex Lennard-Jones potential to render the resulting force, interactions are modeled through this more simple spring model with realistic cutoffs (2.5 Angstroms). As the atomic distance computation is already optimized to take into account only surface and opposite atoms, the refresh rate is sufficient and allows a very precise rendering of the contacts, allowing users to feel holes and bumps at the surface. Hence computation speed and consistent feedback constraints are observed ensuring a biological relevance. Current research aims to determine how the size of the proteins affects computing time. It will also be interesting to compare this atomic clashes-based approach with the geometric one which could provide faster computation.

4.3 Audio rendering

Sonification is the use of non-speech audio to convey information. Due to the high temporal resolution and wide bandwidth, the use of auditory stimuli seems highly suitable for time-varying parameters (very high temporal definition when compared to other modalities such as video and haptics), concurrent streams (overlapping of multiple audio renderings for various parameters is possible and easily understandable if these are properly designed), and spatial information (lower definition if compared to visual stimuli, but perceptible over the 360 degree sphere, therefore allowing true 3D rendering).

A large variety of sonification techniques exist and are used in various applications [Walker & Lane (1994)]. One sonification technique is referred to as "parameter mapping" [Hermann & Ritter (1999)], and it is this technique we used to study protein interaction. Parameter mapping sonification is based on creating a link between the data to be rendered and the parameters of a synthesizer (or of any other device which generates or plays back sound). In this particular sonification typology, three elements need to be carefully considered [Walker & Lane (1994)]:

- The nature of the mapping: which data dimension (*i.e.* temperature, pressure, velocity...) is mapped onto, or represented by, each acoustic parameter (*i.e.* frequency, loudness, tempo...). As an example, for a sonification task the temperature might be linked with the frequency of a sound, therefore as the temperature increases, the frequency of the corresponding sonification increases.
- Mapping polarity: in the event of an increase in the sonified data, the sonification parameter can decrease or increase. In the case of temperature-frequency mapping, it

is common to use an increasing-TO-increasing (up-up) polarity. An alternate example could be the size of an object being mapped to frequency: the polarity would likely be increasing-TO-decreasing such that large objects are linked to low sounds and vice versa.

- Mapping scale: in response to a specific increase of the data to be sonified, how much should the sonification parameter increase or decrease. One must take into account the possible range of the data, and the percentage of the usable audible range which is to be exploited. Human hearing is more sensitive to small frequency changes at low frequencies, rather than higher, following an exponential scale. In the case of temperature-frequency mapping the temperature could be exponentially linked to the frequency.

In our application, sound spatialization is used in two different ways: firstly, for local parameters the sonification is spatialised in the specific position where the parameter is calculated, in accordance with visual or haptic rendering, to provide additional information in the protein coordinate system (*i.e.* if the task is to sonify the collision between two different atoms on both proteins, sonification is spatialised at the position of the collision). Then, multiple concurrent sonifications can be spatially distributed in order to give a better intelligibility of the sonifications themselves (*i.e.* stream segregation, selective attention in auditory perception, cocktail party effect studied by Moore (2003)). In 2007, a set up a test for the validation of different sonification methods for object manipulation. Within this test, the subject was asked to change the orientation of a simplified 3D chemical compound in order to be the same as that of a given reference. To do this, the subject used an orientation tracking device. Three approaches for data parameter sonification were tested to improve the speed and accuracy of this manipulation: manipulation speed, angular distance from the reference configuration, and guidance towards the reference position.

Regarding the protein-protein docking task, the following biophysical information has been selected for the sonification:

4.3.1 Molecular surface collision and complementarity

Atomic surface complementarity is estimated essentially as a calculation of the variance of the inter-atomic distances on the two proteins surfaces. This parameter is used to control the variance of a randomly applied pitch to different grains of a granular synthesis process. Granular synthesis has been applied using a spoken word as audio sample (for this particular application, the french word "complementaire" has been recorded and used), repeated cyclically within the granular engine. In this instance, the word is unintelligible if the geometrical complementarity parameter is low, becoming more intelligible as the parameter increases. The rendered audio stream is doubled and associated to each of the two proteins, in preparation for further processing.

The collision parameter represents the number of collisions computed between the two surfaces. The employed method for atomic collision sonification uses a modulation of the phase of a sinusoidal wave whose parameters (carrier and modulator) are controlled by the global number of collisions. Starting with a continuous 400 Hz sinusoidal wave modulated by a 1 Hz signal, the frequency of the modulation increases as the global collision score gets higher, and with it the number of modulating waves, going from 1 to 4, when the two proteins are completely superposed. A second method developed is based on the individual association of every collision with a broadband noise processed with subtractive synthesis (the result is similar to wind noise). The noise is specifically filtered for every collision, adding a controlled randomization of the filtering parameters, so that every "noise generator" sounds different from the others, and spatialised according to its proper position in space. Both of these sonification methods are based on the principle that the signal produced becomes more and more annoying as the number of collisions increases, encouraging the user to change the position and distance of the proteins in order to reduce the number of collisions, and as such stopping the annoying sound. Regarding the second sonification method, sound spatialization helps the listener to localize the part of the protein surface where the collision is taking place, and to guide him/her towards an orientation of the protein for which no collisions are present.

4.3.2 Electrostatic energy

This electrostatic parameter is computed from electrostatic interaction energies between charged particles. Electrostatic energy sonification is performed through the alternation of two sounds, generated using additive synthesis, whose pitch and timbre vary as a function of the global value of this specific force (scalar value). The electrostatic force value is highly variable, and there is not a direct linear relationship between this parameter and a quality judgement of it being good or bad for the docking condition. The link between the parameter and the quality of its specific value has therefore been traced in a two dimensional Cartesian diagram, with the value of the parameter on the X axis, and the quality (being good or bad) on the Y axis. At a given electrostatic force value, the correspondent value on the Y axis has been sonified with the method previously described. For good values, the frequencies of the two sounds are coincident, and their spectra are perfectly harmonic, whilst as the value worsens, the two frequencies become more distant, and the spectra more inharmonic.

4.3.3 Steric energy

This parameter is computed from van der Waals interaction energies between particles. The sonification of the van der Waals energy is based on the principle of the beatings between two sounds frequentially close. As with the electrostatic force, for the van der Waals force value there is not a linear relationship between the parameter and a quality judgement (being good or bad). A mapping similar to the one described for the previous sonification method (electrostatic force) has been employed, with the Y axis value being sonified. Two intermittent sinusoidal pulses are played back simultaneously: if the quality value for the van der Waals force is good, then the two waves have the same frequency, whilst as it becomes worse, one of the two pulses reduces in frequency by up to 20 Hz from the other. This processing results in the creation of beatings between the two frequencies. If there are no beatings, then the score can be considered to be good. In contrast, if the beatings become more frequent (more rapid beat frequency indicates greater frequency separation between the two pulses) the score is becoming worse.

4.3.4 Hotspots at the interface

The number of "hotspots" at the complex interface refers to the list of amino acids present within the current interface region, previously identified using experimental methods as being important actors for protein-protein interaction. Finding hotspots at the protein-protein interface is an important part in judging the quality of solutions. In a second stage, the two audio streams are processed with a low-pass filter with the cutoff frequency controlled by the percentage of protein hotspots which are situated on the interface region. If none of the hotspots are present on the interface the low-pass filter frequency is set at 200 Hz, making the sound nearly inaudible. The cutoff frequency of the filter increases with the number of hotspots present at the interface, making the sound clearer and brighter until, in the optimal position, the frequency filtering is completely deactivated. The two audio streams are rendered stereophonically, associating the left and right channels respectively to the first and second protein.

5. Coupling simulation and interaction codes

Molecular simulation engines previously described provide time-dependent atomic or particle positions, velocities and system energies according to biophysical models at different scale. These models can now compute a molecular dynamics trajectory of interesting biological systems in interactive time. This progress allows to control and visualise a molecular simulation in progress. We have developed a generic library, called *MDDriver*, in order to facilitate the implementation of such interactive simulations. It allows to easily create a network connection between a molecular user interface and a physically-based simulation. We use this library in order to study a real biomolecular system, simulated by various interaction-enabled molecular engines and models. We use a classical molecular visualisation tool and a haptic device to control the dynamic behavior of the molecule. This approach provides encouraging results for interacting with a biomolecule and understanding its dynamics. Our goal is to extend IMS approach to a broader range of simulation engines, as the use of a specific simulation sofware or model often depends on the studied biological system. We have thus developed a generic and independent library, called *MDDriver*, which allows us to easily interface molecular simulation engines with molecular visualisation tools through a network connection. As a first step, we have rendered the calculation modules easily interchangeable while keeping the existing *VMD* user interface as MUI.

5.1 *MDDriver* : a library to coupling molecular simulations codes and molecular user interfaces

In the *VMD/NAMD* architecture, the IMD network protocol [Stone et al. (2001)] was developed in order to interface the Molecular User Interface (MUI) with the MD engine. However, the use of a specific simulation engine and MUI strongly depends on the studied biological system and on user habits. Adding IMD capabilities to other simulation engines and molecular models as well as to a variety of MUIs in addition to *VMD* and *NAMD* enables a whole range of new possibilities in interactive molecular simulations. This approach allows us to address a larger user community working on molecular modeling and simulations, sometimes based on their own home-made simulation engines. Following these motivations, we developed a generic and independent library, called *MDDriver*, inspired by the *VMD/NAMD* approach.

5.1.1 Software architecture

We have thus encapsulated the IMD protocol in the *MDDriver* library, allowing a developer to easily adapt MUI code and MD code in order to extend them with IMD features. This interface provides functions for the exchange of specific data structures over a network: atom positions and system energies, computed for each simulation step by the MD engine (server part), and user-applied forces on a selected atom set.

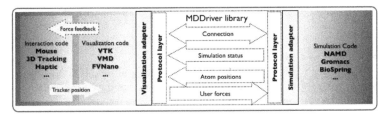

Fig. 12. *MDDriver* library for interfacing a Molecular Dynamics simulation with a Molecular User Interface (Interaction and Visualization code)

5.1.2 Molecular simulation *MDDriver* wrapper

This approach was tested, applied and improved by integrating calls to the *MDDriver* library into the *GROMACS* simulation engine [Hess et al. (2008)], thus rendering the simulation interactive via a MUI. We have used *VMD* as MUI in order to study the molecular behavior of Guanylate Kinase (GK) using an all-atom model and a coarse-grained representation [Baaden & Lavery (2007)] with *GROMACS*. Then we have tested a home-made simulation engine dedicated to molecular docking, which was also IMD-enabled.

MDDriver module offers a simple, modular and generic solution to combine any coordinates-based calculation code with various visualization programs. IMD simulation, this powerful tool for exploration of biomolecules structure in large biological system, is now accessible in a easier way to desktop or virtual reality computational environment. We insist on the fact that the *MDDriver* library was designed for easy integration into any molecular simulation engine providing time series of particle positions. Indeed there are many approaches capable of simulating the dynamic behavior of biomolecules, such as lattice simulations, elastic networks, coarse grain models or even quantum mechanical and semi-empirical methods.

5.1.3 Performances

We will only briefly comment here the desktop use performances obtained for the *MDDriver* library implementation to the *GROMACS* code. The data (coordinates, status and forces) transfer rate between calculation and visualization modules essentially depends on the size of the simulated system.Force application component alters slightly more IMD performances for small systems, depending essentially on selected/total particles ratio (increasing data exchange). In the context of large computing infrastructure deployment for *GROMACS* IMD using *MDDriver*, similar performances have been observed. This confirming robustness of the *MDDriver* library coupled to parallelized applications, performance of the display/interaction installation being the main limitation for IMD simulations of large molecular systems.

6. Applications

We propose in this last section to illustrate the previous simulation, interaction and rendering concepts especially designed for IMS, with several applications. In the first application, these concepts was used to designed new approach and methodology for docking. In the three next ones, these concepts was used in a research context to study some biostructural phenomena. In the two last one, we present two cutting edge scientific sofware that used and included all the innovative concepts presented in this chapter.

6.1 CoRSAIRe : a multimodal and immersive molecular docking project

6.1.1 Main focus

The main focus of the *CoRSAIRe* project [?] is to design a new methodology in that field based on advanced interaction and rendering possibilities, that Virtual Reality (VR) technologies may offer. With respect to other works on docking, we are specifically studying multi-sensorimotor rendering during an interactive docking task.

Usually user participation during the computationnal docking process was very limited, since it only involved configuring docking scripts and choosing one result amongst the computer-generated solutions to the studied problem. Indeed classic approaches to docking provide large numbers of complex configuration based on 3D data describing partner proteins. These algorithms take a long time to produce results, since they test all possible geometric configurations to dock the two proteins. These configurations are then filtered according to energy and physicochemical criteria. Finally, the scientist selects, in this set of results, a smaller set of possible solutions that can be tested against each other experimentally.

Relying on user expertise before applying automatic docking algorithms interactive context allows the user to use natural abilities for the detection of surface complementarity, as well as prior implicit or literature based knowledge regarding for example the nature of the protein-protein interface, what hotspots are present, etc.

It seems thus relevant to develop complementary or alternative approaches to docking. In project *CoRSAIRe* (Combination of Sensorimotor Renderings for the Immersive Analysis of Results) our hypothesis is that using Virtual Reality (VR) technologies and related interactions, which rely on multiple sensory and motor channels, may help experts in this docking task. There are several reasons for this. Firstly, stereoscopy, especially when it is adaptative, may improve perception of 3D protein models. Furthermore, direct manipulation of several proteins at the same time, as afforded by peripherals commonly used today for such tasks (*e.g.* 3D mouse, force-feedback interfaces, etc.) may be more intuitive and efficient than traditional, desktop WIMP[1]-type interfaces. Additionally, multimodal management of sensorimotor feedbacks (based on an approach aiming to dynamically specify adaptation of visual, haptic and audio renderings to the characteristics of the information in use) is one possible answer to the problems related to the simultaneous presentation of large amounts of data. Finally, a strongly interactive approach of VR docking allows the docking expert to be placed on the forefront of the work, rather than giving an automatic algorithm a complete control over the generation of possible sets of solutions. We believe our approach, which combines benefits of multimodal interaction with the capitalization of docking experts' occupational skills (in biology, crystallography, bioinformatics) in modelling will allow improvements in the speed of predictions for the structure of protein-protein complexes, as well as in overall search efficiency and in the quality of results obtained when analyzing possible solutions.

6.1.2 Discussion and results

This project allow us to define the multimodal allocation space ("modal allocation" [André (2000)]) that refers to the specific use of one or more sensory modalities to display an information. It is preferable for users to use optimal modal allocation considering both technical constraint, task (*e.g.* characteristics of information relevant to scientists) and

---

[1] Acronym of Window, Icon, Menu, and Pointing device.

	Visual	Auditive	Haptic
Surface representation	ok		
Surface collisions	ok	ok	ok
Surface complementarity	ok	ok	ok
Electrostatic interactions			ok
Electrostatic energy	ok	ok	ok
Steric interactions			ok
Steric energy	ok	ok	ok
Hydrophobic patchs	ok		
Hotspots at the interface	ok	ok	

Table 1. Restricting the modal allocation space

operator-related constraints (*e.g.* characteristics of perception, of expertise, etc.). The results of the project allow us to formulate the following principles for the design of a multimodal application for molecular docking, summed up in table 1.

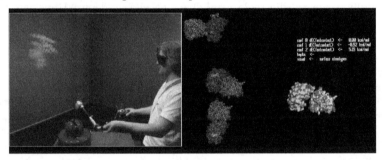

Fig. 13. A user immersed in the docking application (left). On the right, a sample screen capture following selection of three conformations by the user.

The interactive process dedicated to protein protein docking designed into the CoRSAIre project, allows significant reduction of the number of configurations to be tested by algorithms used afterwards, and we maximize the use of the user expertise. Our approach could also be reused in the design of future docking software, integrating factors such as protein flexibility, based on the premise that many docking problems involve flexible partners. Furthermore, this work should also focus on defining future situations of use of such tools. Indeed, our interactions with future users identified several possible avenues for the use of docking tools, *e.g.* teaching, scientific discovery, collaborative work, etc.

6.2 Interactively locating ion binding sites by steering particles into electrostatic potential maps

Interactively locating ion binding sites by steering particles into electrostatic potential maps Metal ions drive important parts of biology, yet it remains experimentally challenging to locate their binding sites into biomolecules (protein, DNA). With the MyPal method (Molecular scrutinY of PotentiALs), implemented in the *BioSpring* program, we use interactive steering of charged ions in an electrostatic potential map in order to identify their potential binding sites [Delalande et al. (2010)]. We use this method in order to facilitate the discovery of new relevant ion binding sites by successfully retrieving the location of cation binding sites in DNase I enzyme and assessing their selectivity, combining atomic and coarse-grained resolutions.

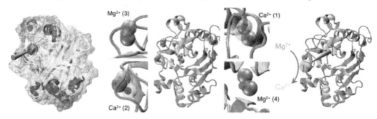

Fig. 14. Visual summary of the interactive experiments. On the left, interactive potential exploration of the DNase I enzyme using the MyPal method. On the middle, results of experiments for detecting a priori unknown ion binding sites. The reference position of each binding pocket is shown as a red sphere and MyPal predictions for a potential map with (orange) or without (green) ionic strength are displayed by transparent spheres. On the right, an ion substitution experiment ("molecular-billiard") at site 2 is depicted. Such an experiment probes the selectivity of a given ionic pocket for different ions

We interactively scanned the electrostatic potential of DNase I by using Na+, Ca2+ and Mg2+ as ionic probes. For the binding sites detection, Mg2+ cation was chosen as its double charge facilitates long-range electrostatic steering towards the binding pockets and its small size increases the accuracy for sensing the rough and detailed molecular surface at atomic resolution. Taking into account ionic strength for calculating the electrostatic potential leads to more accurate maps. However, without considering ionic strength we achieved comparable predictions and more easily detect binding sites thanks long-range driving forces (Figure 16). All four ion binding sites identified were retrieved by the MyPal approach. To assess the selectivity of identified binding sites, we start with different ions (Na+, Ca2+, Mg2+ and Cl-) at a given site and tried to interactively substitute this initial ion by another. Figure 16 and Table 2 illustrate and summarize the results for these ion substitution "molecular-billiard" simulations. As might be expected, chloride as an anion cannot be stabilized within any of the four cation binding pockets, nor can it displace a bound cation. Sites that are magnesium selective are well characterized by our approach. Less efficient substitution experiments may be related to the simplicity of our model in which selectivity depends on the shape of the pocket itself and the pathway for accessing it. Generally speaking, buried and narrow sites are unreachable for large ions, whereas sites localized at the enzyme surface are readily subject to ion exchange. In the latter case, haptic feedback helps the user to distinguish between favourable and unfavourable substitutions.

The current implementation of MyPal/BioSpring was not designed in order to provide precise quantitative binding affinity estimates, but to be capable of distinguishing in real time between non-existing, weak and strong ion binding sites and assess the relative selectivity of significantly different ionic probes. Despite the approximations made in the choice of the model representation it should remain possible to quantify the strength of binding by calculating the work required by the user to extract an ion from its binding site.

6.3 Interactive study of Guanylate Kynase opening and closure

In this study, we have worked on an intensive studied biomolecular system, the Guanylate Kinase (GK) enzyme. Structures for this molecule are provided by experimental methods

Probe (Site)	Ca^{2+} (1)	Ca^{2+} (2)	Mg^{2+} or Ca^{2+}(3)	Mg^{2+} (4)
Mg^{2+}/Ca^{2+}	$Ca \mapsto Mg$	$Ca \mapsto Mg$	$Mg \dashv Ca$	$Mg \dashv Ca$
	$Mg \dashv Ca$	$Mg \mapsto Ca$	$Ca \mapsto Mg$	-
Na^{+}	$Ca \dashv Na$	$Ca \mapsto Na$	$Mg \dashv Na$	$Mg \dashv Na$
	$Na \mapsto Ca$	$Na \mapsto Ca$	$Na \mapsto Mg$	-
Cl^{-}	$Ca \dashv Cl$	$Ca \dashv Cl$	$Mg \dashv Cl$	$Mg \dashv Cl$
	-	-	-	-

Table 2. **Ion substitution interactive simulation results.** The table indicates whether exchange from X to Y is possible ($\mapsto$) or impossible ($\dashv$). For instance, $Ca \dashv Cl$ means that Ca^{2+} cannot be displaced by Cl^{-}. A minus sign indicates that initial positioning of the chosen probe ion at the given binding pocket was not possible *via* our approach.

such as Nuclear Magnetic Resonance or X Ray cristallography. The molecule has a U shape with either a closed or an open conformation (see 15). The closure mechanism of GK consists in increasing the proximity of two substrate binding sites, for GMP and ATP, both essential for the enzymatic reaction. The goal of our study is to understand which parts of this system are involved in the closure mechanism. This mechanism has been investigated using our *MDDriver* framework (VMD/MDDriver/GROMACS) at two levels of detail. The first level corresponds to an all-atom model (18098 atoms), the second to a lower resolution coarse-grain model (1900 beads), and the third to a augmented spring network model. Prospective tests using coarse-grain simulations allowed for the efficient exploration of a broad range of possibilities to close the enzyme, trying to reach a closed conformation similar to the available experimental structures.

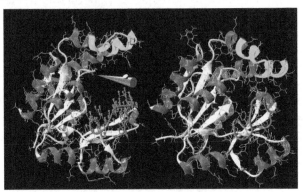

Fig. 15. Haptic control (red arrows in the red loop) of Guanylate Kinase closure. Secondary structure cartoon representation of the open state (left) and the closed state (right)

Figure 15 shows a secondary structure representation of the protein, considering specific architectural units such as the loops (white tubes), the helices (purple ribbons) and the beta sheets (yellow arrows). The crucial role of one loop (highlighted in red in Figure 15) in the initiation of GK's closure could thus be identified. It was then confirmed in a second phase using more detailed all-atom simulations. Understanding the features of this early intermediate state occurring as an impulse for the closure mechanism allows us to propose a novel mechanistic hypothesis. The loop move could be initiated by GMP docking, which may drive this loop via long range electrostatic interactions. When the loop draws closer to the other side of the enzyme, conformational changes could be triggered, subsequently inducing a global closure of the enzyme. The interactive exploration of the simulation using the haptic modality lead us to this theoretical hypothesis. It also suggests that electrostatic interactions could be the main driving force for closure.

6.3.1 Modelling a transient stage of DNA repair by flexible docking of double stranded DNA to RecA nucleoprotein filaments

Homologous recombination is a fundamental process enabling the repair of double-strand breaks with a high degree of fidelity. In prokaryotes, it is carried out by RecA nucleofilaments formed on single-stranded DNA (ssDNA). These filaments incorporate genomic sequences that are homologous to the ssDNA and exchange the homologous strands. Due to the highly dynamic character of this process and its rapid propagation along the filament, the sequence recognition and strand exchange mechanism remains unknown at the structural level. By the interactive and flexible approach available from the BioSpring program, we investigated the possible geometries of association of the early encounter complex between RecA/ssDNA filament and double-stranded DNA (dsDNA) [Saladin et al. (2010)]. Due to the huge size of the system and its dense packing, we used a reduced representation for both protein and DNA. In this study, a systematic docking was also performed to associate dsDNA and the RecA/ssDNA complex, but this approach didn't enable the consideration of flexible regions of the nucleofilament RecA. BioSpring approach promoted to easily build a hybrid rigid-flexible representation of the molecular system by combining an Augmented Spring Network model (ASN) and a static molecular shape, and finally enabled to include very flexible L2 loops in the structure of RecA/ssDNA receptor. These flexible L2 loops constituted the only interactively controlled protein region, the rest of the RecA nucleofilament and the ssDNA were considered as static. Incoming dsDNA (ligand) was the second molecular fragment described by a flexible ASN model. During the interactive docking simulation, L2 loops moves were obtained by pulling user-selected atoms, while position and orientation of the dsDNA were controlled by acting on a fixed particles group (central nucleobases). Each single interactive simulation consisted in (i) moving L2 loops and simultaneously (ii) pulling the dsDNA toward the ssDNA, then (iii) allowing the relaxation of the system and finally (iv) saving the ligand and receptor positions.

Docking of curved dsDNA structures permitted to reach a more stable molecular complex than the one obtained from B-type DNA ligands. These simulations also demonstrate that it is possible for the double-stranded DNA to access the RecA-bound ssDNA while initially retaining its Watson-Crick pairing and emphasize the importance of RecA L2 loop mobility for both recognition and strand exchange.

6.4 ePMV : embedding molecular modeling software directly inside of professional 3D animation applications

ePMV, the Embedded Python Molecular Viewer [Johnson & Autin (2011)] is an open-source plug-in, that runs the molecular modeling software PMV [Sanner (1999)] directly inside of numerous professional 3D animation applications (hosts), to provide seamless access the capabilities of both systems and to simultaneously link the host software to other scientific algorithms. ePMV currently plugs into Maxon Cinema4D, Autodesk Maya, and Blender. Uniting host and scientific algorithms into a single interface allows users from

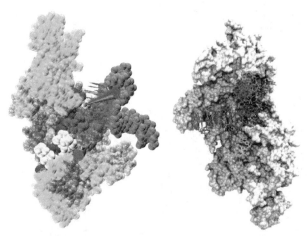

Fig. 16. Visual summary of the interactive experiments. On the left, interactive and flexible low-resolution docking of dsDNA to the RecA/ssDNA complex, using BioSpring. The two trackers enable the user to move at the same time (i) protein L2 loops (yellow or red, if selected) and (ii) dsDNA (pink and red, for selected nucleobases). Static fragments of RecA protein and ssDNA are shown in green/brown and purple spheres, respectively. On the right, all-atom model obtained after reconstruction from one of the best BioSpring prediction.

varied backgrounds to assemble professional quality visuals and to perform computational experiments with relative ease. The hybrid provides:

- high quality rendering with shadows, global illumination, ambient occlusion, etc
- intuitive GUI workflows that help users set up animations ranging from easy turntable rotations to sophisticated mechanism-of-action movies
- mesoscale modeling that allows users to illustrate or animate complex cell events in molecular detail by positioning objects with intuitive controls
- a common Python Platform that allows users to initiate sophisticated algorithms like molecular dynamics or docking energy calculations on the fly and to interoperate these algorithms with each other and with the host

The Interactive Molecular Driver [Stone et al. (2001)] and the callback action from Modeller [Eswar & Sali (2008)] enable real-time interactive molecular simulations with additional forces provided by the user. This interactive steering can operate at different levels, from selected atoms or residues, to selected curve points associated with molecular backbones. Mouse gestures and animated key frames can transmit forces or new coordinates to the simulation calculator that is linked to the host GUI via ePMV. Sophisticated host algorithms like inverse kinematics and efficient collision detection algorithms can operate on the same data as well. With this setup, a ligand can be hand-guided into a binding site with real-time docking scores provided by the Python modules of Autodock [Huey et al. (2007)]. Host-provided physics shortcuts (e.g., soft-body springs for bonds) enable interactive flexible docking with real-time scoring. At the cutting edge of molecular Augmented Reality, a user can interact with data via handheld markers tracked by a camera [Gillet et al. (2005)] to perform an interactive Rigid-body docking with intuitive midair hand gestures (see Figure 17 and http://epmv.scripps.edu/videos/structure2011).

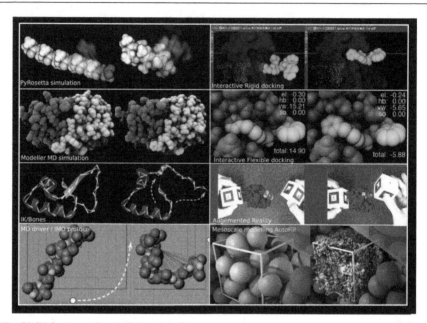

Fig. 17. ePMV features through 8 examples

6.5 FvNano: A virtual laboratory to manipulate and visualize molecular systems

The main goal of FvNano is to provide an easy to use program to manipulate molecular structures in "real time" on regular or high-performance computing (HPC) platforms. The idea is to combine molecular dynamics (MD) software with modern human-computer interaction (HCI) peripherals and GPU rendering. As previously told, combining MD with user interaction is crucial for a better understanding of the molecular motions inside a protein structure when a particular solvent is used or with an important number of active compounds. MD simulations require a lot of computing power. Hence, the ability to use high-performance computing platforms is mandatory when studying complex macromolecular systems. However, the difference between regular and massively parallel architectures can make the program hard to optimize for both platforms. This is solved by using a modular architecture based on the Flow-VR middleware ?, http://flowvr.sourceforge.net/. In that case, MD simulation, interaction and visualization can be represented as modular blocks linked together by Flow-VR. Each of these blocks can then run on single or multiple threads according to the user's choice. Manipulating objects in a 3D environment with a 2D screen can be challenging, for that purpose, FvNano currently implements two types of HCI peripherals: SpaceBalls and haptic arms. SpaceBalls are used to move the viewpoint and haptic arms to manipulate the molecular structures with force feedback support. The visualization part is also modular, as of now two renderers are available: VMD and OpenGL. The OpenGL renderer uses the HyperBalls GPU shaders previously described. FvNano can also be used to visualize molecular trajectories computed by MD softwares with a simple user interface inspired by video players. Related work within the VMD software has recently been discussed [Stone et al. (2001)].

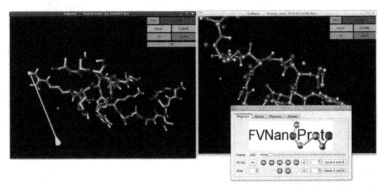

Fig. 18. Screenshot of the interactive molecular dynamics application (left). The cyan cone is the haptic arm avatar and the line shows the atom movement in progress. On the right a screenshot of the molecular trajectory reader.

7. Conclusion

Protein interactions are now routinely studied via computer simulation to understand aspects that cannot currently be studied by experiments. Recently, the Fold It! project [Cooper et al. (2010)] - a 3D-puzzle desktop game, in which the user's task is to fold proteins interactively and without any knowledge prerequisites - showed that using interactivity and insight of human minds can lead to more accurate results than pure computation. Removing false-positive results is done implicitly by users that intuitively avoid erroneous ways of molecule assembly using their experience and logic mind. More generally, molecular simulations can now benefit from this approach to reduce computation and analysis time.

In this document, we presented an interactive approach to assist scientists in their study of protein-protein docking phenomena using some advanced interaction and rendering features offered by a Virtual Reality or advanced human-computer-interaction environment. In such a context, it is important to take into account existing practices of domain experts used in their everyday work. By formalizing user needs and tasks in order to propose a limited set of design principles leading towards an appropriate tool such practices can be further improved, whilst leaving some room for them to evolve in new directions.

Through different examples, we have seen in this chapter that this goal requires efforts from many scientific domains. Experimental biologists describe the needs and validate the results that bioinformaticians extract from the analysis of simulations. Computer science experts are needed to provide efficient codes and graphics. Also, cognitive science helps to design suitable interaction paradigms and user interfaces. Interaction can be used in many applications, from rigid-body docking to accurate atomistic simulations, allowing the user to obtain a wide range of results. The novelty of our approach is that it strives to ensure continuous user participation in the process through direct manipulation of the protein models. In proposing such an approach in which users are involved both upstream and downstream from automatic docking procedures, we hope to maximize the use of their expertise. Hence, the interactive approach efficiently introduces a human element in the process and benefits from the user's experience and insight. The resemblance of this kind of applications with video games should not delude scientists to underestimate the scientific value of such techniques.

8. References

Anderson, J., Lorenz, C. & Travesset, A. (2008). General purpose molecular dynamics simulations fully implemented on graphics processing units, *J. Comp. Phys.* 227: 5342.

André, E. (2000). *Handbook of natural language processing*, chapter The generation of multimedia presentations, pp. 305–327.

Andrusier, N., Nussinov, R. & Wolfson, H. J. (2007). Firedock: fast interaction refinement in molecular docking, *Proteins* 69(1): 139–59.

Baaden, M. & Lavery, R. (2007). There's plenty of room in the middle: Multiscale Modelling of Biological Systems, *Recent Advances in Structural Bioinformatics* pp. 173–196.

Baker, N. A., Sept, D., Joseph, S., Holst, M. J. & McCammon, J. A. (2001). Electrostatics of nanosystems: Application to microtubules and the ribosome, *Proceedings of the National Academy of Science*, Vol. 98, pp. 10037–10041.

Barass, S. & Zehner, B. (2000). Responsive sonification of well-logs, *Proceedings of the International Conference on Auditory Display (ICAD'00)*.

Berman, H., Westbrook, J., Feng, Z., Gilliland, G., Bhat, T., Weissig, H., Shindyalov, I. & Bourne, P. (2000). The protein data bank, *Nucleic Acids Research* 1(28): 235–242.

Borrelli, K., Vitalis, A., Raul Alcantara, R. & Guallar, V. (2005). Pele: Protein energy landscape exploration. a novel monte carlo based technique, *Journal of Chemical Theory and Computation* 6(1): 1304–1311.

Bourdot, P. & Touraine, D. (2002). Polyvalent display framework to control virtual navigations by 6DoF tracking, *Proceedings of the IEEE Virtual Reality International Conference (IEEE-VR'02)*, pp. 277–278.

Chavent, M., Lévy, B., Krone, M., Bidmon, K., Nominé, J., Ertl, T. & Baaden, M. (2011). Gpu-powered tools boost molecular visualization, *Briefings in Bioinformatics* .

Chavent, M., Vanel, A., Tek, A., Lévy, B., Robert, S., Raffin, B. & Baaden, M. (2011). Gpu-accelerated atom and dynamic bond visualization using hyperballs: A unified algorithm for balls, sticks, and hyperboloids, *Journal of Computational Chemistry* 32(13): 2924–2935.

Comeau, S., Gatchell, W., Vajda, S. & Camacho, C. (2004). Cluspro: an automated docking and discrimination method for the prediction of protein complexes, *Bioinformatics* 20(1): 45–50.

Connolly, M. (1983). Analytical molecular surface calculation, *Journal of Applied Crystallography* 16: 548–558.

Cooper, S., Khatib, F., Treuille, A., Barbero, J., Lee, J., Beenen, M., Leaver-Fay, A., Baker, D., Popović, Z. & Players, F. (2010). *Nature* 466(7307): 756–60.

Cui, Q. & Bahar, I. (2006).

Delalande, O., Ferey, N., Grasseau, G. & Baaden, M. (2009). Complex molecular assemblies at hand via interactive molecular simulations, *Journal of Computationnal Chemistry* 30(15): 2375–2387.

Delalande, O., Férey, N., Laurent, B., Gueroult, M., Hartmann, B. & Baaden, M. (2010). Multi-resolution approach for interactively locating functionally linked ion binding sites by steering small molecules into electrostatic potential maps using a haptic device, *Pacific Symposium on Biocomputing*, pp. 205–215.

Dominjon, L., Lécuyer, A., Burkhardt, J., Andrade-Barroso, G. & Richir, S. (2005). The "Bubble" Technique: Interacting with Large Virtual Environments Using Haptic Devices with Limited Workspace, *Proceedings of the World Haptics Conference (joint Eurohaptics Conference and Haptics Symposium)*.

Eswar, N., E.-D. W. B. S. M. & Sali, A. (2008). Protein structure modeling with modeller, *Methods Mol. Biol.* 426: 145?159.

Fernandez-Recio, J., Totrov, M. & Abagyan, R. (2003). Icm-disco docking by global energy optimization with fully flexible side-chains, *Cambridge, MA: Bradford Books / MIT Press* 1(52): 113–117.

Gillet, A., Sanner, M., Stoffler, D. & Olson, A. (2005). Tangible interfaces for structural molecular biology, *Structure* 13(3): 483–91.

Gottschalk, S., Lin, M. & Manocha, D. (1996). Obbtree: A hierarchical structure for rapid interference detection, *Proceedings of the 23rd Conference on Computer graphics and interactive techniques*, Vol. 30, pp. 171–180.

Grayson, P., Tajkhorshid, E. & Schulten, K. (n.d.).

Hermann, T. & Ritter, H. (1999). *Listen to your Data: Model-Based Sonification for Data Analysis*, pp. 189–194.

Hess, B., Kutzner, C. Vanderspoel, D. & Lindahl, E. (2008). Gromacs 4: Algorithms for highly efficient, load-balanced, and scalable molecular simulation., *Journal of Chemical Theory and Computation* 4(3): 435–447.

Hinsen, K. (2000). The molecular modeling toolkit: A new approach to molecular simulation, *Journal of Computational Chemistry* 21: 79–85.

Huey, R., Morris, G. M., Olson, A. J. & Goodsell, D. S. (2007). A semiempirical free energy force field with charge-based desolvation, *Journal of computational chemistry* 28(6): 1145–52.

Isralewitz, B., Baudry, J., Gullingsrud, J., Kosztin, D. & Schulten, K. (2001). Steered molecular dynamics investigations of protein function, *J. Mol. Graph.* 19: 13–25.

Johnson, D. & Willemsen, P. (2003). Six degree-of-freedom haptic rendering of complex polygonal models, *Proceedings of the 11th Symposium on Haptic Interfaces for Virtual Environment and Teleoperator Systems (HAPTICS'03)*.

Johnson, G. & Autin, L., G. D. S. M. O. A. (2011). ePMV Embeds Molecular Modeling into Professional Animation Software Environments, *Structure* 19: 293–303.

Kitagawa, M., Dokko, D., Okamura, A. & Yuh, D. (2005). Effect of sensory substitution on suture-manipulation forces for robotic surgical systems, *Journal of Thoracic and Cardiovascular Surgery* 129(1): 151–158.

Knoll, P. & Mirzaei, S. (2003). Development of an interactive molecular dynamics simulation software package, *Rev. Sci. Instrum.* 74: 2483–2487.

LaViola, J. & Katzourin, M. (2007). An exploration of non-isomorphic 3d rotation in surround screen virtual environments, *Proceedings of the IEEE International Conference on 3D User Interfaces (IEEE-3DUI'07)*, pp. 49–54.

Leech, J., Prins, J. F. & Hermans, J. (1996). SMD : visual steering of molecular dynamics for protein design, *Computational Science and Engineering* 3: 38–45.

Leech, J., Prins, J. F. & Hermans, J. (1997). , *Physica A.* 240: 246–254.

Matthey, T., Cickovski, T., Hampton, S., Ko, A., Ma, Q., Nyerges, M., Raeder, T., Slabach, T. & Izaguirre, J. A. (2004). Protomol, an object-oriented framework for prototyping novel algorithms for molecular dynamics, *ACM Trans. Math. Softw.* 30-265(3): 237–265.

Moore, B. C. J. (2003). *An Introduction to the Psychology of Hearing.*

Nelson, M., Humphrey, W., Kufrin, R., Gursoy, A., Dalke, A., Kale, L., Skeel, R. & Schulten, K. (1995). MDscope - a visual computing environment for structural biology, *Comp. Phys. Comm.* 91: 111–133.

Phillips, J. C., Braun, R., Wang, W., Gumbart, J., Tajkhorshid, E., Villa, E., Chipot, C., Skeel, R. D., Kale, L. & Schulten, K. (2005). Scalable Molecular Dynamics with NAMD, *Journal of Computational Chemistry* 26: 1781–1802.

Pipe, S. (2008). Recombinant clotting factors, *Thromb Haemost* 99(5): 840–850.

Plimpton, S. (1995). Fast parallel algorithms for short-range molecular dynamics, *J. Comp. Phys.* 117: 1–19.

Prins, J. F., Hermans, J., Mann, G., Nyland, L. S. & Simons, M. (1999). A virtual environment for steered molecular dynamics., *Fut. Gen. Comp. Sys.* 15: 485–495.

Richard, P., Chamaret, D., Inglese, F.-X., Lucidarme, P. & Ferrier, J.-L. (2006). Human-scale haptic virtual environment for product design: Effect of sensory substitution, *International Journal of Virtual Reality* 5(2): 37–44.

Ritchie, D. (2003). Evaluation of protein docking predictions using hex 3.1 in capri rounds 1 and 2, *Proteins* 52(1): 98–106.

Saladin, A., Amourda, C., Poulain, P., Ferey, N., Baaden, M., Zacharias, M., Delalande, O. & Prevost, C. (2010). Modeling the early stage of dna sequence recognition within reca nucleoprotein filaments, *Nucleic Acid Research* 38(19): 6313–6323.

Sanner, M. F. (1999). Python: a programming language for software integration and development, *J. Mol. Graph. Model.* 17(1): 57–61.

Sanner, M., Olson, A. & Spehner, J.-C. (1996). Reduced surface: An efficient way to compute molecular surfaces, *Biopolymers* 38: 305–320.

Seeger, A. & Chen, J. (1997). Controlling force feedback over a network, *Proceedings of the Second PHANToM User's Group Workshop*.

Shoemake, K. (1985). Animating rotation with quaternion curves, *Proceedings of the 12th annual conference on Computer graphics and interactive techniques (SIGGRAPH'85)*, pp. 245–254.

Stam, J. (1999). Stable fluids, *In SIGGRAPH 99 Conference Proceedings, Annual Conference Series* 38: 121–128.

Stone, J. E., Gullingsrud, J. & Schulten, K. (2001). A system for interactive molecular dynamics simulation, *Proceedings of Interactive 3D Graphics* pp. 191–194.

Surles, M. C., Richardson, J. S., Richardson, D. C. & Brooks-JR., F. P. (1994). Sculpting proteins interactively : Continual energy minimization embedded in a graphical modelling system, *Protein Science* 3: 198–210.

Villoutreix, B., Bastard, K., Sperandio, O., Fahraeus, R., Poyet, J., Calvo, F., Deprez, B. & Miteva, M. (2008). In silico-in vitro screening of protein-protein interactions: towards the next generation of therapeutics, *Current Pharmaceutical Biotechnology* 9(2): 103–22.

Vormoor, O. (2001). Quick and easy interactive molecular dynamics using Java3D, *Comp. Sci. Eng.* 3: 98–104.

Walker, B. N. & Lane, D. M. (1994). *Auditory Display: Sonification, Audification, and Auditory Interfaces*, Westview Press.

Wang, R., Lu, Y. & Wang, S. (2003). Comparative evaluation of 11 scoring functions for molecular docking, *Journal of Medicinal Chemistry* 46: 2287–2303.

Wu, X. & Wang, S. (2002). Direct observation of the folding and unfolding of b-hairpin in explicit water through computer simulation, *J. Am. Chem. Soc.* 124: 5282–5283.

Zacharias, M. (2005). Attract: protein-protein docking in capri using a reduced protein model., *Proteins* 60(2): 252–6.

2

Computational Methods for Prediction of Protein-Protein Interaction Sites

Aleksey Porollo and Jaroslaw Meller
University of Cincinnati
USA

1. Introduction

Studies of protein-protein interactions play a central role in understanding protein function in biological systems, closing the gap between large-scale sequencing efforts and medically relevant outcomes. Increasingly, protein interaction interfaces that mediate communication between proteins are becoming targets for therapeutics, offering a possibility to disrupt critical interactions and specifically attenuate function (Fletcher and Hamilton, 2007; Fry, 2006).

Efforts to catalog, characterize, and link protein interactions with disease states and other phenotypes are ongoing, building on improvements in experimental techniques, such as high throughput two-hybrid assays or chip-based proteomics. Significant progress has also been achieved in structural genomics, providing detailed information for a growing number of macromolecular complexes and interaction interfaces by means of X-ray crystallography, NMR spectroscopy and other methods.(Aloy et al., 2005; Slabinski et al., 2007)

Despite impressive progress, existing experimental methods for mapping protein interactions suffer from many limitations. High throughput methods, such as two-hybrid or chip-based assays, are characterized by high rates of false positives and false negatives (Bader and Chant, 2006; Han et al., 2005), requiring further validation and detailed characterization of individual interactions. Obtaining detailed high-resolution information about protein interaction interfaces can also be challenging in many instances.

For example, some complexes may not crystallize, or crystallize in a different than biologically relevant conformation. X-ray crystallography may also fail when multiple and incompletely mapped interactions or membrane domains are involved.(Lacapere et al., 2007) This is exacerbated by the fact that each protein has been estimated to have around 9 distinct interacting partners (and some are estimated to have hundreds interactants), with majority of the implied complexes unlikely to be resolved experimentally in the foreseeable future.(Aloy and Russell, 2004; Ritchie, 2008)

Limitations of experimental techniques and attempts to circumvent the problem by focusing directly on protein interactions create an opportunity for computational approaches to complement and facilitate experimental efforts in that regard. In particular, statistical and machine learning-based approaches are being increasingly used to facilitate identification of protein interfaces. There are a growing number of methods for protein interaction sites prediction that vary in terms of principles of the recognition of interaction interfaces, descriptors used to identify interacting sites (feature space) and learning algorithms used.

From the point of view of a representation used to capture characteristics of interaction interfaces, one may distinguish two main groups of methods. The first group attempts to predict interaction sites using sequence information only.(Gallet et al., 2000; Ofran and Rost, 2007) The second group of methods, takes available structural information into account (Fariselli et al., 2002; Lichtarge et al., 1996), typically involving the identification of sites on the surface of a monomeric structure that are either evolutionarily conserved (as for example in the pioneering evolutionary trace method by Lichtarge and colleagues (Lichtarge et al., 1996)), or have a propensity for interaction interfaces (see, e.g., (Jones and Thornton, 1997)).

Although evolutionary trace methods are relatively insensitive to structural detail and can identify conserved "hot spots", their overall accuracy is limited.(Caffrey et al., 2004; Porollo and Meller, 2007) On the other hand, detailed structural information can be used to characterize patches on the surface of a protein in terms of their geometric and other properties (see, e.g., (Bordner and Abagyan, 2005; Koike and Takagi, 2004; Neuvirth et al., 2004)). Structural conservation can also be taken into account when multiple structures within families are available.(Chung et al., 2006; Ma et al., 2003)

While structural information improves prediction accuracies (with the risk of increasing the sensitivity to the choice of a specific structure), challenges remain and new insights are required to improve state-of-the-art in the field.(de Vries and Bonvin, 2008; Zhou and Qin, 2007) Further progress also requires continued systematic evaluation of new methods. In this regard, the lack of standard definitions and consistent evaluation criteria adds to the challenge and often makes direct comparison of existing methods impossible.

One problem that contributes to the difficulty of fair evaluation and objective comparison of different methods is related to the uncertainty concerning the definition of the negative class. The assignment to the "non-interacting" class is at best tentative, given the incompleteness of information regarding all possible interactions and interacting partners. Despite the growing number of resolved structures of protein-protein complexes, another challenge is the relative paucity of carefully curated and properly stratified (to represent different types of complexes) benchmarks.

This chapter reviews computational methods for the prediction of protein interaction sites, with a primary focus on structure-based approaches. The goal is to help the reader better understand the underlying concepts and limitations pertaining to current methods in the field. A number of methodological issues related to the training and validation of such methods are discussed as well. The benchmarks and assessment included in this chapter should also help making an informed decision as to when computational predictions can be regarded as sufficiently confident for a particular system of interest to warrant further experimental validation.

2. Definition of protein-protein interaction site

The recognition of protein-protein interaction sites can be cast as a classification problem, i.e., each amino acid residue is assigned to one of the two classes: interacting or non-interacting residues. Consequently, the problem may be solved using statistical and machine learning techniques, such as neural networks (Ofran and Rost, 2003b; Zhou and Shan, 2001) or Support Vector Machines (Bock and Gough, 2001; Yan et al., 2004).

A clear definition of interacting residues is obviously required in order to predict whether a given amino acid residue is involved in protein-protein interactions. However, many alternative definitions are being used in the field. As the definition of an interaction site varies from one prediction method to another, it becomes difficult to directly compare their performance.

2.1 Commonly used definitions

If available, high resolution structural data readily provides a basis for atom or residue based definition of interaction sites. In fact, prediction methods discussed in this chapter primarily use information from resolved protein complexes to define the positive ("interacting") and negative ("non-interacting") classes. Protein quaternary structures are typically resolved by X-ray crystallography, and less frequently by NMR-spectroscopy or other techniques (Protein Data Bank, PDB - http://www.pdb.org/). While providing a high resolution structure, crystallographic data often remains inconclusive regarding the nature of the observed intermolecular contacts between protein chains. In particular, some of the observed contacts (and the resulting putative interaction interfaces) may be the result of crystal packing, rather than representing biologically relevant interactions.

A number of methods have been introduced to facilitate the process of filtering out crystal packing artefacts. Here, we used the approach adopted by the PISA server (http://www.ebi.ac.uk/msd-srv/prot_int/pistart.html). PISA discriminates crystal packing contacts from the functional protein–protein interaction using the size of solvent exposed area buried during association, as well as the number of residues constituting the interface, the number of salt and disulphide bridges at the interface, and the difference in approximate solvation energy upon complex formation.(Henrick and Thornton, 1998; Krissinel and Henrick, 2007)

Two different approaches are commonly used to define an interaction site based on 3D structural data: (i) interatomic distance and (ii) change in accessible surface area (ASA) upon complex formation. Following the first approach, interaction sites can be defined based on the distance between non-hydrogen atoms of different protein chains. For example, distance cutoffs of 4Å (Bordner and Abagyan, 2005); 4.5Å (Hamer et al., 2010); 5Å (Chen and Zhou, 2005); or 6Å (Ofran and Rost, 2003b) are used. This way of defining interaction sites is likely to miss some interchain contacts when water molecules are involved. A polar solvent, such as water, may bridge the interaction between two charged groups of amino acids that are too far apart to form a direct hydrogen bond.(Janin, 1999) In this regard, Neuvirth *et al.* introduced the Connolly interface index (CII) that is computed for circles of radius 10 Å around anchoring dots on the surface of monomeric structures. Atoms with CII above certain threshold are assigned to be interaction sites.(Neuvirth et al., 2004)

The second approach defines an interaction interface by using the concept of solvent accessibility or ASA. Specifically, ASA or the solvent accessibility of an amino acid residue in an unbound protein chain is contrasted with the corresponding ASA value for the same residue in a complex. Residues with a significant difference in ASA between the isolated chain and complex structures are then classified as "interacting". The following cutoffs for ASA were used: the loss of > 99% ASA for a given atom (Bradford and Westhead, 2005); a residue loses > 1Å^2 ASA in the complex (Chen and Jeong, 2009; Jones and Thornton, 1995; Liang et al., 2006); a residue ASA change by more than 20Å^2 (Kufareva et al., 2007); relative solvent accessibility (RSA) of a given residue decreased by more than 4% and its ASA decreased greater than 5Å^2 (Porollo and Meller, 2007). The latter definition uses relative ASA to address the considerable difference in size of amino acids, e.g. between glycine and tryptophan.

Both approaches require high resolution structural data. However, the interatomic distance based approach seems to be more sensitive to problems with missing atoms or atoms with multiple occupancies. Table 1 illustrates the difference in the protein interface recognition resulting from alternative definitions. As can be seen from the table, the same protein quaternary structure may yield different subsets of residues deemed to be interaction sites, therefore leading to different prediction models and their reported performances.

In what follows, we will refer to protein interfaces derived using our own ASA-based definition, dRSA > 4% and dASA > 5Å^2 (Porollo and Meller, 2007), unless stated otherwise. This definition takes into account both relative and absolute change in ASA, and it attempts to filter out noise related to variation in RSA observed in structures resolved under different conditions, or for closely related homologs.

Definition	Chain	Residues at the interface	Interface ASA, Å^2
dASA > 1Å^2	I	Y35 T41 C42 H57 C58 D60 R61 N95 T96 D97 D98 V99 A99A L143 L151 W172 T175 C191 Q192 G193 S195 T213 S214 F215 V216 S217 R217A L218 K224	830
	E	I18 I19 L20 I21 R22 C23 A24 M25 L26 N27 P29 R31 E46 G47 S48 C49 A52 C53 F54	994
dRSA > 4% and dASA > 5Å^2	I	Y35 T41 H57 D60 R61 T96 D97 V99 A99A L143 L151 W172 T175 C191 Q192 G193 S195 S214 F215 V216 S217 R217A L218	810
	E	I18 I19 L20 I21 R22 C23 A24 M25 L26 N27 P29 R31 E46 G47 S48 C49 A52 F54	989
dASA > 20Å^2	I	Y35 H57 R61 T96 D97 V99 W172 Q192 S195 F215 V216 R217A L218	692
	E	I19 L20 I21 R22 C23 A24 M25 L26 N27 P29 R31 E46 S48 C49	938

Table 1. The effects of using alternative definitions of protein interaction interfaces for a specific hetero-dimeric complex (PDB ID 1fle); dASA is the total loss of ASA for a given protein chain upon complex formation.

It should be noted that information on protein interaction sites may be also derived from the alanine scanning mutagenesis (ASM). Systematic replacement of the residues at the protein interface with alanine enables the evaluation of individual contribution of each interaction site to the binding energy. In this regard, the Alanine Scanning Energetics database (ASEdb, http://www.asedb.org/) provides ASM data on a number of protein-protein, as well as on some protein-DNA and protein-ligand interactions (Thorn and Bogan, 2001)

However, ASM approach is very costly and laborious, thus considerably limiting the number of comprehensively studied proteins. A protein interface needs to be approximately defined beforehand to limit the number of alanine mutants to evaluate. Results of ASM may not necessarily indicate the contribution to the binding energy, as some alanine mutants may cause an adverse protein conformational change and therefore indirectly decrease the efficacy of the protein-protein binding. Moreover, some protein-protein interactions are allosterically regulated, and ASM may not reflect the actual driving forces for a given protein complex. Nevertheless, such data is of great value and may be used as an additional validation of prediction methods. For example, it was used to evaluate ability of the methods ISIS (Ofran and Rost, 2007) and APIS (Xia et al., 2010) to identify hot spots.

2.2 Mapping interaction sites

Methods that do not require information about the interacting partner(s) are the primary focus of this chapter. These methods aim at the recognition of either individual residues, surface patches, or whole interaction interfaces using only sequence, structure and other information about an individual target protein, assuming that it is involved in some sufficiently stable interactions.

In light of the above, an important part of defining the residues as interaction sites is to retrieve as much information as possible on physical interactions for a given protein. Published studies on methods for the prediction of protein-protein interaction quite often ignore the fact that most proteins have multiple interaction partners that are mediated by alternative or overlapping interfaces. Therefore, using just one particular complex to identify the interaction interface and to derive the corresponding definition of the positive class, while ignoring all other complexes and interactions involving the same target protein chain (or its close homolog), may result in highly biased estimates of both false positive and false negative rates.

With the significant growth of structural data, the problem can be addressed by taking into account interaction sites from alternative complexes that contain the same protein chain or its close homologs. Interaction sites identified in such homologs can be mapped to a representative sequence in order to enable more sensitive prediction and perform its fair accuracy evaluation. Figure 1 illustrates this issue for two proteins resolved in complexes with different partners.

The protein shown in the left panel, caspase-9, utilizes overlapping interfaces for homo-oligomerization (PDB ID 1jxq), and for its interaction with ecotin (PDB ID 1nw9). However, the former protein-protein interaction involves many more residues than the latter interaction (affected ASA 1954Å^2 and 1019Å^2, respectively). If the definition of the positive ("interacting") class in caspase-9 were to be derived from the complex with ecotin (1nw9), the accuracy of any method predicting correctly also the more extensive interface would have been wrongly underestimated. This problem can be addressed by mapping the interface from the homooligomer into the target structure, leading to the union of homo-dimerization and caspase-9/ecotin interfaces to be taken as the true positive class.

The second example on the right illustrates the mapping of the known interfaces into the beta subunit of *E. coli* DNA polymerase III. In addition to homodimerization interface (PDB ID 2pol), physical interactions with the delta subunit of the gamma complex (PDB IDs 1jqj, 1jql) and DNA polymerase Pol IV (PDB ID 1unn) are mapped. Again, without this additional mapping step, prediction of these alternative interfaces would be considered as false positives during the evaluation process.

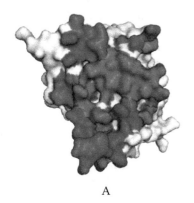 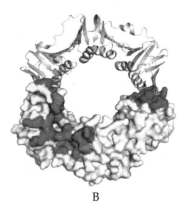

A   B

Fig. 1. Mapping interfaces from alternative protein complexes: A. Interaction interfaces in caspase-9, derived from the complex with ecotin (PDB ID 1nw9, chains B-A, shown in red) and caspase-9 homooligomer (PDB ID 1jxq, chains A-B), which includes both red and blue patches; B. Interaction interfaces mapped into DNA Pol III from the homodimer of the beta subunit of DNA Pol III (PDB ID 2pol, blue), delta subunit (PDB IDs 1jqj and 1jql, red), and DNA Pol IV (PDB ID 1unn, yellow), with the overlap of the latter two shown in magenta. Interfaces identified by using the SPPIDER server (http://sppider.cchmc.org/) and mapped into the target structure by using POLYVIEW-3D (http://polyview.cchmc.org/polyview3d.html).

The mapping, though, needs to be performed carefully, keeping in mind some important caveats. Sequence homology-based approach assumes that similar protein sequences adopt the same 3D fold and carry the same function, which is not always true. For example, paralogs may evolve to have distinct interaction partners and therefore perform different functions while having high sequence homology. Mapping interaction sites from such homologs might then result in incorrect expansion of the positive class to include patches utilized by other proteins with sequence similarity but distinct functions. In this context, one should comment that many methods for the prediction of interaction sites incorporate information about evolutionary profiles of protein families (e.g., obtained using PSI-BLAST to generate PSSM (Altschul et al., 1997)). Therefore, at least in some cases such methods arguably identify sites with a propensity to interact within the whole family, rather than just for the target protein.

Interactions specific to only some (or even only one) family members may require the identification of distinct interaction patches, rather than considering the problem of predicting the union of alternative interaction interfaces. Thus, mapping interaction interfaces might not be appropriate for evaluation of methods that attempt to predict such individual interaction patches. On the other hand, if ANY interaction patch that corresponds to a stable protein complex is to be found, then the union of all known interfaces constitutes the best approximation of the positive class and should be used for evaluation of the overall accuracy. As indicated above, this issue is often ignored altogether, even though it highlights the difficulty with a proper definition of a classification problem that best captures biologically relevant information while providing sufficiently "accurate" predictions.

Conversely, some protein domains with conserved 3D structure and specific function may be very divergent in terms of amino acid sequence, and only structure alignment might be able to detect such distant similarity. For example, PB1 domain displays low sequence homology between proteins, but it has a highly conserved secondary structure pattern and the overall 3D fold.(Lamark et al., 2003) While having just a few conserved residues playing a role of hot spots, this domain is widely utilized in various biological systems for interactions between the PB1-containing proteins to conduct cell signaling.(Moscat et al., 2006)

A PDB-wide structure alignment remains a computationally challenging task when it comes to a large protein set compiled for training or benchmarking a method for protein-protein interaction prediction. However, some current efforts, including for example the Dali database (http://ekhidna.biocenter.helsinki.fi/dali/start) (Holm et al., 2008), provide valuable resources in this regard. There have been also a number of studies published on the structure-based mapping of interaction sites, utilizing different schemes of hit weighting and homology recognition.(Albou et al., 2011; Oldfield, 2002; Park et al., 2001; Xu and Dunbrack, 2011)

However, it remains to be seen how structure-based mapping methods can deal with situations when a protein undergoes a significant conformational change upon complex formation (e.g., in case of calmodulin), and a structure alignment is likely to fail to identify similarity between apo- and holo-forms. Most likely, the future methods will utilize a balanced combination of sequence- and structure-based homology in order to more accurately map interaction sites from the known physical interactions. In this work, in order to test the effects of mapping interaction sites from multiple resolved complexes, we used a sequence homology-based mapping with conservative thresholds for homology hits: 70 or 90% of sequence identity. The interaction sites mapping process was automated through the SCORPPION web-server (http://scorppion.cchmc.org/).

3. Types of protein complexes

Biological diversity is very well represented at molecular level, in particular showing broad versatility in protein-protein interactions. Protein complexes can be classified into a number of broad categories, for example as homo- and hetero-oligomers; transient and obligatory (permanent), rigid and flexible complexes. Homo-oligomers are complexes consisting of two or more protein chains with identical amino acid sequence. Accordingly, assemblies of chains with different sequences are hetero-oligomeres. The number of chains participating in the assembly dictates the distinction on dimers, trimers, tetramers, and so forth.

Obligatory complexes (sometimes called obligomers) are considered to be protein assemblies that perform function only in the coupled state, whereas transient complexes are formed by proteins that were found to exists as monomers and to function separately as well. Rigid complexes may be considered as products of interaction between stable rigid-body domains. Flexible complexes, on the other hand, are formed when one or more constituting proteins undergo significant conformational changes.

Systematic analysis of the known protein complexes by several studies resulted in a number of observations that have significantly influenced the field of protein-protein interaction sites prediction. Ofran and Rost suggested that there are at least 6 types of contacts in proteins that display distinct amino acids compositions and contact preferences.(Ofran and Rost, 2003a) Thus, methods utilizing statistical contact propensities in their prediction models have to take into account different types of interactions. Another study found that even within a single interface the composition of amino acids varies depending on where the interacting amino acids are located, in the core of the interface or at its rim.(Chakrabarti and Janin, 2002)

A closer look at transient complexes was presented in (Nooren and Thornton, 2003). The study distinguished "weak" and "strong" homodimers, and it found that weak transient homodimers demonstrate smaller, more planar and polar interfaces compared to permanent homodimers, whereas strong transient homodimers undergo large conformational changes upon complex formation, and demonstrate larger, less planar, and more hydrophobic interfaces. Interestingly, only weak transient homodimers were found to have residues at interfaces more conserved than other surface residues, whereas other proteins with different oligomeric states showed no pronounced amino acid conservation.

These findings were further supported by the study on a larger set of protein complexes.(Caffrey et al., 2004) Comparing the conservation scores derived from multiple sequence alignments to orthologs vs. paralogs, the study demonstrated that residues at the interfaces are rarely more conserved than other residues on the protein surface. This observation implies that prediction models solely based on evolutionary profiles are likely to have limited overall accuracy.

Another large scale study has recently reported the results of PDB-wide analysis of protein-protein interactions. Both sequence and structure based characteristics of protein interfaces were characterized, with special focus on proteins with multiple interaction partners.(Kim et al., 2006) This analysis showed that, while there are ancient interfaces conserved across archea, bacteria, and eukaryotes (attributed primarily to symmetric homodimers), by and large interfaces are not conserved and vary in shape and amino acid composition due to broad diversity of interactions and interaction partners. The suggested classification introduced as many as 6000 different types of interfaces that are available for search and matching from the SCOPPI database (http://www.scoppi.org/).

4. Benchmarks of protein complexes

Benchmarks specifically designed for the training and evaluation of methods for the recognition of protein-protein interaction sites are critical for further progress in the field. Such benchmarks should allow an unbiased and fair evaluation of prediction methods. Consequently, benchmark sets used for comparison of different methods should comprise a diverse representative set of protein-protein interactions and contain no redundancy to the training sets used by individual methods.

The uncertainty of the negative class assignment further complicates the choice of appropriate benchmarks. Designing a dataset that includes only carefully curated and well-studied proteins, or their domains, with all known physical interactions mapped, may result in a very limited number of data points for training and validation. As a more feasible alternative one could consider assembling several diverse and non-redundant training and validation data sets that include complexes of different type and are characterized by some level of completeness of information regarding interactions and interaction sites.

As a result of these difficulties, there is no established gold standard in the field. Most of the published methods refer to their own compilation of protein complexes derived from PDB. Here, we consider three protein sets used in the literature. The first compilation of protein complexes is a benchmark set for protein-protein docking, current version 3.(Hwang et al., 2008) For this set, proteins in bound and unbound state were retrieved from PDB in a semi-automated manner. Current version contains the total of 124 test cases; among those 88 are rigid-body cases, 19 of medium difficulty, and 17 difficult cases, which are classified by the degree of conformational change at the interface upon complex formation.

While the primary purpose of Hwang *et al.* benchmark was to evaluate the protein docking methods, many protein interface prediction methods used it for their own and comparative evaluation.(de Vries and Bonvin, 2011; de Vries et al., 2006; Fiorucci and Zacharias, 2010; Guharoy and Chakrabarti, 2010; Li et al., 2008; Liu and Zhou, 2009; Qin and Zhou, 2007; Zhou and Qin, 2007) However, a thorough analysis of this benchmark set led us to conclusion that it is not suitable for evaluation of the methods predicting protein-protein interaction sites. For example, it contains 25 antibody-antigen cases (PDB IDs: 1fc2, 1ahw, 1bvk, 1dqj, 1e6j, 1jps, 1mlc, 1vfb, 1wej, 2fd6, 2i25, 2vis, 1bj1, 1fsk, 1i9r, 1iqd, 1k4c, 1kxq, 1nca, 1nsn, 1qfw, 2jel, 1bgx, 1e4k, 2hmi), which are asymmetrical functional protein-protein interactions, i.e. while one partner (in general: antibody, protease, or major histocompatibility complex) is evolved to bind its substrate, the second partner is not (except for the protease inhibitors).

Therefore, all antibody-antigen complexes were removed from the set. In addition, protein chains no longer available in PDB (PDBID_ChainID: 1cd8_B, 1ml0_B, 2pab_C, 2pab_D, 2viu_C, 2viu_E, 1aly_B, 1aly_C, 1jb1_B, 1jb1_C), difficult to interpret in terms of protein chains (1hia_A, 1hia_B, 1n8o_B, 1n8o_C) or too short (1n8o_A, 1k74_B, 1mzn_B, 1zgy_B) were removed. Finally, before using this benchmark set for evaluation of protein interface prediction methods, redundant chains were also removed.

The second benchmark set represents 85 cases of proteins found in PDB both in bound and unbound state.(Albou et al., 2009) No complexes with asymmetrical function are included, such as antibody-antigen cases and others listed above. This set represents diverse protein-protein interactions and allows the evaluators to estimate the role of conformational change on the accuracy of the methods, when predictions using bound structures *versus* unbound are compared. However, the set contains two cases, when only α-carbon coordinates are available (PDBID_ChainID: 3dpa_A and 2tld_I). These cases may be challenging to prediction methods that rely on high resolution data with all atoms resolved.

The last benchmark set to be used in this work is the control set of the SPPIDER method.(Porollo and Meller, 2007) It was compiled based on the protein complexes deposited in PDB after the compilation of the training set for the same prediction method. This manually curated and non-redundant (to the training set and within itself) set includes 149 protein chains, deemed to be sufficiently diverse and representative enough to be used for cross-validation studies. The only update to the set involved replacing the chain 1r72_A by 1xcb_A, as the PDB entry 1xcb now supersedes 1r72. In what follows, this set is referred to as SPPIDER149.

Table 2 and Figure 2 summarize the three datasets described above, after removing problematic cases from the first set, and redundant proteins from the first two sets. Redundancy was defined in terms of sequence homology: BLAST e-value < 0.001 when the alignment covers at least 70% of the query sequence (derived from the ATOM section of a PDB file). 150 chains derived from complexes in the first set and 78 chains in the second set were found non-redundant, and these (sub-) sets will be referred to as Hwang150B and Albou78B, respectively. The corresponding sets of chains that were retrieved from their unbound structures will be referred to as Hwang150U and Albou78U, respectively.

Dataset	Total chains	Families	Domains
Hwang150B	150	42	107
Albou78B	78	16	44
SPPIDER149	149	76	75

Table 2. Protein families and domains represented in non-redundant chains of the three benchmark sets used in this work. Families and domains defined according to the Pfam database (http://pfam.sanger.ac.uk/) (Finn et al., 2008) and mapped using sequence based search as implemented in SCORPPION (http://scorppion.cchmc.org/).

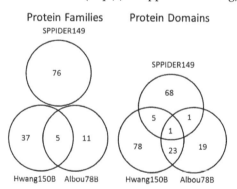

Fig. 2. Overlap between protein families (left) and domains (right) identified within the three benchmark sets used here.

Low to no overlap between the datasets discussed here is observed in terms of protein families and domains, suggesting a broad coverage of protein-protein interactions. This bodes well for estimates of the performance on different types of protein interfaces. On the other hand, the training sets for tested methods might partially overlap with the benchmark sets used here, leading to potentially overestimated accuracy.

Mapping of known interaction interfaces from alternative complexes was performed for each set using different approaches discussed in Section 2.2. Table 3 shows the number and fraction of interacting residues for each protein set. Interaction sites were derived from (i) asymmetric units defined in the original PDB files, (ii) biological units (BUs) as defined by Protein Quaternary Structure (PQS) database, and (iii) BUs as defined by the PISA database. In addition, interaction sites were mapped from the PISA-based BUs of their close homologs using sequence identity 90 and 70% as a cutoff (Table 4). The estimates of accuracy for methods compared here were overall quite similar, and only the results for the latter threshold are reported in the following sections of the chapter.

PDB also provides its own definition of biological units that differs from PISA.(Xu and Dunbrack, 2011) PDB defines biological units as separate models in the same PDB file. In addition, both PISA and PDB may rename chain labels starting from 'A' within each BU. This all makes it difficult sometimes to trace back the chains from the asymmetrical unit in automated manner. To be consistent, we will map interaction sites from BUs as defined by PISA. However, when no information can be mapped for a given chain, due to technical difficulties or inconsistency in BU definition, we will use a PDB-based asymmetric unit for the mapping of interaction sites.

Dataset	Total residues / On the surface	PDB-based, %	PQS-based, %	PISA-based, %
Hwang150B	31208 / 24687	19	21	19
Albou78B	17412 / 13375	16	16	15
SPPIDER149	25883 / 20885	29	28	28

Table 3. Summary of the benchmarks used in this work with regards to the total number of residues, residues on the surface, and percentage of the surface residues found to be at protein interfaces derived from the asymmetric unit (PDB-based), and biological units (PQS-based and PISA-based), respectively.

Dataset	Total / Surface residues	SI70	SI90
Hwang150B	31208 / 24687	10011	9674
Hwang150U	32471 / 24595	10201	9661
Albou78B	17412 / 13375	5819	5506
Albou78U	16838 / 12342	5572	5294
SPPIDER149	25883 / 20885	7863	7668

Table 4. Summary of the benchmarks used in this work with regards to the total number of residues, residues on the surface, and interacting residues on the surface mapped to representative protein chains using BUs derived from the PISA database and 70 or 90% sequence identity cutoffs (SI70 and SI90), respectively.

5. Prediction methods

All prediction methods can be broadly classified by the type of data they use as an input. Sequence-based methods rely on some combination of the following protein features: amino acid hydrophobicity, evolutionary profile (e.g., similarity scores or Shannon entropy), amino acid composition or propensity to be at the interface, predicted structural features (e.g., secondary structure, solvent accessibility, order/disorder region, etc.), or their derivatives like mean or weighted average over a sequence window.

The structure-based methods, on the other hand, also utilize features derived from a 3D protein structure, such as solvent accessibility and secondary structure states, local topology (e.g., protrusions and cavities), hydrophobic and polar surface patches, temperature or B-factors (for X-ray based structures), etc. In addition, there are a number of methods built using a consensus of the individual predictors with reportedly improved accuracy.(de Vries and Bonvin, 2011; Huang and Schroeder, 2008; Qin and Zhou, 2007) However, consensus-based methods are not discussed here in detail, as the goal is to evaluate the discriminating power of the underlying principal features for each representative method.

Described below are selected structure-based methods with at least somewhat orthogonal feature spaces that were available as web-servers at the time of data preparation for this work. Methods are listed in the order of the publication year of the original work.

Evolutionary trace (ET) method (Lichtarge et al., 1996) identifies evolutionary conserved residues and maps them onto a protein 3D structure. Conserved residues in the core of a protein are deemed to be structurally important, whereas those on the surface are assumed to be functionally important. The method starts from constructing a multiple sequence alignments, and partitions the aligned sequences into groups by using their mutual sequence similarity. For each group, a consensus sequence is defined highlighting the positions with invariant amino acids. Consensus sequences are further aligned to identify (i) conserved residues across the entire protein family; (ii) class-specific residues that are invariant in some groups; and (iii) neutral residues that are not preserved in any single sequence group. Conserved and class-specific residues are then mapped onto 3D structure. Clusters of such residues on the surface of a protein structure are predicted to be functional. The ET method is available at http://mammoth.bcm.tmc.edu/ETserver.html

ConSurf (Glaser et al., 2003) follows a similar approach by mapping the evolutionary conserved residues on 3D protein structure. The difference lies in computing the conservation scores that are relative with respect to other residues in a given protein. In addition, the outcome of the method is sensitive to the quality of multiple sequence alignment and to the overall length of a query sequence. For example, two 3D structures of the same protein, but with different sequence length representing its resolved part, may result in different location of the most conserved residues. The ConSurf method is available at http://consurf.tau.ac.il/, whereas its pre-computed results for the PDB deposited proteins are available from the ConSurfDB database (http://consurfdb.tau.ac.il/).

It should be noted that the two methods described above were not designed to identify specifically protein-protein interaction sites, but rather to reveal any functional residues, e.g. involved in protein-DNA or protein-ligand interactions. However, since the authors of these methods refer to identification of protein interfaces as examples in their original publications, we chose these methods to serve as a separate group of predictors that rely primarily on evolutionary information, and can be contrasted with structure-based methods.

PROMATE (Neuvirth et al., 2004) considers residues on the surface of a protein structure within 10Å circles around a given point. Spatially neighboring residues provide the following descriptors: (i) statistically derived chemical composition of binding sites, such as propensity of individual amino acids, atom types, pairs of amino acids, and collective chemical properties (positively and negatively charged, polar, hydrophobic, and aromatic residues); (ii) evolutionary conservation in terms of diagonal elements of the PSI-BLAST-derived position specific scoring matrix (PSSM); (iii) distance in the sequence between residues in the circle; (iv) secondary structure states, including extent of the loops. Additionally, temperature factors (B-factors) and bound waters are incorporated into the model whenever available. These descriptors are combined to yield a cumulative score that allows the circles to be classified as Interface, Non-interface, or Boundary. The neighboring circles are further clustered to define predicted interface patches. PROMATE is available at http://bioinfo.weizmann.ac.il/promate/

Cons-PPISP (Chen and Zhou, 2005) employs a consensus of neural networks trained on (i) the position specific similarity scores derived from the PSI-BLAST multiple sequence alignment and (ii) observed (in the target structure provided as input) solvent accessibility for spatially neighboring residues. In addition to validation on crystal structures, cons-PPISP was shown to provide accurate prediction of protein interfaces for a set of 8 NMR-derived complexes, non-redundant to its training set. The web-server is available at http://pipe.scs.fsu.edu/ppisp.html

WHISCY (de Vries et al., 2006) introduces prediction scores that are based on evolutionary and structural information. Conservation of residues on the surface is computed as the corrected sum of similarity scores between amino acids at a given position by pairwise comparison of a query sequence and sequences from a multiple alignment. Similarity scores are taken from the Dayhoff mutation matrix. ASA is the only structural information used. WHISCY is available at http://nmr.chem.uu.nl/Software/whiscy/index.html

PIER (Kufareva et al., 2007) combines (i) statistically derived interatomic contact potentials, (ii) physical descriptors, such as observed solvent accessibility for separate atomic groups within amino acids, and (iii) sequence alignment based features, in particular, three different conservation scores (frequency-based, similarity matrix-based, and entropy-based). The surface of a protein structure is divided on individual patches. Using the descriptors listed above, all patches obtain a set of cumulative scores that further fed to a partial least squares (PLS) based regression model to predict protein interfaces. Since the PIER scoring heavily relies on atomic resolution, it may have difficulties with incomplete or of low resolution crystal structures. The corresponding prediction server is available at http://abagyan.ucsd.edu/PIER/

SPPIDER (Porollo and Meller, 2007) is a neural network-based method that uses the difference between predicted from sequence and observed in an unbound structure RSA of amino acid residue as a novel and highly informative signal of interaction sites. Solvent accessibility prediction methods tend to predict residues at protein interfaces as buried, which is consistent with the fact that they are indeed getting buried upon complex formation, even though they are exposed in an unbound structure. The SABLE (Adamczak et al., 2004) method for RSA prediction was used to generate the input for SPPIDER. Additional features include averaged over spatially neighboring residues of (i) RSA predicted by SABLE; (ii) evolutionary conservation (in terms of Shannon entropy) of amino acid type, charge, hydrophobicity, and side chain size; (iii) amino acid contact numbers and hydropathy constants. The server is available at http://sppider.cchmc.org/

6. Evaluation

6.1 Accuracy measures

Prediction of protein interaction sites is typically cast as a classification problem. Therefore, a number of commonly used measures for two class classification problems can be employed to evaluate the accuracy. These measures include the two-class classification accuracy (Q_2), recall or sensitivity (R), and precision or specificity (P), all expressed as percentage.

$$Q_2 = \frac{TP+TN}{TP+TN+FP+FN} \cdot 100\% \tag{1}$$

$$R = \frac{TP}{TP+FN} \cdot 100\% \tag{2}$$

$$P = \frac{TP}{TP+FP} \cdot 100\%, \tag{3}$$

where TP are true positives, TN – true negatives, FP – false positives, and FN – false negatives.

However, since the number of interaction sites can be much smaller than the number of non-interacting residues, the classification problem at hand may be highly unbalanced. As a result, the measures listed above may be difficult to interpret and compare for different benchmarks. For example, with 90% of data points assigned to the negative class, a baseline classifier that predicts all residues as non-interacting achieves numerically high 90% classification accuracy. To provide a measure that balances sensitivity and specificity of predictions, the Matthews correlation coefficient (MCC) is often used (4) together with other measures. MCC ranges from -1, indicating an inverse prediction, through 0, which corresponds to a random classifier, to +1 for perfect prediction.

$$MCC = \frac{TP \cdot TN - FP \cdot FN}{\sqrt{(TP+FN)(TP+FP)(TN+FP)(TN+FN)}} \tag{4}$$

Other measures that can be used to assess and compare classification methods are area under the receiver operating characteristic (ROC) curve and F-measure.

6.2 Performance of selected methods

The performance of several representative methods discussed in the previous section is assessed here in order to compare more systematically individual methods, and to quantify the effects of mapping additional interaction interfaces and using truly unbound structures. Different aspects of the performance are evaluated using benchmark datasets described in section 4 (SPPIDER149, Hwang150B/U, and Albou78B/U).

For all evaluations, only residues with RSA of at least 5% were considered, thus excluding all fully buried residues in a given protein conformation. For methods providing a real valued score, multiple thresholds were tested as a basis for projection into two classes. The results for the best performing threshold in terms of MCC are reported in Tables 5 through 9. The following values were found to be optimal for each method: ET with residues being ranked 1 (out of top 1, 5, and 10 rankings evaluated), ConSurf with evolutionary rank ≥ 5 (5,

7, 9 evaluated), WHISCY with threshold ≥ 0 (0, 0.18 evaluated), PIER with threshold ≥ 15 (0, 15, 30 evaluated), and SPPIDER with threshold ≥ 0.3 (0.3, 0.5, 0.7 evaluated).

Method	SPPIDER149	Hwang150B	Albou78B
ET	0.08	0.04	0.01
ConSurf	0.12	0.07	0.02
PROMATE	0.10	0.10	0.09
Cons-PPISP	0.30	0.22	0.17
WHISCY	0.19	0.11	0.08
PIER	0.37	0.27	0.22
SPPIDER	0.41	0.28	0.20

Table 5. The performance of representative methods measured using MCC on three different sets, with only the original PDB complexes used to define the positive class.

As can be seen from Table 5, the overall accuracy of the methods evaluated here is rather limited. The two best performing methods, i.e., PIER and SPPIDER achieve MCC of about 0.4 for SPPIDER149 set, 0.3 for Hwang150B, and 0.2 for Albou78B, respectively. Similar relative drop in accuracy is also observed for other methods, indicating that Hwang150B and Albou78B sets are more difficult to classify. This can be explained in part due to a larger imbalance between positive and negative classes in these benchmarks, especially in the Albou78B dataset (see Table 3).

Method	SPPIDER149		Hwang150B		Albou78B	
	R, %	P, %	R, %	P, %	R, %	P, %
ET	7.03	43.92	3.99	28.18	2.84	17.55
	6.39	51.73	3.44	48.89	3.57	60.60
ConSurf	65.27	32.87	61.42	22.18	55.17	16.40
	63.00	40.97	55.91	41.07	53.19	41.66
PROMATE	3.91	60.71	4.06	48.98	3.69	43.43
	3.22	64.29	2.56	63.78	1.85	58.29
Cons-PPISP	33.40	60.59	26.25	42.42	22.46	34.80
	29.39	69.12	19.35	67.62	15.33	64.40
WHISCY	29.38	45.42	21.15	29.77	20.49	21.83
	26.66	54.32	17.21	51.71	16.53	48.38
PIER	61.10	52.62	49.66	37.46	45.43	30.61
	54.38	60.31	38.64	60.86	31.20	56.99
SPPIDER	80.36	48.47	63.15	34.11	56.22	26.49
	73.14	56.81	53.04	59.82	43.48	55.52

Table 6. The effect of mapping interaction sites from homologous protein complexes on recall (R) and precision (P): the first line in each row shows R and P using original PDB complexes, whereas the second line indicates accuracy derived after mapping interaction sites using PISA BUs and homologous chains with 70% sequence identity.

It should be noted that due to a sufficiently large number of data points (surface residues, see Table 3) included in each benchmarks, each of the correlation coefficients reported above is statistically significantly different from 0 with a p-value < 0.05. Nevertheless, practical applicability of methods that achieve correlations of 0.2 and lower has to be judged using also other criteria and specific examples. In particular, evolutionary methods achieve very limited accuracy in this test, even though they may provide biologically valuable insights, as discussed later.

The effects of mapping interaction residues from alternative complexes are illustrated in Table 6 using measures of sensitivity and specificity. The accuracy using the assignment of the positive class (interaction sites) derived from the original complexes is compared to the accuracy obtained re-labeling the "non-interacting" residues in mapped interfaces as "interacting" sites. Due to largely canceling effects of decreased rates of false positives and increased rates of false negatives, the mapping of interaction sites from PISA biological units does not affect significantly the performance of the prediction methods in terms of MCC, although a systematic small drop in accuracy is observed in most cases (data not shown).

However, as can be seen from Table 6, all methods show a drop in recall while precision improves when mapping is applied. These results also allow one to trace how the trade-off between sensitivity and specificity was optimized for different methods. One striking example is ConSurf vs. ET comparison. On the other hand, most structure-based methods provide fairly well balanced predictions. In particular, precision improves considerably, with only a relatively limited drop in recall for the best performing SPPIDER method, followed by PIER and Cons-PPISP. The observed ranking could reflect the fact that SPPIDER was trained (although on a different set without homology to SPPIDER149 set) using mapping from alternative complexes to reduce the noise in learning from data and to provide a more balanced classification problem.

Method	Hwang150B SI70	Hwang150U SI70	Albou78B SI70	Albou78U SI70
ET	0.03	0.00	0.06	0.08
ConSurf	0.03	0.05	0.00	0.00
PROMATE	0.06	0.05	0.04	0.01
Cons-PPISP	0.20	0.18	0.14	0.13
WHISCY	0.09	0.16	0.06	0.08
PIER	0.24	0.23	0.15	0.11
SPPIDER	0.29	0.29	0.17	0.14

Table 7. The effect of the bound versus unbound state of the protein structures used as an input in terms of MCC. In all cases, interacting residues were mapped using homology to PISA BUs with 70% sequence identity.

The impact of conformational change and the use of structures in bound as opposed to unbound state as an input is assessed in Table 7. For that purpose, the overall accuracy in terms of MCC is compared using two pairs of sets of bound (taken from a complex by simply ignoring other chains) and truly unbound structures: Hwang150B vs. Hwang150U and Albou78B vs. Albou78U, respectively. Slight decrease in performance is observed for all but one structure-based method, the exception being WHISCY. The latter method starts from a low level, though. In addition, the WHISCY server did not generate results for a number of more difficult cases, suggesting that this trend might not hold on other data sets.

While the drop in accuracy is limited for other methods tested, it should be emphasized that benchmarks included here sample relatively small conformational changes due to induced fit. Therefore, further systematic studies will be required to better delineate the range of applicability of structure-based method for the recognition of protein interaction sites.

Table 8 demonstrates how the performance estimates can be inflated when accuracy measures are computed based on all residues as opposed to computing the accuracy for each protein and then averaging over all proteins. Per protein averages, together with measures of variance (here we report standard deviations), allow one to assess better the range of expected accuracies for individual proteins. As can be seen from Table 8, the observed large standard deviations suggest large protein to protein variation and indicate that all tested methods fail dramatically for at least some proteins. It should be also noted that using per protein measures PIER is the top performing method, followed by SPPIDER and Cons-PPISP.

Method	MCC	Q_2, %	R, %	P, %
ET	0.06±0.12	65.64±17.83	9.60±16.07	29.35±35.01
	0.08	71.21	7.03	43.92
ConSurf	0.12±0.15	54.44±8.16	64.54±14.06	39.61±22.69
	0.12	52.80	65.27	32.87
PROMATE	0.07±0.13	64.01±19.63	5.72±8.93	28.30±39.31
	0.10	71.16	3.91	60.71
Cons-PPISP	0.23±0.23	69.52±13.23	37.50±22.11	58.99±29.71
	0.30	74.15	33.40	60.59
WHISCY	0.14±0.20	67.39±13.14	26.58±19.79	42.64±28.00
	0.19	71.03	29.38	45.42
PIER	0.30±0.23	71.18±11.47	58.73±24.80	55.22±27.09
	0.37	72.54	61.10	52.62
SPPIDER	0.29±0.20	66.94±13.82	79.16±24.79	49.19±21.69
	0.41	69.39	80.36	48.47

Table 8. Comparison of the accuracy measures calculated per residue by merging data from all chains (the bottom line in each row) and per protein averages and standard deviations (the top line in each row), using the SPPIDER149 set (similar effect is observed on other benchmarks).

Not all web-based implementations of the methods are reliable. While requesting and retrieving predictions from the evaluated servers, we faced multiple failures. Table 9 illustrates the reliability of the corresponding servers from the user's point of view by presenting the numbers of proteins failed to be processes within each benchmark set. The most reliable web-servers appear to be PIER and SPPIDER, whereas ET, ConSurf, and WHISCY are quite unreliable, which makes it more difficult to evaluate servers on a large scale.

Prediction methods that seemingly perform poorly according to some evaluation criteria can still greatly facilitate further experimental and computational studies on protein interactions. One might argue that predicting possible interaction interfaces should be directed at the recognition of the sites that contribute most to the binding energy. Such hot spots also represent the most natural target for further validation, e.g., using mutagenesis, or as targets for therapeutics.

Method	SPPIDER149	Hwang150B	Hwang150U	Albou78B	Albou78U
ET	14	14	28	8	8
ConSurf	21	12	13	4	4
PROMATE	1	3	8	3	12
Cons-PPISP	7	3	1	0	4
WHISCY	34	15	17	8	9
PIER	0	0	0	0	0
SPPIDER	0	0	0	0	0

Table 9. The number of proteins not included in each benchmark due to problems with the retrieval of the results as an indicator of the reliability of web-servers tested.

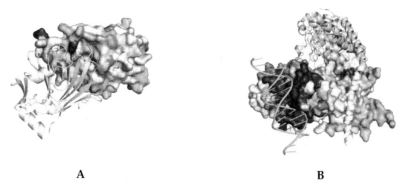

A   **B**

Fig. 3. Examples of protein interaction sites predicted by ConSurf: **A.** A successful identification of the protein interface for the homodimer of phosphoglucose isomerase (PDB ID 1qxr, chain A); **B.** A multi-interface protein (CSL transcription factor) illustrates possible confusion with DNA binding sites that are the most slowly evolving residues at the surface of the protein in this case (PDB ID 2fo1, chain A). Residues in magenta are the most conserved, whereas variable sites are colored using cyan (see the ConSurf documentation).

In this context, a special note needs to be made on the performance of evolutionary methods, such as ET and ConSurf. As we mentioned before, these methods were not designed specifically to predict protein-protein interaction sites, but rather to identify evolutionary conserved residues. Therefore, these methods may not able to discriminate between protein-protein, protein-ligand (e.g., co-factor or substrate), and protein-DNA/RNA binding sites. An example of such a case is shown in Figure 3.

On the other hand, highly conserved residues that are exposed on the surface of a protein are very likely functionally relevant, irrespective of the actual involvement in interaction. Despite all the limitations, evolutionary methods for the prediction of interaction sites have significantly contributed to the mapping of protein interactions and other functional annotations, see e.g., (Kniazeff et al., 2002; Shenoy et al., 2006) and (He et al., 2003; Lietha et al., 2007), for ET and ConSurf, respectively.

7. Discussion and conclusions

Protein-protein interactions are essential for enzymatic functions, signal transduction, cell cycle regulation and other fundamental biological processes. In addition to addressing the fundamental questions of molecular biology, identification of residues involved in protein-protein interactions has important medical relevance. Combined with recent advances in genome sequencing it facilitates delineating natural functional variants from pathological mutants, and conducting 'molecular diagnostics' as part of personalized medicine.(Su et al., 2011) Detailed structural information on thousands of protein complexes also stimulates growth in the field of rational drug design by providing a new class of targets that include known protein interaction interfaces.(White et al., 2008)

However, experimental identification and validation of a protein interface remains a challenging task, both in terms of labor and cost. Therefore, efforts to map and characterize protein interactions can considerably benefit from computational biology and structural bioinformatics. In particular, methods that integrate sequence and structure information achieved accuracies that are useful in selecting and prioritizing targets for mutagenesis and other experimental studies.

In this chapter, we reviewed state-of-the-art in the field of computational prediction of protein-protein interaction sites. We evaluated some representative methods using several published benchmarks of protein complexes. The overall accuracy of existing methods, in accord with other recent evaluations, was found to be limited (the Matthews correlation coefficient between the predicted and true class assignment of up to 0.4). Therefore, further concerted efforts will be required to improve state-of-the-art in the field. To that end, we discussed the need for standard definition of protein interaction sites, developing more comprehensive benchmark protein sets, and appropriate ways of measuring/reporting the accuracy of predictions.

We quantified the effects of taking into account multiple interaction interfaces and using as an input unbound structures that were resolved without interacting partners. Both of these issues are often ignored when evaluating the performance of interaction sites prediction methods. Yet, they are shown to impact significantly the estimates of performance. These two issues also highlight more fundamental difficulties with the definition of the negative class and current attempts to cast the problem in a computationally feasible way.

Casting the prediction of interaction sites in terms of a two-class classification problem requires that examples of the negative ("non-interacting") class be used for the training. With data points representing both "interacting" and "non-interacting" residues, a decision boundary separating the two classes can be optimized. These negative examples are defined in most cases by simply taking the complement of the positive class, i.e., all other (surface exposed) residues that are not known to be involved in interactions.

Consequently, without mapping known interfaces alternative complexes, residues within such interfaces are incorrectly regarded as "non-interacting". This could introduce problems in training, as misclassified vectors from the negative class may coincide with the bulk of the density for the positive class. One strategy to address this issue is to filter out such difficult cases. As an alternative, one could also consider one-class approaches, in which only the positive class examples are used to learn a predictor. On the other hand, if residues from multiple complexes are systematically mapped, as advocated here, the negative class assignment as a source of noise should be gradually reduced with the progress in experimental mapping of interaction sites.

Conformational changes upon complex formation pose another problem for the methods considered here. Protein flexibility and the induced fit effects upon complex formation are assumed to be limited. Obviously, this assumption does not hold in many instances of protein-protein interactions (and sometimes it breaks spectacularly, e.g., when the co-folding of otherwise disordered interacting domains occurs). Therefore, methods presented here are of limited applicability when large conformational changes or flexible domains are involved.

It should be also stressed that even a limited induced fit can pose significant challenges for structure-based methods. Simply ignoring all but one chain in a protein complex, and thus taking a *de facto* bound conformation as input, may lead to spurious effects in training and overly optimistic estimates of accuracy. For example, low B-factors of surface residues, which can be "locked" in a specific conformation by interactions with a co-factor, may not be a true signal of interaction sites (in many cases the opposite can actually be observed). Features that are capable of identifying interaction sites starting from a truly unbound structure should be emphasized.

Reliable identification of residues that participate in binding to other proteins can help direct and streamline mutagenesis and other experimental studies, and to facilitate efforts to map entire interactomes. It can also reduce the levels of false positives (by assessing compatibility between predicted interfaces), and false negatives (by helping identify novel interactions) observed for experimental approaches that are used to map protein interactions. Another promising application is protein docking, in which predicted interfaces can be used for evaluating and ranking potential complex structures (de Vries and Bonvin, 2011), in analogy to docking methods that utilize limited NMR data. (Dominguez et al., 2003; Kohlbache et al., 2001)

Further progress in the field will require new insights to overcome current limitations, as well as careful assessment of the accuracy in order to address possible biases in training and validation. Constant improvements in experimental techniques and a growing number of resolved macromolecular complexes, from which to learn better predictors, bode well for future efforts in this regard.

8. References

Adamczak, R., Porollo, A., and Meller, J. (2004). Accurate prediction of solvent accessibility using neural networks-based regression. Proteins 56, 753-767.

Albou, L. P., Poch, O., and Moras, D. (2011). M-ORBIS: mapping of molecular binding sites and surfaces. Nucleic Acids Res 39, 30-43.

Albou, L. P., Schwarz, B., Poch, O., Wurtz, J. M., and Moras, D. (2009). Defining and characterizing protein surface using alpha shapes. Proteins 76, 1-12.

Aloy, P., Pichaud, M., and Russell, R. B. (2005). Protein complexes: structure prediction challenges for the 21st century. Curr Opin Struct Biol *15*, 15-22.

Aloy, P., and Russell, R. B. (2004). Ten thousand interactions for the molecular biologist. Nat Biotechnol *22*, 1317-1321.

Altschul, S. F., Madden, T. L., Schaffer, A. A., Zhang, J., Zhang, Z., Miller, W., and Lipman, D. J. (1997). Gapped BLAST and PSI-BLAST: a new generation of protein database search programs. Nucleic Acids Res *25*, 3389-3402.

Bader, J. S., and Chant, J. (2006). Systems biology. When proteomes collide. Science *311*, 187-188.

Berman, H. M., Westbrook, J., Feng, Z., Gilliland, G., Bhat, T. N., Weissig, H., Shindyalov, I. N., and Bourne, P. E. (2000). The Protein Data Bank. Nucleic Acids Res *28*, 235-242.

Bock, J. R., and Gough, D. A. (2001). Predicting protein--protein interactions from primary structure. Bioinformatics *17*, 455-460.

Bordner, A. J., and Abagyan, R. (2005). Statistical analysis and prediction of protein-protein interfaces. Proteins *60*, 353-366.

Bradford, J. R., and Westhead, D. R. (2005). Improved prediction of protein-protein binding sites using a support vector machines approach. Bioinformatics *21*, 1487-1494.

Caffrey, D. R., Somaroo, S., Hughes, J. D., Mintseris, J., and Huang, E. S. (2004). Are protein-protein interfaces more conserved in sequence than the rest of the protein surface? Protein Sci *13*, 190-202.

Chakrabarti, P., and Janin, J. (2002). Dissecting protein-protein recognition sites. Proteins *47*, 334-343.

Chen, H., and Zhou, H. X. (2005). Prediction of interface residues in protein-protein complexes by a consensus neural network method: test against NMR data. Proteins *61*, 21-35.

Chen, X. W., and Jeong, J. C. (2009). Sequence-based prediction of protein interaction sites with an integrative method. Bioinformatics *25*, 585-591.

Chung, J. L., Wang, W., and Bourne, P. E. (2006). Exploiting sequence and structure homologs to identify protein-protein binding sites. Proteins *62*, 630-640.

de Vries, S. J., and Bonvin, A. M. (2011). CPORT: a consensus interface predictor and its performance in prediction-driven docking with HADDOCK. PLoS One *6*, e17695.

de Vries, S. J., and Bonvin, A. M. J. J. (2008). How proteins get in touch: Interface prediction in the study of biomolecular complexes. Curr Protein Pept Sc *9*, 394-406.

de Vries, S. J., van Dijk, A. D., and Bonvin, A. M. (2006). WHISCY: what information does surface conservation yield? Application to data-driven docking. Proteins *63*, 479-489.

Dominguez, C., Boelens, R., and Bonvin, A. M. (2003). HADDOCK: a protein-protein docking approach based on biochemical or biophysical information. J Am Chem Soc *125*, 1731-1737.

Fariselli, P., Pazos, F., Valencia, A., and Casadio, R. (2002). Prediction of protein--protein interaction sites in heterocomplexes with neural networks. Eur J Biochem *269*, 1356-1361.

Finn, R. D., Tate, J., Mistry, J., Coggill, P. C., Sammut, S. J., Hotz, H. R., Ceric, G., Forslund, K., Eddy, S. R., Sonnhammer, E. L., and Bateman, A. (2008). The Pfam protein families database. Nucleic Acids Res *36*, D281-288.

Fiorucci, S., and Zacharias, M. (2010). Prediction of protein-protein interaction sites using electrostatic desolvation profiles. Biophys J *98*, 1921-1930.

Fletcher, S., and Hamilton, A. D. (2007). Protein-protein interaction inhibitors: small molecules from screening techniques. Curr Top Med Chem *7*, 922-927.

Fry, D. C. (2006). Protein-protein interactions as targets for small molecule drug discovery. Biopolymers *84*, 535-552.

Gallet, X., Charloteaux, B., Thomas, A., and Brasseur, R. (2000). A fast method to predict protein interaction sites from sequences. J Mol Biol *302*, 917-926.

Glaser, F., Pupko, T., Paz, I., Bell, R. E., Bechor-Shental, D., Martz, E., and Ben-Tal, N. (2003). ConSurf: identification of functional regions in proteins by surface-mapping of phylogenetic information. Bioinformatics *19*, 163-164.

Guharoy, M., and Chakrabarti, P. (2010). Conserved residue clusters at protein-protein interfaces and their use in binding site identification. BMC Bioinformatics *11*, 286.

Hamer, R., Luo, Q., Armitage, J. P., Reinert, G., and Deane, C. M. (2010). i-Patch: interprotein contact prediction using local network information. Proteins *78*, 2781-2797.

Han, J. D., Dupuy, D., Bertin, N., Cusick, M. E., and Vidal, M. (2005). Effect of sampling on topology predictions of protein-protein interaction networks. Nat Biotechnol *23*, 839-844.

He, X. L., Bazan, J. F., McDermott, G., Park, J. B., Wang, K., Tessier-Lavigne, M., He, Z., and Garcia, K. C. (2003). Structure of the Nogo receptor ectodomain: a recognition module implicated in myelin inhibition. Neuron *38*, 177-185.

Henrick, K., and Thornton, J. M. (1998). PQS: a protein quaternary structure file server. Trends Biochem Sci *23*, 358-361.

Holm, L., Kaariainen, S., Rosenstrom, P., and Schenkel, A. (2008). Searching protein structure databases with DaliLite v.3. Bioinformatics *24*, 2780-2781.

Huang, B., and Schroeder, M. (2008). Using protein binding site prediction to improve protein docking. Gene *422*, 14-21.

Hwang, H., Pierce, B., Mintseris, J., Janin, J., and Weng, Z. (2008). Protein-protein docking benchmark version 3.0. Proteins *73*, 705-709.

Janin, J. (1999). Wet and dry interfaces: the role of solvent in protein-protein and protein-DNA recognition. Structure *7*, R277-279.

Jones, S., and Thornton, J. M. (1995). Protein-protein interactions: a review of protein dimer structures. Prog Biophys Mol Biol *63*, 31-65.

Jones, S., and Thornton, J. M. (1997). Analysis of protein-protein interaction sites using surface patches. J Mol Biol *272*, 121-132.

Kim, W. K., Henschel, A., Winter, C., and Schroeder, M. (2006). The many faces of protein-protein interactions: A compendium of interface geometry. PLoS Comput Biol *2*, e124.

Kniazeff, J., Galvez, T., Labesse, G., and Pin, J. P. (2002). No ligand binding in the GB2 subunit of the GABA(B) receptor is required for activation and allosteric interaction between the subunits. J Neurosci *22*, 7352-7361.

Kohlbache, O., Burchardt, A., Moll, A., Hildebrandt, A., Bayer, P., and Lenhof, H. P. (2001). Structure prediction of protein complexes by an NMR-based protein docking algorithm. J Biomol NMR *20*, 15-21.

Koike, A., and Takagi, T. (2004). Prediction of protein-protein interaction sites using support vector machines. Protein Eng Des Sel *17*, 165-173.

Krissinel, E., and Henrick, K. (2007). Inference of macromolecular assemblies from crystalline state. J Mol Biol 372, 774-797.

Kufareva, I., Budagyan, L., Raush, E., Totrov, M., and Abagyan, R. (2007). PIER: protein interface recognition for structural proteomics. Proteins 67, 400-417.

Lacapere, J. J., Pebay-Peyroula, E., Neumann, J. M., and Etchebest, C. (2007). Determining membrane protein structures: still a challenge! Trends Biochem Sci 32, 259-270.

Lamark, T., Perander, M., Outzen, H., Kristiansen, K., Overvatn, A., Michaelsen, E., Bjorkoy, G., and Johansen, T. (2003). Interaction codes within the family of mammalian Phox and Bem1p domain-containing proteins. J Biol Chem 278, 34568-34581.

Li, N., Sun, Z., and Jiang, F. (2008). Prediction of protein-protein binding site by using core interface residue and support vector machine. BMC Bioinformatics 9, 553.

Liang, S., Zhang, C., Liu, S., and Zhou, Y. (2006). Protein binding site prediction using an empirical scoring function. Nucleic Acids Res 34, 3698-3707.

Lichtarge, O., Bourne, H. R., and Cohen, F. E. (1996). An evolutionary trace method defines binding surfaces common to protein families. J Mol Biol 257, 342-358.

Lietha, D., Cai, X., Ceccarelli, D. F., Li, Y., Schaller, M. D., and Eck, M. J. (2007). Structural basis for the autoinhibition of focal adhesion kinase. Cell 129, 1177-1187.

Liu, R., and Zhou, Y. (2009). Using support vector machine combined with post-processing procedure to improve prediction of interface residues in transient complexes. Protein J 28, 369-374.

Ma, B., Elkayam, T., Wolfson, H., and Nussinov, R. (2003). Protein-protein interactions: structurally conserved residues distinguish between binding sites and exposed protein surfaces. Proc Natl Acad Sci U S A 100, 5772-5777.

Moscat, J., Diaz-Meco, M. T., Albert, A., and Campuzano, S. (2006). Cell signaling and function organized by PB1 domain interactions. Mol Cell 23, 631-640.

Neuvirth, H., Raz, R., and Schreiber, G. (2004). ProMate: a structure based prediction program to identify the location of protein-protein binding sites. J Mol Biol 338, 181-199.

Nooren, I. M., and Thornton, J. M. (2003). Structural characterisation and functional significance of transient protein-protein interactions. J Mol Biol 325, 991-1018.

Ofran, Y., and Rost, B. (2003a). Analysing six types of protein-protein interfaces. J Mol Biol 325, 377-387.

Ofran, Y., and Rost, B. (2003b). Predicted protein-protein interaction sites from local sequence information. FEBS Lett 544, 236-239.

Ofran, Y., and Rost, B. (2007). ISIS: interaction sites identified from sequence. Bioinformatics 23, e13-16.

Oldfield, T. J. (2002). Data mining the protein data bank: residue interactions. Proteins 49, 510-528.

Park, J., Lappe, M., and Teichmann, S. A. (2001). Mapping protein family interactions: intramolecular and intermolecular protein family interaction repertoires in the PDB and yeast. J Mol Biol 307, 929-938.

Porollo, A., and Meller, J. (2007). Prediction-based fingerprints of protein-protein interactions. Proteins 66, 630-645.

Qin, S., and Zhou, H. X. (2007). meta-PPISP: a meta web server for protein-protein interaction site prediction. Bioinformatics 23, 3386-3387.

Ritchie, D. W. (2008). Recent progress and future directions in protein-protein docking. Curr Protein Pept Sci *9*, 1-15.

Shenoy, S. K., Drake, M. T., Nelson, C. D., Houtz, D. A., Xiao, K., Madabushi, S., Reiter, E., Premont, R. T., Lichtarge, O., and Lefkowitz, R. J. (2006). beta-arrestin-dependent, G protein-independent ERK1/2 activation by the beta2 adrenergic receptor. J Biol Chem *281*, 1261-1273.

Slabinski, L., Jaroszewski, L., Rodrigues, A. P., Rychlewski, L., Wilson, I. A., Lesley, S. A., and Godzik, A. (2007). The challenge of protein structure determination--lessons from structural genomics. Protein Sci *16*, 2472-2482.

Su, Z., Ning, B., Fang, H., Hong, H., Perkins, R., Tong, W., and Shi, L. (2011). Next-generation sequencing and its applications in molecular diagnostics. Expert Rev Mol Diagn *11*, 333-343.

Thorn, K. S., and Bogan, A. A. (2001). ASEdb: a database of alanine mutations and their effects on the free energy of binding in protein interactions. Bioinformatics *17*, 284-285.

White, A. W., Westwell, A. D., and Brahemi, G. (2008). Protein-protein interactions as targets for small-molecule therapeutics in cancer. Expert Rev Mol Med *10*, e8.

Xia, J. F., Zhao, X. M., Song, J., and Huang, D. S. (2010). APIS: accurate prediction of hot spots in protein interfaces by combining protrusion index with solvent accessibility. BMC Bioinformatics *11*, 174.

Xu, Q., and Dunbrack, R. L., Jr. (2011). The protein common interface database (ProtCID)--a comprehensive database of interactions of homologous proteins in multiple crystal forms. Nucleic Acids Res *39*, D761-770.

Yan, C., Honavar, V., and Dobbs, D. (2004). Identification of interface residues in protease-inhibitor and antigen-antibody complexes: a support vector machine approach. Neural Comput Appl *13*, 123-129.

Zhou, H. X., and Qin, S. (2007). Interaction-site prediction for protein complexes: a critical assessment. Bioinformatics *23*, 2203-2209.

Zhou, H. X., and Shan, Y. (2001). Prediction of protein interaction sites from sequence profile and residue neighbor list. Proteins *44*, 336-343.

3

Protein Interactome and Its Application to Protein Function Prediction

Woojin Jung[1], Hyun-Hwan Jeong[1,2], and KiYoung Lee[1]

[1]Department of Biomedical Informatics, Ajou University School of Medicine
[2]Department of Computer Engineering, Ajou University
Republic of Korea

1. Introduction

Diverse molecules interact with proteins to produce a biological function. Proteins exhibit many interactions with other molecules including other proteins, nucleic acids, carbohydrates, lipids, minerals, metabolites, and chemical compounds, resulting in diverse roles within and/or between cells. Some of these proteins locate in subcellular organelles, where they modulate biochemical reactions, and some other proteins locate in membranes mediating various stimuli to signaling pathways. Cellular systems can be represented as complex networks. We may consider the molecules as nodes and the associations among the molecules as edges in the network. In this network, all kinds of the molecular interactions can be referred to as an interactome. Even though all kinds of the interactome are important, we here focus on protein-protein interactions (PPIs) since they are fundamental in cellular systems. To function correctly, a protein should interact with other proteins in the context of complex formation, signalling pathways and biochemical reactions. To perform a specific biological function, these interactions need to be specifically formed with proper interacting partners at the right time and locations.

Given the knowledge of genome sequencing on model organisms including human, we have elucidated a large number of unknown molecular structures and interactions within nucleic acids. In the post-genomic era, functional genomics is an emerging area of research that seeks to annotate every bit of information of the genome structure with relevant biological function. Still, many proteins (or genes) remain functionally unannotated (Apweiler et al, 2004; Sharan et al, 2007). These missing links between structures and functions need to be resolved to understand complex biological phenomena including human diseases, development and aging.

Protein function is widely defined in several different ways. It is highly context- and condition-dependent, which means that proteins participate in most biological processes. There have been various attempts to categorize the protein functions (Bork et al, 1998). One of them categorized the protein function into three parts: molecular function, cellular function and phenotypic function. First, the molecular function is defined as biochemical reactions performed by proteins. Second, the cellular function is defined as various pathways associated with proteins. Lastly, the phenotypic function is defined as an integration of all physiological subsystems to environmental stimuli.

Aside from the conceptual definition, many annotation efforts on protein function have been undertaken (Table 1). One of these efforts, the Gene Ontology (GO) consortium (Ashburner et al, 2000), made a standard and multi-labelled hierarchical annotation on proteins in the category of biological process, molecular function and cellular component. The GO consortium is regularly accumulating annotations on proteins according to GO category in open databases. In this chapter, we consider the three kinds of GO terms in annotation of protein function.

Many experimental techniques are available for discovering the protein function, such as gene knockout and transcript knockdown, but these approaches are low-throughput and time-consuming. In recent decades novel high-throughput techniques have been developed, and we are now able to analysis genome-wide data, which is broadening our biological insights. Computational methods are necessary for analysing the massive quantity of data and they are complementary with the low- and high-throughput experimental methods.

In this chapter, we first introduce PPI data available through public databases and compare the contents of major databases. We also describe PPI detection methods by experimental and computational approaches. Next, network- and non-network-based computational methods for the identification of protein function are described. Finally, computational prediction methods of protein subcellular localization, especially by exploiting PPI data, are shown.

Databases	Description
GO	The Gene Ontology project/consortium
COGs	Clusters of Orthologous Groups of proteins
ENZYME	A repository of information relative to the nomenclature of enzymes
Pfam	A database of protein families that includes their annotations and multiple sequence alignments
PROSITE	Database of protein domains, families and functional sites
HAMAP	High-quality Automated and Manual Annotation of microbial Proteomes
UniProt	The Universal Protein Resource
FunCat	MIPS (Munich Information Center for Protein Sequences) Functional Catalogue
DAVID	The Database for Annotation, Visualization and Integrated Discovery
FANTOM	A database for functional annotation of the mammalian genome
ANNOVAR	Functional annotation of genetic variants from high-throughput sequencing data
EFICAz	A genome-wide enzyme function annotation database
KEGG	Kyoto Encyclopedia of Genes and Genomes

Table 1. Databases for the functional annotation of genes and proteins.

2. PPI data

PPI can be considered as one kind of protein interactome. Proteins mutually interact in the biological context for specific functions. Given the knowledge of a single gene, expressing distinct transcripts and protein isoforms, a protein also interacts with other proteins including itself to give specific function. PPIs are defined as physical interactions between protein pairs (Bonetta, 2010). There are also non-physical interactions such as genetic and functional interactions. Genetic interaction is typically defined as when two genes are simultaneously perturbed, with the quantitative phenotype being more or less than expected (Mani et al, 2008). Functional interaction between two proteins is a much broader concept than other experiment-derived interactions. It may include any functionally associated gene/protein pairs which are integrated and predicted from heterogeneous data. We will explain these computational prediction methods later in this section.

The physical interactions between protein pairs also can be either direct or indirect. Binary interaction is an example of a direct interaction while indirect interaction includes subunits of protein complex. To give a specific function, proteins often form a large complex including direct and indirect interaction among the participant proteins. These interactions are also separable according to their binding lifetime. Some interactions between protein pairs are transient, with the interactions associating and dissociating under particular physiological conditions. On the other hand, some of proteins form stable complexes where the participants in the complexes permanently interact with each other. Various PPI types are defined in standard and annotated across many PPI databases (Cote et al, 2010; Kerrien et al, 2007).

2.1 PPI databases

Currently, there are 132 PPI databases indexed by the Pathguide (Bader et al, 2006; accessed 23 Dec 2011). The quantity of physical interactions to date is 386,495 across all species when integrated among major 11 databases by the iRefWeb (Turner et al, 2010; accessed 23 Dec 2011). The PPI data derived from both high- and low-throughput experiments are altogether deposited into any of primary databases which manually curate experimental results. These primary databases include not only physical interactions but also genetic interactions and annotate standard minimal information about a molecular interaction (MIMIx) (Orchard et al, 2007). There is an inconsistency problem related to the literature curation across different databases (Turinsky et al, 2010). Turinsky et al. confirmed that the agreement between curated interactions from 15,471 papers shared across nine databases was only 42% for interactions and 62% for proteins. This result was averaged between any two databases curated from the same publication. Some of the primary databases altogether formed a consortium called IMEx (The International Molecular Exchange) to enhance the quality of literature curation efforts.

Since we have plenty of primary databases, comprehensive integration of those primary databases has become an intriguing research field. Such meta-databases minimize redundancy and inconsistency that are limitations of the primary databases (Turinsky et al, 2010). Moreover, functional interaction databases consist of both experimentally-detected and computationally-predicted data. Sometimes, these predicted and experimental PPIs need to be distinguished for the degree of confidence. They both give useful information but should be separated according to the relevant evidence codes. There are also species-specific functional interaction databases (Lee et al, 2011; Lee et al, 2010a).

Type	Name	Description	URL
Primary databases	BioGRID	Physical and genetic interaction	http://thebiogrid.org
	MINT	Physical interaction	http://mint.bio.uniroma2.it
	IntAct	Physical interaction	http://www.ebi.ac.uk/intact
	DIP	Physical interaction	http://dip.doe-mbi.ucla.edu
	BIND	Physical and genetic interaction	http://bond.unleashedinformatics.com
	Phospho-POINT	A human kinase interactome resource	http://kinase.bioinformatics.tw
	PIG	Host-Pathogen interactome	http://pig.vbi.vt.edu
	SPIKE	A database of highly curated human signaling pathways	http://www.cs.tau.ac.il/~spike
	MPPI	The MIPS mammalian PPI database	http://mips.helmholtz-muenchen.de/proj/ppi
	HPRD	Human physical interaction	http://www.hprd.org
	CORUM	Mammalian protein complexes	http://mips.helmholtz-muenchen.de /proj/corum
Meta-databases	APID	Agile Protein Interaction DataAnalyzer	http://bioinfow.dep.usal.es/apid
	MiMi	Michigan Molecular Interactions	http://mimi.ncibi.org
	UniHI	Unified Human Interactome	http://www.mdc-berlin.de/unihi
	iRefWeb	Interaction Reference Index	http://wodaklab.org/iRefWeb
	DASMI	Distributed Annotation System for Molecular Interactions	http://dasmi.de/dasmiweb.php
	HIPPIE	Human Integrated Protein-Protein Interaction rEference	http://cbdm.mdc-berlin.de/tools/hippie
	HAPPI	Human Annotated and Predicted Protein Interaction database	http://bio.informatics.iupui.edu/HAPPI
Functional databases	STRING	Search Tool for the Retrieval of Interacting Genes/Proteins	http://string-db.org
	Gene-MANIA	Multiple Association Network Integration Algorithm	http://genemania.org
	Functional-Net	Species-specific functional gene networks	http://www.functionalnet.org

Table 2. List of PPI databases.

Contents	BIND	BioGRID	DIP	HPRD	IntAct	MINT	MPPI
Biological role (PSI-MI)					O	O	
Experimental role (PSI-MI)					O	O	
Taxonomy ID					O	O	
Interaction category					O	O	
Interaction title							
Interaction type (Text)		O					
Interaction type (PSI-MI)			O		O	O	
Interactor type (peptide, protein)					O	O	
Detection method (Text)				O			O
Detection method (PSI-MI)		O	O		O	O	
Evidence (PMID or doi-number)						O	
PubMed ID	O	O	O	O	O		O
BioGRID ID		O					
HPRD ID				O			
NCBI Gene ID	O	O		O			
Protein ID					O	O	O
ID type					O	O	O
Protein accession number	O			O			
UniProt ID			O				
Link to source ID	O	O	O			O	
Description	O				O		
Confidence score		O				O	

Table 3. Contents of primary PPI databases. Available contents are colored in grey with "O" shape.

We have listed some of the major primary databases, meta-databases, and functional databases in Table 2. Comparisons among the primary databases are shown in Table 3. We compared various features including interaction types, detection methods, references, and biological and experimental roles. This information would be valuable for researchers when they need to select and integrate various PPI data bases.

2.2 Methods for PPI identification

There are two major ways to determine PPIs. One is an experimental detection and the other is computational prediction. The former method is more reliable and well-established in both small and large scales while the latter method is based on the characteristics of accumulated protein interactions. In this section, we will briefly describe both approaches.

2.2.1 Experimental detection methods for PPIs

Experimental detection of interactions between protein pairs is achieved by various methods. Here, we describe only two representative methods: yeast two-hybrid (Y2H) (Suter et al, 2008) and mass spectrometry (MS) (Berggard et al, 2007). These methods both detect physical PPIs but the type of PPIs is different. As previously stated, direct and binary PPIs are distinct from protein groups in a complex and this type of PPI is detected only by Y2H method. This method uses a transcription factor found in yeast which consists of two other domains. Y2H method relies on an artificial insertion of a protein coding sequence to one of the domains and another protein inserted on the other domain using a plasmid. PPI can be assessed by confirming phenotype of the target gene of the transcription factor. The Y2H method can detect PPIs in large-scale and the sensitivity is high, enabling detection of even weak transient PPIs. But, since the experiment is done only in the nucleus, the real location information of such PPIs is hard to annotate, which obscures the detailed biological interpretation. Moreover, Y2H detects only binary interactions and results in a high rate of false positive, which are noteworthy limitations.

Another method in this category is based on mass-spectrometry (MS). The MS analyzes the mass of molecules rapidly and accurately. If the weight of all proteins in question is known, this information can be linked to the specific protein. This method is powerful when protein co-complexes are examined. Although it cannot provide details on the direct-level of interactions, the grouping of the proteins in a complex can be revealed. For this method, one protein ("bait") and all of interacting partners in a complex are pulled out and separated by electrophoresis. Finally, all the constructs derived from electrophoresis are used for MS. This method yields many false positive results when the sampling strategy is thoroughly different. This sampling might include fake interactions resulting in a high rate of false readings. There are many strategies related to this problem (Bousquet-Dubouch et al, 2011; Gingras et al, 2007). The experimental results obtained with MS-based methods are different from those obtained with binary methods (Y2H). Data derived from co-complex experiments cannot directly assign a binary interpretation. An algorithm is needed to translate group-based observations into pairwise interactions.

2.2.2 Computational prediction methods for PPIs

While recent reviews (Lees et al, 2011; Pitre et al, 2008; Shoemaker & Panchenko, 2007; Skrabanek et al, 2008; Xia et al, 2010) have discussed computational prediction methods for PPIs in details, we here briefly introduce some of approaches that are widely used. Although the amounts of experimental resources of PPIs are growing rapidly, proteome-wide PPIs information is still lacking and mostly limited on several model organisms. Given wide types of indirect but genome-wide resources, we can enhance our understanding of overall protein interactome. Methods in prediction of direct physical PPIs are less investigated than those of functional association between protein pairs. These functional association methods of PPIs can give information of which protein pairs have same biological process and potential physical interactions.

The first data used in these prediction methods is genomic sequences. Co-occurrence-based methods use assumption that if gene pairs are co-inherited across evolutionary processes (i.e. species), they are considered as functionally associated (Barker & Pagel, 2005; Bowers et al, 2004; Pellegrini et al, 1999). These methods applied to microorganisms and successfully discovered novel participants of known pathway (Carlson et al, 2004; Luttgen et al, 2000). Other similar methods based on this genomic sequence use the information of gene fusion events (Marcotte et al, 1999; Reid et al, 2010; Zhang et al, 2006) and gene neighbourhood (Ferrer et al, 2010; Itoh et al, 1999; Koonin et al, 2001). Another type of data used is amino acid (AA) sequences and the interface of interacting protein pairs are composed of specific AA residues (Tuncbag et al, 2008; Tuncbag et al, 2009). This knowledge is reflected in the co-evolution of specific interface residues between interacting proteins and by alignments of multiple sequences, the results are highly correlated with physical PPIs (Pazos et al, 2005). Commonly occurring domain pairs are also considered in this context (Eddy, 2009; Finn et al, 2010; Stein et al, 2009; Yeats et al, 2011) and simple AA sequence such as 3-mers of interacting residues can be used (Ben-Hur & Noble, 2005). Another well-known information is homology of PPIs across different species. Methods on this information simply find PPIs which are conserved across species, called interologs (Matthews et al, 2001). Here, any known PPIs regarded as query to find conserved interactions across species using an ortholog database. There are many algorithms which follow this approach (Kemmer et al, 2005; Persico et al, 2005). Aside from the sequence-level data, structural information is also a valuable resource to predict PPIs, especially a protein 3D structure. (Aloy & Russell, 2003; Ezkurdia et al, 2009; Hosur et al, 2011; Shoemaker et al, 2010; Singh et al, 2010; Zhang et al, 2010). A huge amount of genome-wide gene expression profiles are another useful data to predict PPIs and they are investigated to define gene co-expression patterns of any pairs and consider higher correlation degree as higher probability of PPIs (Grigoriev, 2001; Lukk et al, 2010; Stuart et al, 2003). As shown in the earlier section, there are many literature-curated PPI databases. While those approaches are based on the manual inspection, such PPIs information can be automatically extracted using a text-mining algorithm (Blaschke et al, 2001; Szklarczyk et al, 2011; Tikk et al, 2010).

3. Computational prediction methods for protein function

Even before the prevalence of genome sequencing technologies, typical experimental identification on a protein function has been executed. Such identification has focused on a specific target gene or protein, or a small set of protein complexes. Gene knockout, knockdown of gene expression, and targeted mutations are some methods for protein function identification (Recillas-Targa, 2006; Skarnes et al, 2011). Such low-throughput experiments were replaced by high-throughput experiments including genome sequencing and determination of the protein interactome. Computational methods followed by massively archived data have been developed for better analysis. Based on the assumption that structural similarity correlates with functional similarity, homology-based functional annotation across organisms has now become a trivial approach (Aloy et al, 2001; Gaudet et al, 2011).

3.1 Non-network based approaches

Classical computational methods use features from only a single protein in prediction of protein function (Bork et al, 1998). These approaches use a set of features like amino acid sequences, genome sequences, protein structures (2D and 3D), phylogenetic data, and gene expression data. PSI-BLAST (Altschul et al, 1997) and FASTA (Mount, 2007) are popular sequence alignment tools used to reveal homologous proteins between known and unknown (query) proteins. Proteins with similar sequences are assumed to have similar functions. Moreover, protein folding patterns are also preserved enough to identify homologs (Huynen et al, 1998; Sanchez-Chapado et al, 1997). The comparative genomics across different species is a powerful approach for analysing functional annotation of proteins. In fact, it has been suggested that correlation of sequence-structure is much stronger than that of sequence-function (Smith et al, 2000; Whisstock & Lesk, 2003). So many approaches take the sequence to structure to function route for protein function prediction (Fetrow & Skolnick, 1998).

Likewise, these data are showing only single aspect of functional features conserved during evolution. Data derived from different sources can be inter-connected it should be integrated to analyse simultaneously (Kemmeren & Holstege, 2003). We next show that PPI networks potentially enrich functional relationship between protein pairs that may not be detectable from other genomic data such as primary or higher level sequence structure.

3.2 Network-based approaches

As we mentioned in the Introduction, biological function is never achieved by a single protein. Rather, proteins dynamically interact with each other and the interacting partners adopt similar performances for specific functions. With a plethora of data being generated by high-throughput proteomic experiments, it has become possible to use proteome-wide PPI patterns in protein function prediction. Among a broad type of protein interactome, a PPI network generates well-known data that is invaluable in prediction of protein function. It is possible to annotate the function of undefined proteins according to its neighbours that are functionally annotated. This assumption is based on simple idea called "guilt-by-association", and we consider an association by possible physical interaction in any condition and, sometimes, functional association are given with relevant evidence score.

Here, we review the general network-based approaches in predicting protein functions. These approaches are categorized into two methods for better description. The first one is a straightforward method of inferring protein function based on the topological structure of a PPI network. The other method first identifies distinct sub-networks from a whole PPI network. These sub-networks are also referred to as functional modules since they perform specific biological functions such as protein complexes, and metabolic and signalling pathways. Functional modules are detected by a broad variety of clustering algorithms and, thereafter, each module is annotated with appropriate functional association. In this section, basic concepts and pioneering studies on this corresponding approaches are introduced.

3.2.1 Direct annotation of protein function using PPI network

3.2.1.1 Neighbourhood approaches

Direct functional annotation considers the correlation of the network distance between two proteins, which means the closer the two proteins are in the network the more similar are their functions. One of the earliest studies extrapolated only adjacent neighbours within an entire PPI network. This simple approach used information of the immediate neighbourhood and took the most common functions up to three among its neighbours. In spite of the effectiveness, accuracy was achieved by 72% (Schwikowski et al, 2000). However, this method lacked significance values for each association and the full network topology was not considered in the annotation process. A strategy was proposed to tackle the first problem of assigning statistical significance (Hishigaki et al, 2001). This was done by using χ^2.-like scores and, instead of using the immediate neighbours, the n-neighbourhood of a protein that consists of proteins with distance of k-links to the protein is considered. Simply put, the neighbours of adjacent neighbours are taken into account with the frequencies of all the distance of in this neighbourhood. For an unknown protein, the functional enrichment in its n-neighbourhood in identified with χ^2 test, and the top ranking functions are assigned to the unknown protein. In another approach, the shared neighbourhood of a pair of proteins are considered besides from the neighbourhood of the protein of interest. Chua et al. investigated the correlation between functional similarity and network distance (Chua et al, 2006). They developed a functional similarity score, called the FS-weight measure, which gives different weights to proteins depending on their network distance from the query protein. This approach showed higher accuracy when employing indirect interactions and its functional association.

3.2.1.2 Global optimization approaches

Although the neighbourhood approach is very attractive and effective by its simplicity, shortcomings arise when there is not enough number of protein neighbours and sufficiently annotated proteins. To overcome this issue, several approaches that utilize the entire topology of the network have been proposed. These global approaches attempt to optimize annotation of function-unknown protein using the topology of a whole network. One of the first studies that took this approach used the theory of Markov random fields, which determines the probability of a protein having a certain function (Deng et al, 2004). This theory is then used to determine the joint probability of the whole interaction network regarding to a certain function. This formulation is transformed to that of the conditional probability of a protein having a certain function given the annotations of its interaction partners. After that, the Gibbs sampling technique is iteratively applied to determine the stable values of this probability for each protein. This approach resulted in higher performance than those of neighbourhood-based approaches (Chua et al, 2006; Hishigaki et al, 2001; Schwikowski et al, 2000) when utilized to the yeast PPI data.

Additional attempts according to this approach had been followed. Here, the objective function is defined for the whole network, which is a sum of the following variables (Vazquez et al, 2003).

1.   The number of neighbours of a protein having the same function as itself.
2.   The number of neighbours of a protein having the function under consideration.

Thus, this function estimates the number of pairs of interacting proteins with no common functional annotation. Since a high value of this function is biologically undesirable, it is minimized using a simulated annealing procedure. As expected, this approach outperformed the majority rule-based strategy on the *Saccharomyces cerevisiae* interaction data (Schwikowski et al, 2000), since the latter tried to optimize only the second factor above. An additional advantage of this approach was that multiple annotations of all proteins were obtained in one shot, unlike earlier approaches which ran independent optimization procedures for different functions.

The above discussion shows that a wide variety of approaches based on principles of global optimization have been proposed in the literature and many more are in the pipeline. The most accurate results in the field of function prediction from PPI networks have also been achieved by these approaches, which is intuitively acceptable since they extract the maximum benefit from the knowledge of the structure of the entire network.

3.2.2 Indirect annotation of protein function

This approach uses a protein interaction network, not directly for annotation, but identifies functional modules first and then assigns functions to unknown proteins based on their membership in the functional modules. This is based on the assumption that most biological networks are organized as distinct sub-networks to give specific functions (Hartwell et al, 1999). We assume that proteins in the same module participate in a similar biological process. Modular patterns and dense regions are found in the PPI network (Gavin et al, 2006).

3.2.2.1 Distance-based clustering approaches

To find biologically significant modules, clustering algorithms can be applied efficiently. Clustering is a popular unsupervised learning algorithm that does not use any prior information about the class label. There are two widely-used ways of clustering: topology-based or distance-based. The key procedure in distance-based clustering is to select the similarity measure between two proteins to detect modules. The distance between two proteins (also called as nodes) in a network is usually defined as the number of interactions (also called as edges) on the shortest path between them. However, there is a serious problem in this hierarchical clustering, known as the 'ties in proximity' problem (Arnau et al, 2005). This means that the distance between many protein pairs are identical.

To solve this problem, a network clustering method was developed to identify modules in the biological network based on the fact that each node has a unique pattern of shortest path lengths to every other node. But for a specific module in the network, the nodes/members of the module shared similar pattern of shortest path lengths (Rives & Galitski, 2003). Another study used the hierarchical clustering method with the shortest path length between proteins as a distance measure to overcome the 'ties in proximity'. This was achieved by exploiting equally valid hierarchical clustering solution with a random select when ties are met (Arnau et al, 2005). Although many methods in the similarity measures have been proposed, a single validation for such methods is insufficient. For this, two evaluation schema are suggested, which are based on the depth of a hierarchical tree and width of the ordered adjacency matrix (Lu et al, 2004). Furthermore, there are various types of cellular network with distinct modular patterns, and so network-specific methods should be investigated in the future.

3.2.2.2 Graph-based clustering approaches

Dissecting functional modules in a large PPI network is the same problem of graph partitioning and clustering. One of the pioneering method using this network topology-based concept was the MCODE (molecular complex detection algorithm) (Bader & Hogue, 2003). This method predicts complexes in a large PPI network consisting of three processes. First, the nodes of the network are weighted by their core clustering coefficients (the density of the largest k-core of its adjacent neighbourhood), and then densely connected modules are identified in a greedy fashion. The use of this coefficient instead of a standard clustering coefficient was proposed, as it increases the weights of densely interconnected graph regions while giving small weights to the less connected nodes. The next step is to filter or add proteins based on the connectivity criteria. This method was applied to large-scale PPI networks and given as a plug-in for the Cytoscape (Kohl et al, 2011).

Another similar study to find complexes and functional modules is based on super paramagnetic clustering. This method used an analogy to the physical features of a heterogeneous ferromagnetic model to detect densely connected clusters in a large graph (Spirin & Mirny, 2003). There is also an algorithm called the restricted neighbourhood search clustering (RNSC), which starts with an initial random cluster assignment and then proceeds by reassigning nodes to maximize the partition's score. Here, the score represents an intra-connectivity in the cluster, not an inter-connectivity across other clusters. The RNSC algorithm is known to perform better than the MCODE algorithm (King et al, 2004). The Markov clustering algorithm (MCL) is another fast and scalable clustering algorithm based on simulation of random walks on the underlying graph (Pereira-Leal et al, 2004). This algorithm has an assumption that a random walker in natural clusters (i.e. dense region of the graph) sparsely goes from one to another natural cluster. Such clusters in a whole graph are structurally identified by the MCL algorithm. It starts by measuring the probabilities of random walks through the graph to build a stochastic "Markov" matrix, by alternating two operations: expansion and inflation. The expansion takes the squared power of the matrix while the inflation takes the Hadamard power of a matrix, followed by a re-scaling. Therefore the resulting matrix is remained as stochastic. Clusters are detected by alternation of expansion and inflation until the graph is partitioned into distinct subsets where no paths between these subsets are available. This algorithm can be efficiently implemented to weighted and large dense graphs. Various PPI networks were applied using the MCL algorithm to find functional modules such as protein complex (Krogan et al, 2006).

It is true that a protein might have multiple functions and this characteristics of a protein leads to overlap of different modules. That means graph partitioning in a strict manner might not be reasonable for the PPI network. However, most current methods are based on the hard-partition algorithms, meaning that each protein can belong to only one specific module. To handle this limitation, a clustering algorithm based on the information flow was suggested. This algorithm efficiently identified the overlapping clusters in weighted PPI network by integrating semantic similarity between GO function terms (Cho et al, 2007). Since the common proteins in the overlapping modules are interpreted as a connecting bridge across the different modules, biologically significant and functional sub-networks could be identified. Still, there are few clustering methods identifying such overlapping modules. Novel clustering methods for this theme are required with enhancement of prediction accuracy.

4. Prediction of protein subcellular localization

4.1 Introduction

Proteins should move to specific locations after synthesis to work in our body correctly. Thus, knowing subcellular localization of proteins is important to understand their own functions. Unicellular organisms like budding and fission yeasts can find systematic protein localization by experimental studies. However, such studies could not be performed well in higher eukaryotes such as *Caenorhabditis elegans*, *Drosophila melanogaster*, or mammals because of large-scale proteome sizes and technical difficulties associated with protein tagging.

Therefore, bioinformatical approaches to develop efficient methods are required instead of wet experiments. Actually, many computational methods to predict subcellular localization of protein have been proposed over several decades. A considerable number of computational classification methods have been developed for this purpose. Typically these algorithms input list of features and output subcellular localizations of target proteins. The features contain various characteristics of the proteins. Molecular weight, amino acid content and codon bias can be the features. Input features for prediction of subcellular localization can be broadly categorized into four categories: protein sorting signals, empirically correlated characteristics, sequence homology with known answer sets, and other sources (Imai & Nakai, 2010).

During the training phase, in the methods, learning utilizes a set of gold-standard proteins whose localizations are well known. This set consists of the feature vectors. After the training phase, a model is constructed to recognize those features or patterns of features that are useful and then predicts the subcellular localization of proteins whose localization is unknown. Various algorithms have been used to construct a model for prediction of sub-cellular localization.

In the field of bioinformatics, there are several problems to resolve for predicting subcellular localization of proteins. First, there are generally too many classes (localization). According to Huh et al, 22 distinct localizations exist in budding yeast. Next, one protein may have multiple different localizations (Huh et al, 2003). This is referred to a multi-label classification problem and traditional classification algorithms have a limit on handling the multi-label problem well. Another problem is that there may be a higher dimensional feature space for prediction. More than tens of thousands features exist in some cases.

Another issue is that data for each localization is too imbalanced. All these characteristics make the prediction difficult. More importantly, the localization prediction is sometimes difficult to achieve sufficient performance when we use information of single proteins only. Recently, large-scale protein-protein interaction networks have been elucidated in yeast, fly, worm, and human. To interact physically, two proteins should localize to the same or adjacent subcellular localization. That means we can get useful information of a protein from its interacting neighbours. Thus, we can improve the localization prediction performance particularly using PPI networks.

4.2 Computational prediction of protein subcellular localization

4.2.1 Single-protein feature based localization predictions

Table 4 summarizes previous studies that have used the features of single proteins. The studies for prediction of subcellular localization have the following trends. The first is an increase in the number of predicting localizations. At first, Nakashima & Nishikawa predicted localization of a protein that is inter-cellular or extra-cellular using Amino Acid (AA) and Pair coupled Amino Acid (PairAA) (Nakashima & Nishikawa, 1994). After their study, many studies tried to increase the number of distinct localizations to predict. For example, Gardy et al predicted five distinct subcellular localization including 'cytoplasmic', 'inner membrane', 'periplasmic', 'outer membrane' and 'extra-cellular' (Gardy et al, 2003). Nair & Rost predicted ten distinct subcellular localizations (Nair & Rost, 2003). Also, Chou & Cai predicted 22 distinct subcellular localizations that experimentally identified localization of Huh et al. (Chou & Cai, 2003).

The second trend is handling of a multi-label problem. A protein can localize to several sub-cellular locations. However, most of these studies did not consider multiple localization property, but rather assumed that a protein has a single representative localization. Also, the accuracy of prediction is lower when the number of distinct localizations for a protein is increased. Some researchers have been tried to address this issue (Lee et al, 2006).

Another tendency is the development of a classification algorithm for an elaborate and efficient model construction. Least distance algorithm, artificial neural network, a nearest neighbour approach, a Markov model, a Bayesian network approach, and support vector machine (SVM) were used to archive the goal. Some studies mixed several algorithms. Lee et al. developed an algorithm that reflects of property of the prediction task (Lee et al, 2006). They developed an extended Density-induced Support Vector Data Description (D-SVDD) classification algorithm to handle well the issues related to class imbalance, higher dimensionality, multi-label, and many distinct classes. The classical D-SVDD algorithm can handle only one-class classification tasks. Thus, Lee et al. extended it to handle multi-label classification tasks.

4.2.2 Network-based localization prediction

As mentioned earlier, two proteins that localize to same or adjacent subcellular localization have a tendency to interact with each other. That means two proteins can be a tag protein to one other for subcellular localization. Therefore, if a molecular network such as PPIs is available, we may take advantage of the PPI network for the prediction. Several studies tried to predict subcellular localization using network data. This section consists two parts: first one is a brief explanation of the study by Lee et al. (Lee et al, 2008), which is the cornerstone of the network-based approach for location prediction using PPI network. We describe a methodology to generate of feature vectors for a protein in the aforementioned study and introduce a DC-kNN classifier for the prediction. The second part is a summary of the network-based approaches from the work of Lee et al. to the present.

Author(s)	Method(s)	Feature(s)	# Classes	Multi-label	Imbalanced
(Nakai & Kanehisa, 1991)	Expert Systems	SignalMotif	4	X	X
(Nakai & Kanehisa, 1992)	Expert Systems	AA, SingalMotif	14	X	X
(Nakashima & Nishikawa, 1994)	Scoring System	AA, diAA	2	X	X
(Cedano et al, 1997)	LDA using Mahalanobis distance	AA	5	X	X
(Reinhardt & Hubbard, 1998)	ANN Approach	AA	3, 4	X	X
(Chou & Elrod, 1999)	CDA	AA	12	X	X
(Yuan, 1999)	Markov Model	AA	3, 4	X	X
(Nakai & Horton, 1999)	k-NN approach	SignalMotif	11	X	X
(Emanuelsson et al, 2000)	Neural network	SignalMotif	4	X	X
(Drawid & Gerstein, 2000)	CDA	Gene Expression Pattern	8	X	X
(Drawid & Gerstein, 2000)	Bayesian Approach	SignalMotif, HDEL motif	5, 6	X	X
(Cai et al, 2000)	SVM	AA	12	X	X
(Chou, 2000)	Augumented CDA	AA, SOC factor	5, 7, 12	X	X
(Chou, 2001)	LDA using various distance measures	pseuAA	5, 9, 12	X	X
(Hua & Sun, 2001)	SVM	AA	4	X	X
(Chou & Cai, 2002)	SVM	SBASE-FunD	12	X	X
(Nair & Rost, 2002)	Nearest Neighbor Approach	functional annotation	10	X	X
(Cai et al, 2003)	SVM	SBASE-FunD, pseuAA	5	X	X
(Cai & Chou, 2003)	Nearest Neighbor Approach	GO, InterProFunD, pseuAA	3, 4	X	X
(Chou & Cai, 2003)	LDA using various distance measures	pseuAA	14	X	X
(Pan et al, 2003)	Augumented CDA	pseuAA with filler	12	X	X
(Park & Kanehisa, 2003)	SVM	AA, diAA, gapAA	12	X	X
(Zhou & Doctor, 2003)	Covariant discrinant algorithm	AA	4	X	X
(Cai et al, 2003)	SVM	SBASE-FunD, pseuAA	5	X	X

Author(s)	Method(s)	Feature(s)	# Classes	Multi-label	Imbalanced
(Gardy et al, 2003)	SVM, HMM, Baysian	AA, motif, homlogy analysis	5	X	X
(Reczko & Hatzigerrorgiou, 2004)	ANN Approach	AA, SingalMotif	3	X	X
(Huang & Li, 2004)	fuzzy k-NN	diAA	11	X	X
(Cai & Chou, 2004)	Nearest Neighbor Approach	GO, InterProFunD, pseuAA	3, 4	X	X
(Chou & Cai, 2005)	Nearest Neighbor Approach	FunDC(5875D), pseuAA	3, 4	X	X
(Bhasin & Raghava, 2004)	SVM	AA, diAA	4	X	X
(Lee et al, 2006)	PLPD	AA, diAA, gapAA, InterProFunD	22	O	O
(Chou & Shen, 2007)	Nearest Neighbor Approach	GO, InterProFunD, pseuAA	22	O	X
(Shatkay et al, 2007)	SVM	SignalMotif, AA, text-based feature	11	X	X
(Garg et al, 2009)	k-NN, PNN	AA, sequence order, physicochemical properties	11	X	X
(Zhu et al, 2009)	SVM	AA, PSSM	14	O	X
(Shen & Burger, 2010)	SVM	AA, groupedAA, gapAA,, GO	4	X	X
(Mei et al, 2011)	SVM	AA, diAA, gapAA, GO	10	O	X
(Wang et al, 2011)	Frequent Pattern Tree	Motif, Overall-sequence	12	X	X
(Mooney et al, 2011)	N-to-1 Neural Network	BLAST	5	X	O
(Tian et al, 2011)	PCA, WSVM	PesAA	20	X	X
(Pierleoni et al, 2011)	SVM	AA, ChemAA, protein length, GO	3	X	X

Table 4. Summary of previous methods for prediction of protein subcellular location.

4.2.2.1 Generation of feature vectors

Lee et al. used three types of feature to predict the localization and integrated these features (Lee et al, 2008). These are single protein features (S) and two kinds of network neighbourhood features (N and L).

Seven S features were based on a protein's primary sequence and its chemical properties. Amino acid composition frequencies (AA), adjacent pair amino acid frequencies (diAA) and pair-wise amino acid frequencies with a gap which is length of 1 (gapAA) from a protein's primary sequence were used. Also, three kinds of chemical amino acid compositions (chemAA) were generated from normalized hydrophobicity (HPo), hydrophilicity (HPil), or side-chain mass (SCM). Also, they combined these chemical properties into pseudo-amino acid composition (pseuAA), which is another S feature vector. Occurrences of known signalling motifs in the primary protein sequence (Motif) are also used as one of the S features. The last S feature encoded functional annotations of the protein from Gene Ontology (GO) (Ashburner et al, 2000). Figure 1 provides an example.

N network features are summary of S features from neighbourhood of a protein. Knowledge for neighbours of a protein comes from PPI data, which are pooled from various databases such as BioGRID (Stark et al, 2011), DIP (Salwinski et al, 2004) and SGD (Engel et al, 2010). L network features are summary of location distribution of interacting neighbours. Figure 2A shows a relationship among the three PPI databases. It shows that a single protein interaction database covers a different part of the whole reported interactions. The diagonal pattern in Figures 2B-D shows that interacting protein pairs share similar localization information. For example, a protein in an "ER to Golgi" tends to interact with other proteins which localized in the "ER to Golgi" more than other localizations.

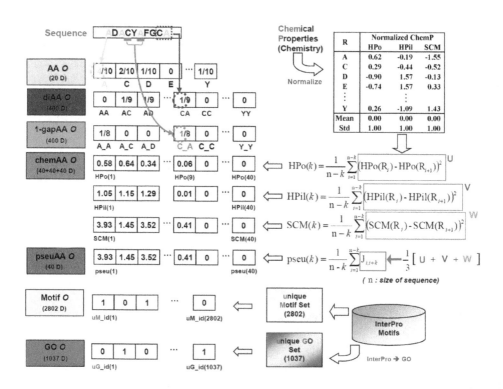

Fig. 1. Summary of feature generation scheme for a single protein (adapted from Lee et al, 2008).

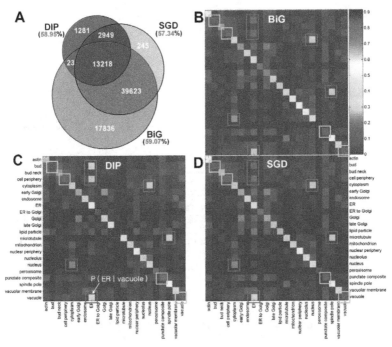

Fig. 2. Correlation between known localizations and protein interactions of yeast proteins. (A) The number of interactions (inside the circles) and the fraction of interactions whose proteins share localization information (outside the circles) of three interaction databases: BiG, DIP and SGD. (B-D) They show that interacting protein pairs have similar localization information in DIP, BiG and SGD (adapted from Lee et al, 2008).

4.2.2.2 Divide-and-Conquer k-Nearest Neighbour (DC-kNN) Classifier

After generating feature vectors, large-scale feature vectors with a high order may generate. A high dimensional feature vectors generally cause some problems like *curse-of-dimensionality*. In other words, data from higher dimensional feature vectors usually require a corresponding amount of inputs and it, sometimes, causes an over-fitting problem to a given dataset (Guyon et al, 2002). Also some feature vectors may be useless in constructing a model for a specific localization. Thus, individual model for different subcellular localizations may require different sets of useful feature sets. Therefore, extraction for feasible feature vectors for individual localizations may be needed to construct robust and reliable prediction models.

To construct a prediction model, Lee et al. proposed a DC-kNN classifier which is a variety of a k-Nearest Neighbours classification algorithm. A DC-kNN classifier tackles high-dimensional features in a divide-and-conquer manner. Briefly mentioning, a DC-kNN has three main steps (Figure 3): dividing, choosing, and synthesizing. In the dividing step, the full feature vector is divided into m meaningful subsets. After the dividing step, the k-nearest neighbours are chosen for each protein and for each subvector. In the synthesizing step, results of kNNs of individual m sets are synthesized to produce confidence scores using an average of Area under the ROC curve (AUC) for each localization. DC-kNN finds a feasible combination of feature sub-vectors for each label (localization) based on a feature forward selection approach.

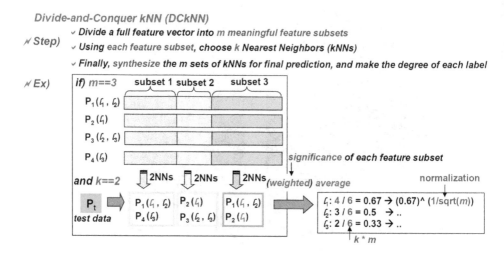

Fig. 3. Brief description of a DC-kNN (adapted from Lee et al, 2008).

4.2.3 Results of location prediction

Lee et al. first compared prediction performance of a DC-kNN for localization prediction with different feature sets: S features only, N features only, L features only, all features together ($S+N+L$), and random guesses. N and L features are generated using DIP (Salwinski et al, 2004). Performance of each case was evaluated by the technique of leave-one-out cross-validation (LOOCV). Proteins of *Saccharomyces cerevisiae* (n=3914) (Huh et al, 2003) were used for the LOOCV. They used three different performance metrics: Top-K, Total, and Balanced. These metrics were used to summarize the results of 3914 LOOCV runs. Top-K measurement considers as correct if at least one of the real localization of a protein is in the top-K predictions. Total measurement counts all the correctly predicted localizations based on the number of real localizations of test data. Balanced measure calculates the averaged fraction of correctly predicted proteins in each localization. As a result, every classifier showed clearly better performance than random guess (Figure 4A), and combination of S, N, and L features showed the highest performance.

Figures 4A and 4B inform that information of neighbourhood acquired from a PPI database improves prediction performance. However, Figure 4C illustrates that acquiring more information does not always contribute to an improvement of performance. On the contrary, additional information can decrease prediction performance. To find the necessary feature vectors for each localization, Lee et al. used a DC-kNN and found feasible subsets using the prepared feature vectors for individual localizations (Figures 4C and 4D). Using the selected features for individual localizations, the average of the AUC values was 0.94.

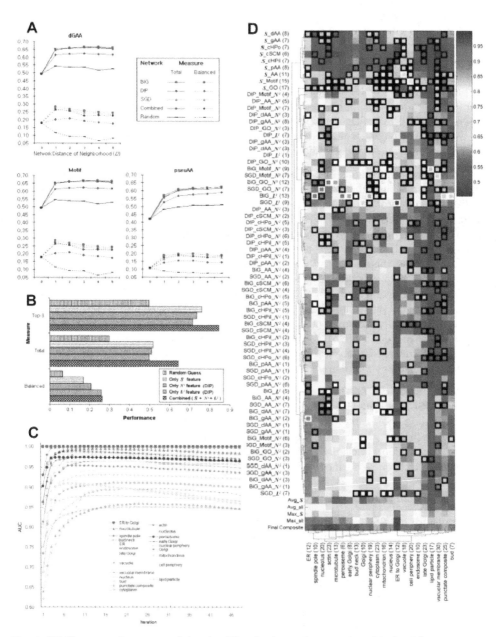

Fig. 4. (A) Shows performance of the classifiers by input from various kinds of feature. (B) shows performance for combination of feature vectors. (C) shows averaged AUC of the classifier for each localization based on feature selection using a DC-kNN. (D) shows selected feature sets for each of 22 localizations in yeast (adapted from Lee et al, 2008).

Based on the methodology, Lee et al. applied their method to the prediction of the localizations of genome-wide yeast proteins. Surprisingly, they also validated novel localizations of 61 proteins. For example, Huh et al. reported that Noc4/Ypr144c and Utp21/Ylr409c were localized in the nucleus (Huh et al, 2003). However, the proposed method developed by Lee et al. predicted the localization of the both proteins as the nucleolus. They revaluated for both proteins using new experiments and finally confirmed the previous results of Huh et al. had errors (Figures 5A and 5B). The correct prediction mainly owes to the fact that Lee et al. combined evidence from multiple interacting partners. For example, Noc4 interacts with many other proteins known to exist in the nucleolus, so we can assume that Noc4 localizes nearby or directly in the nucleolus. They confirmed the assumption by the network neighbours (Lee et al, 2008) (Figure 5C).

The number of localizations and known PPIs for yeast proteins are larger than those for other organisms. In other words, some organisms have less information on known localization and protein interaction, which might make the location prediction difficult based on a PPI network. Lee et al. evaluated their method using yeast data with some random missing information (Lee et al, 2008). As a result relatively robust results were obtained with less information. For example, the average number of neighbours of a protein in yeast is 27 and the number in worm is three. Decrement in the number of neighbours from yeast to worm was 9-fold. However, the average of AUC value decreased from 0.94 (yeast) to 0.87 (worm) (Figure 6). In other words, their method can be easily applied, not only to yeast but to other species with less known localization and/or interaction information. Actually they predicted subcellular localization of fly, human, and Arabidopsis (Lee et al, 2008; Lee et al, 2010b) using protein interactions. The results of both works showed that the prediction worked well for the other organisms and could find real localizations of some unknown proteins (Figures 6-7).

They also compared a DC-kNN with two previous popular methods, ISort (Chou & Cai, 2005) and PSLT2 (Scott et al, 2005). ISort is a comprehensive sequence-based machine learning method. ISort can predict more than 15 compartments. PSLT2 is a previous method that used a protein interaction network to predict subcellular localizations. They compared to DC-kNN with ISort and PSLT2 using both total and balanced measures. As illustrated in Figure 8, DC-kNN outperformed both methods in total and balanced measurement.

4.2.4 Other network-based methods

After the study of Lee et al. in 2008, several studies based on network-based approaches tried to predict subcellular localization. Mintz-Oron et al. used a constraint-based method for predicting subcellular localization of enzymes based on their embedding metabolic network, relying on a parsimony principle of a minimal number of cross-membrane metabolite transporters (Mintz-Oron et al, 2009). They showed that their method outperformed pathway enrichment-base methods. Another group constructed a decision tree-based meta-classifier for identification of essential genes (Acencio & Lemke, 2009). Their method relied on network topological features, cellular localization and biological process information for prediction of essential genes. Tung & Lee integrated various biological data sources to get information of neighbour proteins in a probabilistic gene-network (Tung & Lee, 2009). They predicted the subcellular localization using a Fuzzy k-nearest neighbour classifier. Lee et al. curated IntAct *Arabidopsis thaliana* PPI dataset

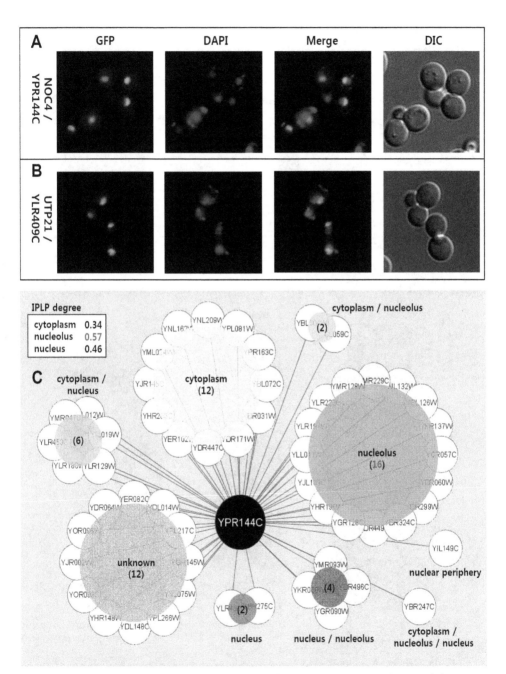

Fig. 5. (A, B) represent results of new experiments for Noc4/Ypr144c and Utp21/Ylr409c. (C) shows the interacting neighbours of Ypr144c (adapted from Lee et al, 2008).

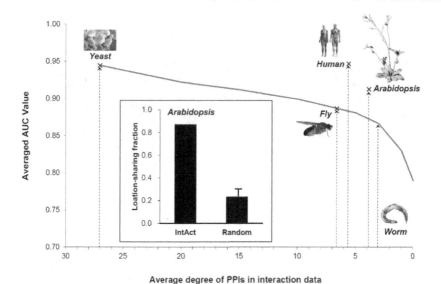

Fig. 6. Averaged AUC values across different organisms (adapted from Lee et al, 2010b).

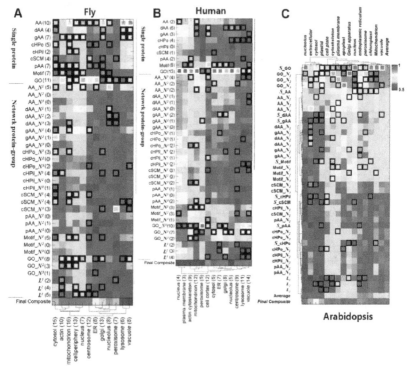

Fig. 7. Generated models for the location prediction for Fly (A), Human (B), and Arabidopsis (C) (adapted from Lee et al, 2008 and Lee et al, 2010b).

Lee et al. curated IntAct *Arabidopsis thaliana* PPI dataset (Aranda et al, 2010) using the DC-kNN method, which was proposed before and which showed good performance (Lee et al, 2010b). They also showed that the DC-kNN is applicable to other organisms. Kourmpetis et al. predicted a function of proteins in *Saccharomyces cerevisiae* based on network data, such as PPI data (Kourmpetis et al, 2010). They took a Bayesian Markov Random field analysis method for prediction and predicted the functions of 1170 un-annotated *Saccharomyces cerevisiae* proteins.

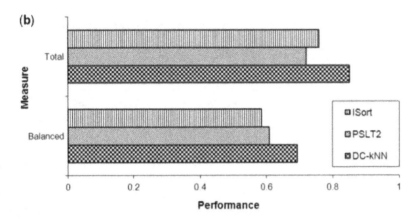

Fig. 8. Performance comparison of Isort, PSLT2 and DC-kNN (adapted from Lee et al, 2008).

5. Conclusions

We reviewed on PPI databases and the methods for detection of PPIs. Then, the computational methods of protein function prediction were briefly reviewed. We finally discussed that the prediction of protein function, especially the subcellular localization, shows outstanding performance when using PPIs data. This is because real biological functions are maintaining through a cascade of PPIs. Moreover, the computational approaches are very much promising when compared to the experimental identification especially for the false reading corrections. Functional genomics is an ongoing field in systems biology and this must be done well to drive further progress. We are facing other issues concerning the lack of conditional protein interactomes. We have identified and accumulated only static information at the molecular level in cells to make a scaffold of cellular systems. Computational methods should be applied to this conditional analysis when sufficient data become available and the next field of utilization would be personalized medicines, such as the early diagnosis with specific markers and treatments with specific drug targets.

6. Acknowledgement

This work was supported by Basic Science Research Programs through the National Research Foundation of Korea (NRF) funded by the Ministry of Education, Science and Technology (2010-0022887 and 2010-0018258).

7. References

Acencio ML, Lemke N (2009) Towards the prediction of essential genes by integration of network topology, cellular localization and biological process information. *BMC bioinformatics* 10: 290

Aloy P, Querol E, Aviles FX, Sternberg MJ (2001) Automated structure-based prediction of functional sites in proteins: applications to assessing the validity of inheriting protein function from homology in genome annotation and to protein docking. *Journal of molecular biology* 311: 395-408

Aloy P, Russell RB (2003) InterPreTS: protein interaction prediction through tertiary structure. *Bioinformatics* 19: 161-162

Altschul SF, Madden TL, Schaffer AA, Zhang J, Zhang Z, Miller W, Lipman DJ (1997) Gapped BLAST and PSI-BLAST: a new generation of protein database search programs. *Nucleic acids research* 25: 3389-3402

Apweiler R, Bairoch A, Wu CH, Barker WC, Boeckmann B, Ferro S, Gasteiger E, Huang H, Lopez R, Magrane M, Martin MJ, Natale DA, O'Donovan C, Redaschi N, Yeh LS (2004) UniProt: the Universal Protein knowledgebase. *Nucleic acids research* 32: D115-119

Aranda B, Achuthan P, Alam-Faruque Y, Armean I, Bridge A, Derow C, Feuermann M, Ghanbarian AT, Kerrien S, Khadake J, Kerssemakers J, Leroy C, Menden M, Michaut M, Montecchi-Palazzi L, Neuhauser SN, Orchard S, Perreau V, Roechert B, van Eijk K et al (2010) The IntAct molecular interaction database in 2010. *Nucleic acids research* 38: D525-531

Arnau V, Mars S, Marin I (2005) Iterative cluster analysis of protein interaction data. *Bioinformatics* 21: 364-378

Ashburner M, Ball CA, Blake JA, Botstein D, Butler H, Cherry JM, Davis AP, Dolinski K, Dwight SS, Eppig JT, Harris MA, Hill DP, Issel-Tarver L, Kasarskis A, Lewis S, Matese JC, Richardson JE, Ringwald M, Rubin GM, Sherlock G (2000) Gene ontology: tool for the unification of biology. The Gene Ontology Consortium. *Nature genetics* 25: 25-29

Bader GD, Cary MP, Sander C (2006) Pathguide: a pathway resource list. *Nucleic acids research* 34: D504-506

Bader GD, Hogue CW (2003) An automated method for finding molecular complexes in large protein interaction networks. *BMC bioinformatics* 4: 2

Barker D, Pagel M (2005) Predicting functional gene links from phylogenetic-statistical analyses of whole genomes. *PLoS computational biology* 1: e3

Ben-Hur A, Noble WS (2005) Kernel methods for predicting protein-protein interactions. *Bioinformatics* 21 Suppl 1: i38-46

Berggard T, Linse S, James P (2007) Methods for the detection and analysis of protein-protein interactions. *Proteomics* 7: 2833-2842

Bhasin M, Raghava GP (2004) ESLpred: SVM-based method for subcellular localization of eukaryotic proteins using dipeptide composition and PSI-BLAST. *Nucleic acids research* 32: W414-419

Blaschke C, Hoffmann R, Oliveros JC, Valencia A (2001) Extracting information automatically from biological literature. *Comparative and functional genomics* 2: 310-313

Bonetta L (2010) Protein-protein interactions: Interactome under construction. *Nature* 468: 851-854

Bork P, Dandekar T, Diaz-Lazcoz Y, Eisenhaber F, Huynen M, Yuan Y (1998) Predicting function: from genes to genomes and back. *Journal of molecular biology* 283: 707-725

Bousquet-Dubouch MP, Fabre B, Monsarrat B, Burlet-Schiltz O (2011) Proteomics to study the diversity and dynamics of proteasome complexes: from fundamentals to the clinic. *Expert review of proteomics* 8: 459-481

Bowers PM, Cokus SJ, Eisenberg D, Yeates TO (2004) Use of logic relationships to decipher protein network organization. *Science* 306: 2246-2249

Cai YD, Chou KC (2003) Nearest neighbour algorithm for predicting protein subcellular location by combining functional domain composition and pseudo-amino acid composition. *Biochemical and biophysical research communications* 305: 407-411

Cai YD, Chou KC (2004) Predicting 22 protein localizations in budding yeast. *Biochemical and biophysical research communications* 323: 425-428

Cai YD, Liu XJ, Xu XB, Chou KC (2000) Support vector machines for prediction of protein subcellular location. *Molecular cell biology research communications : MCBRC* 4: 230-233

Cai YD, Zhou GP, Chou KC (2003) Support vector machines for predicting membrane protein types by using functional domain composition. *Biophysical journal* 84: 3257-3263

Carlson BA, Xu XM, Kryukov GV, Rao M, Berry MJ, Gladyshev VN, Hatfield DL (2004) Identification and characterization of phosphoseryl-tRNA[Ser]Sec kinase. *Proceedings of the National Academy of Sciences of the United States of America* 101: 12848-12853

Cedano J, Aloy P, Perez-Pons JA, Querol E (1997) Relation between amino acid composition and cellular location of proteins. *Journal of molecular biology* 266: 594-600

Cho YR, Hwang W, Ramanathan M, Zhang A (2007) Semantic integration to identify overlapping functional modules in protein interaction networks. *BMC bioinformatics* 8: 265

Chou KC (2000) Prediction of protein subcellular locations by incorporating quasi-sequence-order effect. *Biochemical and biophysical research communications* 278: 477-483

Chou KC (2001) Prediction of protein cellular attributes using pseudo-amino acid composition. *Proteins* 43: 246-255

Chou KC, Cai YD (2002) Using functional domain composition and support vector machines for prediction of protein subcellular location. *The Journal of biological chemistry* 277: 45765-45769

Chou KC, Cai YD (2003) Prediction and classification of protein subcellular location-sequence-order effect and pseudo amino acid composition. *Journal of cellular biochemistry* 90: 1250-1260

Chou KC, Cai YD (2005) Predicting protein localization in budding yeast. *Bioinformatics (Oxford, England)* 21: 944-950

Chou KC, Elrod DW (1999) Protein subcellular location prediction. *Protein engineering* 12: 107-118

Chou KC, Shen HB (2007) Recent progress in protein subcellular location prediction. *Analytical biochemistry* 370: 1-16

Chua HN, Sung WK, Wong L (2006) Exploiting indirect neighbours and topological weight to predict protein function from protein-protein interactions. *Bioinformatics* 22: 1623-1630

Cote R, Reisinger F, Martens L, Barsnes H, Vizcaino JA, Hermjakob H (2010) The Ontology Lookup Service: bigger and better. *Nucleic acids research* 38: W155-160

Deng M, Chen T, Sun F (2004) An integrated probabilistic model for functional prediction of proteins. *Journal of computational biology : a journal of computational molecular cell biology* 11: 463-475

Drawid A, Gerstein M (2000) A Bayesian system integrating expression data with sequence patterns for localizing proteins: comprehensive application to the yeast genome. *Journal of molecular biology* 301: 1059-1075

Eddy SR (2009) A new generation of homology search tools based on probabilistic inference. *Genome informatics International Conference on Genome Informatics* 23: 205-211

Emanuelsson O, Nielsen H, Brunak S, von Heijne G (2000) Predicting subcellular localization of proteins based on their N-terminal amino acid sequence. *Journal of molecular biology* 300: 1005-1016

Engel SR, Balakrishnan R, Binkley G, Christie KR, Costanzo MC, Dwight SS, Fisk DG, Hirschman JE, Hitz BC, Hong EL, Krieger CJ, Livstone MS, Miyasato SR, Nash R, Oughtred R, Park J, Skrzypek MS, Weng S, Wong ED, Dolinski K et al (2010) Saccharomyces Genome Database provides mutant phenotype data. *Nucleic acids research* 38: D433-436

Ezkurdia I, Bartoli L, Fariselli P, Casadio R, Valencia A, Tress ML (2009) Progress and challenges in predicting protein-protein interaction sites. *Briefings in bioinformatics* 10: 233-246

Ferrer L, Dale JM, Karp PD (2010) A systematic study of genome context methods: calibration, normalization and combination. *BMC bioinformatics* 11: 493

Fetrow JS, Skolnick J (1998) Method for prediction of protein function from sequence using the sequence-to-structure-to-function paradigm with application to glutaredoxins/thioredoxins and T1 ribonucleases. *Journal of molecular biology* 281: 949-968

Finn RD, Mistry J, Tate J, Coggill P, Heger A, Pollington JE, Gavin OL, Gunasekaran P, Ceric G, Forslund K, Holm L, Sonnhammer EL, Eddy SR, Bateman A (2010) The Pfam protein families database. *Nucleic acids research* 38: D211-222

Gardy JL, Spencer C, Wang K, Ester M, Tusnady GE, Simon I, Hua S, deFays K, Lambert C, Nakai K, Brinkman FS (2003) PSORT-B: Improving protein subcellular localization prediction for Gram-negative bacteria. *Nucleic acids research* 31: 3613-3617

Garg P, Sharma V, Chaudhari P, Roy N (2009) SubCellProt: predicting protein subcellular localization using machine learning approaches. *In silico biology* 9: 35-44

Gaudet P, Livstone MS, Lewis SE, Thomas PD (2011) Phylogenetic-based propagation of functional annotations within the Gene Ontology consortium. *Briefings in bioinformatics* 12: 449-462

Gavin AC, Aloy P, Grandi P, Krause R, Boesche M, Marzioch M, Rau C, Jensen LJ, Bastuck S, Dumpelfeld B, Edelmann A, Heurtier MA, Hoffman V, Hoefert C, Klein K, Hudak M, Michon AM, Schelder M, Schirle M, Remor M et al (2006) Proteome survey reveals modularity of the yeast cell machinery. *Nature* 440: 631-636

Gingras AC, Gstaiger M, Raught B, Aebersold R (2007) Analysis of protein complexes using mass spectrometry. *Nature reviews Molecular cell biology* 8: 645-654

Grigoriev A (2001) A relationship between gene expression and protein interactions on the proteome scale: analysis of the bacteriophage T7 and the yeast Saccharomyces cerevisiae. *Nucleic acids research* 29: 3513-3519

Guyon I, Weston J, Barnhill S, Vapnik V (2002) Gene Selection for Cancer Classification using Support Vector Machines. *Machine Learning* 46: 389-422

Hartwell LH, Hopfield JJ, Leibler S, Murray AW (1999) From molecular to modular cell biology. *Nature* 402: C47-52

Hishigaki H, Nakai K, Ono T, Tanigami A, Takagi T (2001) Assessment of prediction accuracy of protein function from protein--protein interaction data. *Yeast* 18: 523-531

Hosur R, Xu J, Bienkowska J, Berger B (2011) iWRAP: An interface threading approach with application to prediction of cancer-related protein-protein interactions. *Journal of molecular biology* 405: 1295-1310

Hua S, Sun Z (2001) Support vector machine approach for protein subcellular localization prediction. *Bioinformatics (Oxford, England)* 17: 721-728

Huang Y, Li Y (2004) Prediction of protein subcellular locations using fuzzy k-NN method. *Bioinformatics (Oxford, England)* 20: 21-28

Huh WK, Falvo JV, Gerke LC, Carroll AS, Howson RW, Weissman JS, O'Shea EK (2003) Global analysis of protein localization in budding yeast. *Nature* 425: 686-691

Huynen M, Doerks T, Eisenhaber F, Orengo C, Sunyaev S, Yuan Y, Bork P (1998) Homology-based fold predictions for Mycoplasma genitalium proteins. *Journal of molecular biology* 280: 323-326

Imai K, Nakai K (2010) Prediction of subcellular locations of proteins: where to proceed? *Proteomics* 10: 3970-3983

Itoh T, Takemoto K, Mori H, Gojobori T (1999) Evolutionary instability of operon structures disclosed by sequence comparisons of complete microbial genomes. *Molecular biology and evolution* 16: 332-346

Kemmer D, Huang Y, Shah SP, Lim J, Brumm J, Yuen MM, Ling J, Xu T, Wasserman WW, Ouellette BF (2005) Ulysses - an application for the projection of molecular interactions across species. *Genome biology* 6: R106

Kemmeren P, Holstege FC (2003) Integrating functional genomics data. *Biochemical Society transactions* 31: 1484-1487

Kerrien S, Orchard S, Montecchi-Palazzi L, Aranda B, Quinn AF, Vinod N, Bader GD, Xenarios I, Wojcik J, Sherman D, Tyers M, Salama JJ, Moore S, Ceol A, Chatr-Aryamontri A, Oesterheld M, Stumpflen V, Salwinski L, Nerothin J, Cerami E et al (2007) Broadening the horizon--level 2.5 of the HUPO-PSI format for molecular interactions. *BMC biology* 5: 44

King AD, Przulj N, Jurisica I (2004) Protein complex prediction via cost-based clustering. *Bioinformatics* 20: 3013-3020

Kohl M, Wiese S, Warscheid B (2011) Cytoscape: software for visualization and analysis of biological networks. *Methods Mol Biol* 696: 291-303

Koonin EV, Wolf YI, Aravind L (2001) Prediction of the archaeal exosome and its connections with the proteasome and the translation and transcription machineries by a comparative-genomic approach. *Genome research* 11: 240-252

Kourmpetis YA, van Dijk AD, Bink MC, van Ham RC, ter Braak CJ (2010) Bayesian Markov Random Field analysis for protein function prediction based on network data. *PloS one* 5: e9293

Krogan NJ, Cagney G, Yu H, Zhong G, Guo X, Ignatchenko A, Li J, Pu S, Datta N, Tikuisis AP, Punna T, Peregrin-Alvarez JM, Shales M, Zhang X, Davey M, Robinson MD, Paccanaro A, Bray JE, Sheung A, Beattie B et al (2006) Global landscape of protein complexes in the yeast Saccharomyces cerevisiae. *Nature* 440: 637-643

Lee I, Blom UM, Wang PI, Shim JE, Marcotte EM (2011) Prioritizing candidate disease genes by network-based boosting of genome-wide association data. *Genome research* 21: 1109-1121

Lee I, Lehner B, Vavouri T, Shin J, Fraser AG, Marcotte EM (2010a) Predicting genetic modifier loci using functional gene networks. *Genome research* 20: 1143-1153

Lee K, Chuang HY, Beyer A, Sung MK, Huh WK, Lee B, Ideker T (2008) Protein networks markedly improve prediction of subcellular localization in multiple eukaryotic species. *Nucleic acids research* 36: e136

Lee K, Kim DW, Na D, Lee KH, Lee D (2006) PLPD: reliable protein localization prediction from imbalanced and overlapped datasets. *Nucleic acids research* 34: 4655-4666

Lee K, Thorneycroft D, Achuthan P, Hermjakob H, Ideker T (2010b) Mapping plant interactomes using literature curated and predicted protein-protein interaction data sets. *The Plant cell* 22: 997-1005

Lees JG, Heriche JK, Morilla I, Ranea JA, Orengo CA (2011) Systematic computational prediction of protein interaction networks. *Physical biology* 8: 035008

Lu H, Zhu X, Liu H, Skogerbo G, Zhang J, Zhang Y, Cai L, Zhao Y, Sun S, Xu J, Bu D, Chen R (2004) The interactome as a tree--an attempt to visualize the protein-protein interaction network in yeast. *Nucleic acids research* 32: 4804-4811

Lukk M, Kapushesky M, Nikkila J, Parkinson H, Goncalves A, Huber W, Ukkonen E, Brazma A (2010) A global map of human gene expression. *Nature biotechnology* 28: 322-324

Luttgen H, Rohdich F, Herz S, Wungsintaweekul J, Hecht S, Schuhr CA, Fellermeier M, Sagner S, Zenk MH, Bacher A, Eisenreich W (2000) Biosynthesis of terpenoids: YchB protein of Escherichia coli phosphorylates the 2-hydroxy group of 4-diphosphocytidyl-2C-methyl-D-erythritol. *Proceedings of the National Academy of Sciences of the United States of America* 97: 1062-1067

Mani R, St Onge RP, Hartman JLt, Giaever G, Roth FP (2008) Defining genetic interaction. *Proceedings of the National Academy of Sciences of the United States of America* 105: 3461-3466

Marcotte EM, Pellegrini M, Ng HL, Rice DW, Yeates TO, Eisenberg D (1999) Detecting protein function and protein-protein interactions from genome sequences. *Science* 285: 751-753

Matthews LR, Vaglio P, Reboul J, Ge H, Davis BP, Garrels J, Vincent S, Vidal M (2001) Identification of potential interaction networks using sequence-based searches for conserved protein-protein interactions or "interologs". *Genome research* 11: 2120-2126

Mei S, Fei W (2010) Amino acid classification based spectrum kernel fusion for protein subnuclear localization. *BMC bioinformatics* 11 Suppl 1: S17

Mei S, Fei W, Zhou S (2011) Gene ontology based transfer learning for protein subcellular localization. *BMC bioinformatics* 12: 44

Mintz-Oron S, Aharoni A, Ruppin E, Shlomi T (2009) Network-based prediction of metabolic enzymes' subcellular localization. *Bioinformatics (Oxford, England)* 25: i247-252

Mooney C, Wang YH, Pollastri G (2011) SCLpred: protein subcellular localization prediction by N-to-1 neural networks. *Bioinformatics (Oxford, England)* 27: 2812-2819

Mount DW (2007) Using a FASTA Sequence Database Similarity Search. *CSH protocols* 2007: pdb top16

Nair R, Rost B (2002) Inferring sub-cellular localization through automated lexical analysis. *Bioinformatics (Oxford, England)* 18 Suppl 1: S78-86

Nair R, Rost B (2003) Better prediction of sub-cellular localization by combining evolutionary and structural information. *Proteins* 53: 917-930

Nakai K, Horton P (1999) PSORT: a program for detecting sorting signals in proteins and predicting their subcellular localization. *Trends in biochemical sciences* 24: 34-36

Nakai K, Kanehisa M (1991) Expert system for predicting protein localization sites in gram-negative bacteria. *Proteins* 11: 95-110

Nakai K, Kanehisa M (1992) A knowledge base for predicting protein localization sites in eukaryotic cells. *Genomics* 14: 897-911

Nakashima H, Nishikawa K (1994) Discrimination of intracellular and extracellular proteins using amino acid composition and residue-pair frequencies. *Journal of molecular biology* 238: 54-61

Orchard S, Salwinski L, Kerrien S, Montecchi-Palazzi L, Oesterheld M, Stumpflen V, Ceol A, Chatr-aryamontri A, Armstrong J, Woollard P, Salama JJ, Moore S, Wojcik J, Bader GD, Vidal M, Cusick ME, Gerstein M, Gavin AC, Superti-Furga G, Greenblatt J et al (2007) The minimum information required for reporting a molecular interaction experiment (MIMIx). *Nature biotechnology* 25: 894-898

Pan YX, Zhang ZZ, Guo ZM, Feng GY, Huang ZD, He L (2003) Application of pseudo amino acid composition for predicting protein subcellular location: stochastic signal processing approach. *Journal of protein chemistry* 22: 395-402

Park KJ, Kanehisa M (2003) Prediction of protein subcellular locations by support vector machines using compositions of amino acids and amino acid pairs. *Bioinformatics (Oxford, England)* 19: 1656-1663

Pazos F, Ranea JA, Juan D, Sternberg MJ (2005) Assessing protein co-evolution in the context of the tree of life assists in the prediction of the interactome. *Journal of molecular biology* 352: 1002-1015

Pellegrini M, Marcotte EM, Thompson MJ, Eisenberg D, Yeates TO (1999) Assigning protein functions by comparative genome analysis: protein phylogenetic profiles. *Proceedings of the National Academy of Sciences of the United States of America* 96: 4285-4288

Pereira-Leal JB, Enright AJ, Ouzounis CA (2004) Detection of functional modules from protein interaction networks. *Proteins* 54: 49-57

Persico M, Ceol A, Gavrila C, Hoffmann R, Florio A, Cesareni G (2005) HomoMINT: an inferred human network based on orthology mapping of protein interactions discovered in model organisms. *BMC bioinformatics* 6 Suppl 4: S21

Pierleoni A, Martelli PL, Casadio R (2011) MemLoci: predicting subcellular localization of membrane proteins in eukaryotes. *Bioinformatics (Oxford, England)* 27: 1224-1230

Pitre S, Alamgir M, Green JR, Dumontier M, Dehne F, Golshani A (2008) Computational methods for predicting protein-protein interactions. *Advances in biochemical engineering/biotechnology* 110: 247-267

Recillas-Targa F (2006) Multiple strategies for gene transfer, expression, knockdown, and chromatin influence in mammalian cell lines and transgenic animals. *Molecular biotechnology* 34: 337-354

Reczko M, Hatzigerrorgiou A (2004) Prediction of the subcellular localization of eukaryotic proteins using sequence signals and composition. *Proteomics* 4: 1591-1596

Reid AJ, Ranea JA, Clegg AB, Orengo CA (2010) CODA: accurate detection of functional associations between proteins in eukaryotic genomes using domain fusion. *PloS one* 5: e10908

Reinhardt A, Hubbard T (1998) Using neural networks for prediction of the subcellular location of proteins. *Nucleic acids research* 26: 2230-2236

Rives AW, Galitski T (2003) Modular organization of cellular networks. *Proceedings of the National Academy of Sciences of the United States of America* 100: 1128-1133

Salwinski L, Miller CS, Smith AJ, Pettit FK, Bowie JU, Eisenberg D (2004) The Database of Interacting Proteins: 2004 update. *Nucleic acids research* 32: D449-451

Sanchez-Chapado M, Angulo JC, Ibarburen C, Aguado F, Ruiz A, Viano J, Garcia-Segura JM, Gonzalez-Esteban J, Rodriquez-Vallejo JM (1997) Comparison of digital rectal examination, transrectal ultrasonography, and multicoil magnetic resonance imaging for preoperative evaluation of prostate cancer. *European urology* 32: 140-149

Schwikowski B, Uetz P, Fields S (2000) A network of protein-protein interactions in yeast. *Nature biotechnology* 18: 1257-1261

Scott MS, Calafell SJ, Thomas DY, Hallett MT (2005) Refining protein subcellular localization. *PLoS computational biology* 1: e66

Sharan R, Ulitsky I, Shamir R (2007) Network-based prediction of protein function. *Molecular systems biology* 3: 88

Shatkay H, Hoglund A, Brady S, Blum T, Donnes P, Kohlbacher O (2007) SherLoc: high-accuracy prediction of protein subcellular localization by integrating text and protein sequence data. *Bioinformatics (Oxford, England)* 23: 1410-1417

Shen YQ, Burger G (2010) TESTLoc: protein subcellular localization prediction from EST data. *BMC bioinformatics* 11: 563

Shoemaker BA, Panchenko AR (2007) Deciphering protein-protein interactions. Part II. Computational methods to predict protein and domain interaction partners. *PLoS computational biology* 3: e43

Shoemaker BA, Zhang D, Thangudu RR, Tyagi M, Fong JH, Marchler-Bauer A, Bryant SH, Madej T, Panchenko AR (2010) Inferred Biomolecular Interaction Server--a web server to analyze and predict protein interacting partners and binding sites. *Nucleic acids research* 38: D518-524

Singh R, Park D, Xu J, Hosur R, Berger B (2010) Struct2Net: a web service to predict protein-protein interactions using a structure-based approach. *Nucleic acids research* 38: W508-515

Skarnes WC, Rosen B, West AP, Koutsourakis M, Bushell W, Iyer V, Mujica AO, Thomas M, Harrow J, Cox T, Jackson D, Severin J, Biggs P, Fu J, Nefedov M, de Jong PJ, Stewart AF, Bradley A (2011) A conditional knockout resource for the genome-wide study of mouse gene function. *Nature* 474: 337-342

Skrabanek L, Saini HK, Bader GD, Enright AJ (2008) Computational prediction of protein-protein interactions. *Molecular biotechnology* 38: 1-17

Smith KL, DeVos V, Bryden H, Price LB, Hugh-Jones ME, Keim P (2000) Bacillus anthracis diversity in Kruger National Park. *Journal of clinical microbiology* 38: 3780-3784

Spirin V, Mirny LA (2003) Protein complexes and functional modules in molecular networks. *Proceedings of the National Academy of Sciences of the United States of America* 100: 12123-12128

Stark C, Breitkreutz BJ, Chatr-Aryamontri A, Boucher L, Oughtred R, Livstone MS, Nixon J, Van Auken K, Wang X, Shi X, Reguly T, Rust JM, Winter A, Dolinski K, Tyers M (2011) The BioGRID Interaction Database: 2011 update. *Nucleic acids research* 39: D698-704

Stein A, Panjkovich A, Aloy P (2009) 3did Update: domain-domain and peptide-mediated interactions of known 3D structure. *Nucleic acids research* 37: D300-304

Stuart JM, Segal E, Koller D, Kim SK (2003) A gene-coexpression network for global discovery of conserved genetic modules. *Science* 302: 249-255

Suter B, Kittanakom S, Stagljar I (2008) Two-hybrid technologies in proteomics research. *Current opinion in biotechnology* 19: 316-323

Szklarczyk D, Franceschini A, Kuhn M, Simonovic M, Roth A, Minguez P, Doerks T, Stark M, Muller J, Bork P, Jensen LJ, von Mering C (2011) The STRING database in 2011: functional interaction networks of proteins, globally integrated and scored. *Nucleic acids research* 39: D561-568

Tian J, Gu H, Liu W, Gao C (2011) Robust prediction of protein subcellular localization combining PCA and WSVMs. *Computers in biology and medicine* 41: 648-652

Tikk D, Thomas P, Palaga P, Hakenberg J, Leser U (2010) A comprehensive benchmark of kernel methods to extract protein-protein interactions from literature. *PLoS computational biology* 6: e1000837

Tuncbag N, Gursoy A, Guney E, Nussinov R, Keskin O (2008) Architectures and functional coverage of protein-protein interfaces. *Journal of molecular biology* 381: 785-802

Tuncbag N, Kar G, Keskin O, Gursoy A, Nussinov R (2009) A survey of available tools and web servers for analysis of protein-protein interactions and interfaces. *Briefings in bioinformatics* 10: 217-232

Tung TQ, Lee D (2009) A method to improve protein subcellular localization prediction by integrating various biological data sources. *BMC bioinformatics* 10 Suppl 1: S43

Turinsky AL, Razick S, Turner B, Donaldson IM, Wodak SJ (2010) Literature curation of protein interactions: measuring agreement across major public databases. *Database : the journal of biological databases and curation* 2010: baq026

Turner B, Razick S, Turinsky AL, Vlasblom J, Crowdy EK, Cho E, Morrison K, Donaldson IM, Wodak SJ (2010) iRefWeb: interactive analysis of consolidated protein interaction data and their supporting evidence. *Database : the journal of biological databases and curation* 2010: baq023

Vazquez A, Flammini A, Maritan A, Vespignani A (2003) Global protein function prediction from protein-protein interaction networks. *Nature biotechnology* 21: 697-700

Wang J, Li C, Wang E, Wang X (2011) An FPT approach for predicting protein localization from yeast genomic data. *PloS one* 6: e14449

Whisstock JC, Lesk AM (2003) Prediction of protein function from protein sequence and structure. *Quarterly reviews of biophysics* 36: 307-340

Xia JF, Wang SL, Lei YK (2010) Computational methods for the prediction of protein-protein interactions. *Protein and peptide letters* 17: 1069-1078

Yeats C, Lees J, Carter P, Sillitoe I, Orengo C (2011) The Gene3D Web Services: a platform for identifying, annotating and comparing structural domains in protein sequences. *Nucleic acids research* 39: W546-550

Yuan Z (1999) Prediction of protein subcellular locations using Markov chain models. *FEBS letters* 451: 23-26

Zhang QC, Petrey D, Norel R, Honig BH (2010) Protein interface conservation across structure space. *Proceedings of the National Academy of Sciences of the United States of America* 107: 10896-10901

Zhang Z, Sun H, Zhang Y, Zhao Y, Shi B, Sun S, Lu H, Bu D, Ling L, Chen R (2006) Genome-wide analysis of mammalian DNA segment fusion/fission. *Journal of theoretical biology* 240: 200-208

Zhou GP, Doctor K (2003) Subcellular location prediction of apoptosis proteins. *Proteins* 50: 44-48

Zhu L, Yang J, Shen HB (2009) Multi label learning for prediction of human protein subcellular localizations. *The protein journal* 28: 384-390

4

Integrative Approach for Detection of Functional Modules from Protein-Protein Interaction Networks

Zelmina Lubovac-Pilav
University of Skövde, Systems Biology Research Centre
Sweden

1. Introduction

Advances in large scale technologies in proteomics, such as yeast two-hybrid (Y2H) screening and mass spectrometry (MS) have enabled us to generate large protein-protein interaction (PPI) networks. The structure of such networks has been frequently analysed to identify the modules, which constitute the basic "building blocks" of molecular networks. One of the challenges that systems biology is facing consists of explaining biological organisation in the light of the existence of modules in networks (Han et al., 2004; Pereira-Leal et al., 2004; Petti and Church, 2005; Rives and Galitski, 2003). A series of studies attempting to reveal the modules in cellular networks, ranging from metabolic (Ravasz et al., 2002), to protein networks (Spirin and Mirny, 2003; Yook et al., 2004), support the proposal that modular architecture is one of the principles underlying biological organisation.

Several key issues are being addressed in current research in systems biology, as a result of our post-genomic view that has expanded the role of the protein into an element of a network in which it has contextual functions within functional modules (Eisenberg et al., 2000; Jeong et al., 2001). How do modules interact to achieve a certain functionality (Han et al., 2004; Rives and Galitski, 2003)? How can we evaluate the biological relevance of modules (Pereira-Leal et al., 2004; Poyatos and Hurst, 2004)? Answering those questions may contribute to better understanding of the relationships between structure, function and regulation of molecular networks, which is an important aim of systems biology (Qi and Ge, 2006; Stelling et al., 2002).

From the structural perspective, modules are often associated with highly connected clusters of proteins. Many efforts in this area have been directed towards analysing structural properties of the protein interaction graph, measured by clustering coefficient and shortest path distance for example, to derive modular formations. The main focus presented in this chapter is on defining similarity between protein interactions based on an integrated score that takes into consideration topology of PPI network along with the functional knowledge determined by semantic similarity. An important reason for considering knowledge represented in annotations a valuable complement to topological characteristics is encompassed in the concept of functional modules themselves. A functional module consists of proteins that cooperate towards achieving a particular function or participate in similar processes. Hence, considering annotation that describes molecular functions and biological processes should enrich the protein-protein interactions. Functional information can be retrieved from Gene Ontology (GO), which is a structured vocabulary used to annotate proteins with information about their molecular function, participation in biological processes or localization in cellular components. A module-identifying algorithm proposed earlier (Lubovac et al., 2006), SWEMODE (Semantic WEights for MODule Elucidation), that relies on an integrated measure, called semantic cohesiveness, corresponds to one of the successful approaches that contributes to achieve the important aims of systems biology. This method will be the focus of attention in this chapter.

2. Background

Molecular biology is becoming a highly modular science where functional modules are considered to be a critical level of biological organization. The term "module", as understood in molecular biology, was originally defined as a discrete unit with a function that is separable from those of other modules (Hartwell et al., 1999). Furthermore, modularity refers to clusters of elements that work in a co-operative fashion to achieve some defined function. Protein complexes constitute one example type of module, since the proteins within a complex interact functionally and physically to form a robust unit, which in its turn carries out some biological function (Yook et al., 2004).

One of the key issues to be solved with help of bioinformatics is the deciphering of the complex architecture of biological networks.

2.1 Climbing life's complexity pyramid

Biological networks are often modular and compound, and involve connections between groups of genes and proteins as well as between individual elements. A simple complexity pyramid (see Fig. 1) suggested by Oltvai and Barabasi (2002), illustrates different levels of cellular organisation.

Living systems are organised at both logical and physical levels. The individual nucleotides are elementary building blocks of DNA and RNA molecules, which, in turn, are organised into higher level structures such as regulatory elements, and genes. DNA is physically organised into larger structures such as chromatin and chromosomes. Groups of genes, proteins, RNAs (the bottom level of the pyramid in Fig. 1) may be organised into pathways in metabolism, and motifs in genetic regulatory networks (see level 2). Regulatory motifs may in turn serve as building blocks of functional modules (level 3). There is a growing body of evidence that the modules are then organised in a hierarchical manner (Barabasi and Oltvai, 2004; Oltvai and Barabasi, 2002; Ravasz et al., 2002), defining the large-scale functional organisation of the cell (level 4 in Fig. 1).

The way these various structures interact with each other determines the machinery of a cell. Cells and the extracellular matrix, which surrounds and supports cells, build up the tissues that in turn are organised into organs, and so forth.

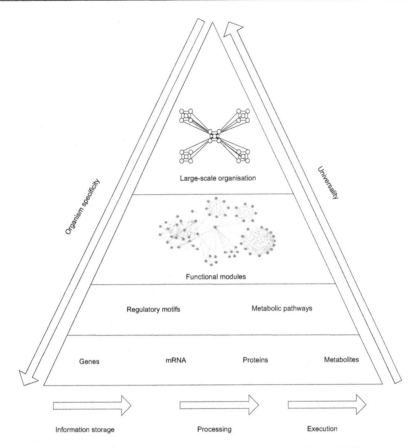

Fig. 1. Life's complexity pyramid redrawn from (Oltvai and Barabasi, 2002).

The integration of different layers in the pyramid to achieve a better understanding of system-level rules that govern cell function is one of the challenges in systems biology. Computational analysis tools and methods are needed at each level but also across different levels. Here, the integrative approach for deriving modules at the third level in the pyramid is described, which also make it possible to climb to the top, and provide means for revealing large-scale organisation.

2.2 Modularity in cellular networks

"Modularity is a fundamental design principle whereby components are partitioned according to common physical, regulatory, or functional properties" (Petti and Church, 2005). Modules can be found in many systems, for example, food webs, networks of web pages describing related subjects (Flake et al., 2002), networks of friends in sociology (Newman, 2003), or scientific collaboration networks (Newman, 2001). A usual synonym for the term module in other scientific disciplines, like sociology for example, is community or community structure. In a study by Flake et al., (2002), the term web community is for example defined as "a collection of web pages such that each member page has more hyperlinks within the community than outside of the community". This definition may be adjusted further, according to Flake et al., (2002), to identify communities of varying sizes and levels of cohesiveness (clustering).

Furthermore, modularity involves groups of elements that work in a co-operative fashion to achieve some well-defined function. In a general network representation, a module appears as a highly interconnected group of nodes (Barabasi and Oltvai, 2004). Modules can be interpreted as separated substructures of a network or pathway, e.g. a protein complex is a module of a protein interaction network. Protein complexes are well-defined examples of modularity since they consist of proteins that interact functionally and physically to form a tightly connected unit, which, in turn, carries out some biological function (Yook et al., 2004). Another example of modular organisation can be found in genetic regulatory networks where several transcription factor binding sites, organised into functional units, i.e. modules, play a crucial role in gene transcription.

The members that constitute modules are more strongly related to each other than to members of other modules, which is reflected in the network topology. The modular nature of PPI networks is reflected by a high degree of clustering, measured by the clustering coefficient. The clustering coefficient measures the local cohesiveness around a node, and it is defined, for any node i, as the fraction of neighbours of i that are connected to each other (Watts and Strogatz, 1998). Simply stated, the clustering coefficient c_i measures the presence of 'triangles' which have a corner at i (see the triangles with dashed sides in Fig. 2). The high degree of clustering is based on local sub-graphs with a high density of internal connections, while being less tightly connected to the rest of the network (Uhrig, 2006).

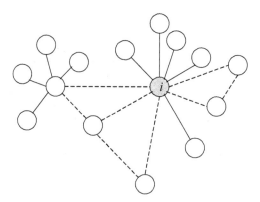

Fig. 2. Example of a protein sub-graph with triangle-forming proteins.

As pointed out by Barabasi and Oltvai (2004), each module may be reduced to a set of triangles, and a high density of such triangles is highly characteristic for PPI networks, pointing at the modular nature of such networks. By averaging the clustering coefficient over all nodes we can obtain a global measure of the cohesiveness of the network, where a high average clustering coefficient indicates the presence of modularity. It has been confirmed in many studies that most real large-scale networks tend to contain dense clusters, in the sense that the average clustering coefficient of such networks is much greater than for random networks. In contrast, if modularity is absent in the network, the average clustering coefficient is comparable to that of a randomised network.

The exact meaning of modularity in biological networks depends on the network under consideration. For example, modules in protein networks are often seen as static molecular complexes (such as the ribosome) or as dynamic signalling pathways (such as the MAPK cascade). There are also examples of large modular molecule complexes that are in turn organised in modules. One of such complexes is yeast Mediator, which transmits regulatory signals from DNA-binding transcription factors to RNA polymerase II. The Mediator complex is thought to be composed of 24 subunits organised in four modules, named the head, middle, tail and Cdk8 modules. In gene regulatory networks, modules are often seen as sets of genes controlled by the same set of transcription factors under certain conditions (Segal et al., 2003).

Modules should not be seen as isolated components, since it has been shown that some crosstalk and overlap exists between them (Han et al., 2004; Schwikowski et al., 2000). Instead, modules should be considered as components that have dense intra-connectivity but sparse inter-connectivity. In a study analysing protein interaction networks in the yeast *Saccharomyces cerevisiae*, Schwikowski et al., (2000) reported global patterns of interactions of proteins within functional classes or subcellular compartments, as well as many possible cross-connections. It is further pointed out by Qi and Ge (2006) that the existence of the links between modules emphasises the coordination of the cellular processes. For example, Petti and Church (2005) investigated possible transcriptional coordination between glycolysis and lipid metabolism modules.

A growing body of work supports the idea that such modules underlie much of cellular functioning (Gavin et al., 2006; Han et al., 2004; Pereira-Leal et al., 2004; Qi and Ge, 2006; Rives and Galitski, 2003), and that functional modules are the most relevant organisational units of a cell from the perspective of systems biology (Hartwell et al., 1999).

2.3 Integrating functional knowledge in module discovery

Although topology-based network measures, such as clustering coefficient, play an important role in module discovery, there are some reasons why we should integrate functional knowledge as well when deriving modular formations. High-throughput protein interaction data that is often used to identify modules is very noisy (Titz et al., 2004). Technologies such as Y2H often result in many false positives that may cause false conclusions in the analysis. A possible approach to decrease the number of false interactions may be to focus on the "high confidence" data sets, where all interactions have been confirmed by several experiments. However, in this way the majority of the existing interactions would be discarded from further analysis. A better approach should imply incorporating the functional knowledge associated with available interactions into the analysis. This has also been pointed out in previous studies that focus on deriving protein complexes by using topological information. In (Przulj et al., 2004), it has been observed that the increasing size of PPI networks (by including medium and low confidence interactions) has resulted in a decreasing number of highly connected sub-graphs or clusters which may correspond to protein complexes. As Przulj, et al., (2004) state, the reason for this may be the increasing noise in the data, and a possible solution to this problem is the integration of PPI networks with annotation or gene expression data. In sub-chapter 2.4 a possible general framework for such integrative approach for module identification is described.

2.4 A general framework for integrative module identification

There are many ways of measuring similarity between proteins. The main proposal presented here considers protein similarity based on an integrated score that takes into consideration protein interaction data (as a topology source) and functional information based on semantic similarity. As pointed out previously, an ideal approach should take into consideration both temporal and spatial data, to be able to reflect the true dynamics of the cellular networks. It is therefore worthwhile to discuss how the methods presented here may be generalised to cope with several sources of information. Our module-identifying framework may be generalised by:

1.   considering several sources of topological information
2.   considering several sources of functional information

Topological information may refer to, for example, protein-protein interactions obtained from different experimental sources, such as Y2H and MS. However, this information may also be derived from different topological properties like clustering coefficient, edge betweenness, etc.

Besides semantic similarity values based on protein GO terms that we used in this work, there are many other sources of functional information that may be useful for predicting membership in protein complexes. One of the most prominent sources is gene expression data generated using various high-throughput platforms, such as microarrays. Expression profile correlation coefficients may, for example, be used to assign similarity scores to pairwise interactions. Other sources of functional information are essentiality, phylogenetic profiles, localisation, the MIPS functional catalogue, etc.

In this study, as in the majority of others, protein interactions are treated as binary, i.e. the edges in a network are either present or absent. Bearing in mind the fact that large-scale methods, although offering vast improvements in efficiency, still have much higher error rates than small-scale methods, a step towards generalisation of the proposed algorithms would be to treat protein interaction networks probabilistically. By treating the edges as binary (indicating presence/absence of interaction), we cannot distinguish edges supported by multiple evidence types, from edges supported by evidence of differing quality. There are several ways of assigning probabilities to individual pairs of proteins based on the amount and type of supporting evidence (Asthana et al., 2004; Jansen et al., 2002; Jansen et al., 2003). When dealing with several data sources that need to be combined in order to improve the prediction, a usual way of combining these consists of overlapping different interactomes. This approach, in turn, gives rise to the question whether it is more beneficial to consider the union of the disparate datasets or their intersection. One of the extremes that may be envisaged is that each one of the networks that are to be integrated has a low rate of false positives (FP) but a high rate of false negatives (FN). In this case, the union of the two sets of interactions would be advantageous. At the other extreme, when dealing with networks with high FP rates and low FN rates, the intersection between the different networks is preferable.

The problem of finding an optimal combination of unions and intersections among the different networks may be defined, as described in (Jansen et al., 2002), as finding a trade-off between the highest possible coverage (TP/(TP+FN)) and the lowest possible error rate (FP/(TP+FP)). Determining the error rate is still an open question, as pointed out in (Jansen et al., 2002).

A hypothetical example of integrating different data sources that may be useful in generalising the proposed approaches is given in Fig. 3. The top part of the figure shows four possible data sources that may be useful for module identification. Two of them are topological sources, denoted as t_1 and t_2, and are usually treated as binary networks. The other two sources, denoted as f_1 and f_2, may be used to assign functional weights to the edges. For example, when using gene expression as a possible source for weighting the edges, the probability of finding two proteins in a complex, given a certain correlation between their expression profiles, may be a possible way to assign weights (Jansen et al., 2002). Gene ontology sub-graphs as a possible source of functional information is visualised in the third square in Fig. 3, where semantic similarity between ontology terms may be used to reflect the functional similarity between the proteins, as assumed in this work. These functional weights may also be transformed into binary values, by setting different thresholds, where the level of the threshold determines the sensitivity and specificity of the experiment.

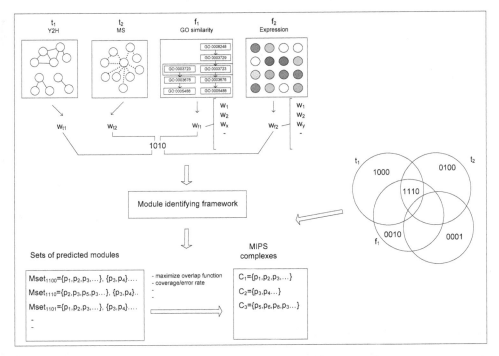

Fig. 3. Hypothetical integration of four data sources for module identification.

The bottom part of Fig. 3 shows the hypothetical module sets generated with different combinations of data sets. The Venn diagram to the right in the figure shows binary subset profiles, where profile 1110 includes all data points that are present in data sets t_1, t_2, and f_1. Mset1110, for example, denotes the set of modules derived from the combination of MS, Y2H, and GO semantic similarity weights, where p_x denotes a protein x belonging the module.

3. Module identification based on an integrated approach

The algorithm described in previous work (Lubovac et al., 2006), SWEMODE (Semantic WEeights for MODule Elucidation), is an example of a method that employs an integrated approach for deriving functional modules, based on the functional and topological cohesiveness of the sub-graphs. Here, an integrated weighting score, called weighted clustering coefficient, that forms the bases for this method will be described. The reason for focusing on description of the integrative score here is that it can be applied as a part of node weighting procedure in other methods for deriving modules of PPI networks.

3.1 Weighted clustering coefficient

As depicted in earlier work, the separate edge weights do not provide an overall picture of the network's complexity. Therefore, we here consider the sum of all weights between a particular node and its neighbours, also referred to as the node strength. The strength s_i of the node i is defined as:

$$s_i = \sum_{\forall j, j \in N(i)} ss_{ij} \tag{1}$$

Given two proteins, i and j, with T_i and T_j containing m and n terms, respectively, the protein-protein semantic similarity ss_{ij} based on GO terms, is defined as the average inter-set similarity between terms from the given term sets (see Equation 2).

$$ss_{ij} = \frac{1}{m \times n} \sum_{t_k \in T_i, t_l \in T_j} sim(t_k, t_l) \tag{2}$$

Determining the similarity between two proteins i and j, is preceded by calculation of the similarity between the terms belonging to the term sets T_i and T_j that are used to annotate these proteins. Given the ontology terms $t_k \in T_i$ and $t_l \in T_j$, the semantic similarity measure proposed by (Lin, 1998) is defined as:

$$sim(t_k, t_l) = \frac{2 \ln p_{ms}(t_k, t_l)}{\ln p(t_k) + \ln p(t_l)} \tag{3}$$

Where $p(t_x)$ is the probability of term t_x and $p_{ms}(t_k, t_l)$ is the probability of the minimum subsumer of t_k and t_l, which is defined as the lowest probability found among the parent terms shared by t_k and t_l (Lord et al., 2003).

In previous work, some extensions of the topological clustering coefficient have been developed for weighted networks. In (Barrat et al., 2004), two scores that integrate topological and weighted features of the nodes – weighted clustering coefficient c^w and weighted average nearest-neighbours degree nn^w are introduced. These scores have previously been applied to two types of complex weighted networks, namely, the world-wide airport network and the scientist collaboration network. A first attempt to apply these integrated scores on PPI networks was described in (Lubovac et al., 2006). A weighted measure that uses semantic similarity weights was introduced. Weighted clustering coefficient c^w is defined as:

$$c_i^w = \frac{1}{s_i(k_i-1)} \sum_{\forall j,h \mid \{j,h\} \in K(i)} (ss_{ij} + ss_{ih}) \qquad (4)$$

Where s_i is the functional strength of node i (see Equation 1) and ss_{ij} is the semantic similarity reflecting the functional weight of the interaction (see Equation 2). For each triangle formed in the neighbourhood of node i, involving nodes j and h, the semantic similarities ss_{ij} and ss_{ih} are calculated. Hence, not only the number of triangles in the neighbourhood of the node i is considered but also the relative functional similarity between the nodes that form those triangles, with regard to the total functional strength of the node. The normalisation factor $s_i(k_i-1)$ represents the summed weight of all edges connected from node i, multiplied by the maximum possible number of triangles in which each edge may participate. It also ensures that $0 \le c^w \le 1$. This measure can be involve any of the three aspects of Gene Ontology - molecular function, biological process and cellular component, or the combination of these.

4. Comparison with topology-based methods for module identification

The aim of this sub-chapter is to demonstrate the performance of the approach called SWEMODE (Lubovac et al., 2006), based on an integrative score described in 3.1, by comparing it to two purely topological approaches. One of the topology-based method for detecting modules from a PPI networks has been developed by Luo and Scheuerman (2006) and further analysed in (Luo et al., 2007). The module notion proposed was based on the degree definition of the sub-graphs. Unlike the approach described in Section 3, this method is based solely on topological properties of the protein sub-graph.

Modules generated with SWEMODE were also compared with the modules derived in (Przulj et al., 2004), based on HCS (Highly Connected Subgraphs) clustering algorithm (Hartuv and Shamir, 2000). This method aims to find disjoint subsets (clusters) that should satisfy following criteria: homogeneity – members of the same cluster are highly similar to each other; and separation: members of different clusters have low similarity to each other.

4.1 Protein-protein interaction data

For the evaluation purpose, two different PPI networks have been used. The first one was derived from the Database of Interacting Proteins (DIP: http://dip.doe-mbi.ucla.edu), which is a database that stores and organises experimentally determined PPI (Xenarios et al., 2000). There is the subset of PPI from Yeast *S. cerevisiae*, denoted as CORE, which is the result of assessment with the Expression Profile Reliability Index (ERP Index) and the Paralogous Verification Method (PVM) (for further details, see (Deane et al., 2002)). The CORE subset contained 6379 interactions.

The second data set of PPI is obtained from the study by (von Mering et al., 2002). In that study, a quality assessment of large-scale data sets of protein-protein interactions in yeast was performed. A critical evaluation of the accuracy of high-throughput data is needed, because of the high rate of false interactions in these data sets. In (von Mering et al., 2002), data sets from yeast two-hybrid (Y2H) systems, protein complex purification techniques that rely on mass-spectroscopy (TAP and HMS-PCI), correlated mRNA expression profiles, genetic interactions, and *in silico* interaction predictions were analysed. As stated further in this study, each of these methods can be used to predict protein interactions, even though their goals are slightly different.

The authors integrated about 80 000 interactions between yeast proteins and found that only 2 455 were supported by more than one method. This low overlap between sets of protein interactions obtained from different methods may be due to the high fraction of false positives, but may also be caused by the difficulties for some methods to capture certain types of interactions. All interactions are classified by the level of confidence (low, medium, high), based on the evidence that supports them. In our study, we have used the interaction set with high level of confidence, meaning that all interactions are confirmed by several methods. This data set will be referred to as "von Mering". The data set contains 2 455 interactions between 988 proteins.

4.2 Evaluation against MIPS functional categories

The Munich Information Center for Protein Sequences (MIPS) provides high quality curated genome-related information, such as protein-protein interactions, protein complexes, protein functional categories, etc., spanning over several organisms.

The MIPS functional catalogue database consists of different fields, such as functional catalogue (FunCat) number, EC number, GO number, keywords etc. FunCat is an annotation scheme that provides functional descriptions of proteins (Ruepp et al., 2004). There are in total 28 main functional categories that are hierarchically structured. These categories cover functional fields such as metabolism, signal transduction, cellular transport etc.

The MIPS Comprehensive Yeast Genome Database (CYGD) provides information on the molecular structure and functional network of *S. cerevisiae*. The information used here for the evaluation purposes is the protein complex catalogue that contains a manually curated set of protein complexes that serve as an example of a type of module. There is another data set containing protein complexes obtained from (Gavin et al., 2002). This data set was produced by using a single experimental method, whereas the complex data set from MIPS has been derived from experiments from many labs using different techniques. Therefore, MIPS database is more realistic and appropriate to use for evaluation.

To evaluate and compare the performance of SWEMODE with two other methods for module identification, overlap score is used. In previous work, a similar evaluation has been applied to the clustering algorithm MCODE (Bader and Hogue, 2003), with respect to the number of matched complexes, but here slightly different definition of overlap score is used (see Equation 5).

The overlap score Ol (Poyatos and Hurst, 2004), is defined as:

$$Ol_{ij} = \left|M_i \cap M_j\right| \Big/ \sqrt{\left|M_i\right|\left|M_j\right|}$$ (5)

where M_i is the predicted module, and M_j is a module from the MIPS complex data set. The Ol measure assigns a score of 0 to modules that have no intersection with any known protein complex, whereas modules that exactly matches a known complex get the score 1.

4.3 Results

A total of 99 modules were detected in (Luo and Scheuermann, 2006). A new agglomerative algorithm was developed to identify modules from the network by combining the new module definition with the relative edge order generated by the Girvan-Newman algorithm. A JAVA program, MoNet, was developed to implement the algorithm Luo et al. (2007). Applying MoNet to the yeast core protein interaction network from the database of interacting proteins (DIP) identified 86 simple modules with sizes larger than 3 proteins. For convenience, those modules will be referred to as MoNet modules.

Evaluation of the MoNet modules with the overlap score threshold has been performed, and the results are compared with the resulting modules from SWEMODE, generated across approximately 400 different parameter settings (for parameter settings, see (Lubovac et al., 2006). We found that the modules derived from the latter show higher agreement with MIPS complexes (see Fig. 4). This comparison also indicates that introducing knowledge in terms

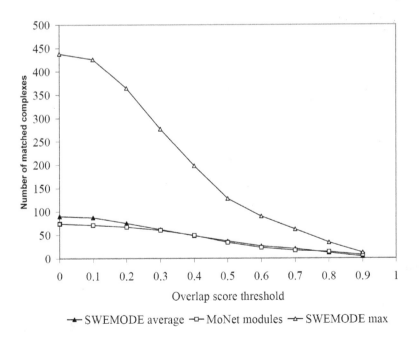

Fig. 4. Comparison between MoNet modules and SWEMODE modules.

of semantic similarity into the network topology seems to be advantageous over using only topology information. Furthermore, this method produces one single partition of the network, which does not seem biologically plausible, as many proteins may be involved in different processes.

We also compared our SWEMODE modules obtained from von Mering data with the modules derived in (Przulj et al., 2004), based on HCS. The modules generated with SWEMODE showed also here higher overlap with MIPS complexes (see Fig. 5). A more detailed analysis shows that both algorithms resulted in 39 identical modules. However, as HCS only discern the complexes that are highly interconnected, it discards many clusters that correspond to known complexes.

Another disadvantage of both methods that are here compared to SWEMODE is that they do not allow any overlap between modules, i.e. they produce disjoint clusters.

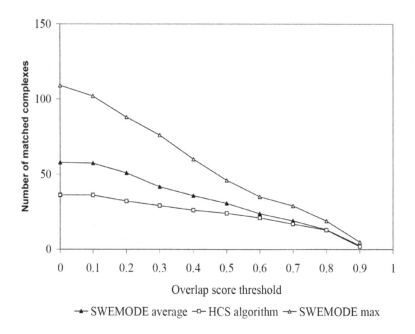

Fig. 5. Comparison between SWEMODE modules and modules generated with HCS clustering method.

5. Conclusion

The focus of attention in this chapter is the knowledge-based method that integrates domain specific knowledge, in this case functional information from Gene Ontology, with topological information, to derive modular structures from PPI networks. There are clear disadvantages with the approaches that only rely on topological information, as previously described. In contrast to these methods that often suffer from lack of biological plausibility, the approach described here takes into consideration the functional knowledge about the experimental interactions, and in this way strengthen the validity of the obtained modular structures. Modules obtained in this way serve as models for studying interconnectivity, which is a step towards reconstruction of the higher order hierarchy of cellular networks.

Three different biological aspects – molecular function, biological process and cellular component, have been employed and tested for their suitability for deriving modules. The identification of protein complexes may become more challenging as additional PPI data becomes available, because the interactions are noisy, and the integration of PPI data with annotation might prove a useful solution to this problem. The integrated approaches contribute to this solution, by increasing the confidence in high-throughput Y2H data. The approach also provides means for an increased understanding of the higher-order structures underlying cellular function. As annotations become more complete, the increased biological relevance of our module predictions with integrated approaches is expected to be even more evident.

One of the biggest issues in this type of study is the difficulty to clearly characterise modules. There is no generally accepted definition of modules. A pioneering work in this area, performed by Hartwell et al. (1999) provides a wide definition, which leaves space for different authors to define different more specific criteria. This is, as also pointed out in (Schlosser and Wagner, 2004), unavoidable, and "retaining a pragmatic pluralism of different modularity concepts is probably a fruitful strategy for broadening our perspective and illuminating the importance of modularity at many different levels of organization".

A possible future application of the method described in this chapter is identification of modules of genes and proteins involved in various diseases, such as cancer. This module-level knowledge can contribute to the understanding of cancer on system-level, which may be useful for developing new drugs. Cancer-related networks for a specific type of cancer may be derived from, for example, gene expression data. Deriving gene networks makes it possible to apply network theoretic approaches on the interconnected genes that are potentially related to cancer development. Furthermore, a comparative analysis of the cancer-related networks derived from different types of cancer could be performed to identify modules that are shared among different types, but also to identify the specific processes that characterize a certain type of cancer.

Modular analysis may also be applied to identify general properties of the interrelated genes that are involved in the origin of cancer cells. A suitable model for this analysis is a gene fusion network in human neoplasia (Hoglund et al., 2006). By investigating topological properties of the cancer nodes in the network, such as node betweenness centrality, the cancer-related genes that act as "bridges" or communication points between various modules that correspond to cancer related processes may be identified.

Explaining the relationships between structure, function and regulation of molecular networks at different levels of the complexity pyramid of life is one of the main goals in systems biology. By integrating the topology, i.e. various structural properties of the networks with the functional knowledge encoded in protein annotations, and also analysing the interconnectivity between modules at different levels of the hierarchy, we aim to contribute to this goal. With the increasing availability of protein interaction data and more fine-grained GO annotations, this will help constructing a more complete view of interconnected modules to better understand the organisation of cells.

6. References

Asthana, S., King, O. D., Gibbons, F. D., & Roth, F. P. (2004). Predicting protein complex membership using probabilistic network reliability. *Genome Res* 14, 1170-1175.

Bader, G. D., & Hogue, C. W. (2003). An automated method for finding molecular complexes in large protein interaction networks. *BMC Bioinformatics* 4, 2.

Barabasi, A. L., & Oltvai, Z. N. (2004). Network biology: understanding the cell's functional organization. *Nat Rev Genet* 5, 101-113.

Barrat, A., Barthelemy, M., Pastor-Satorras, R., & Vespignani, A. (2004). The architecture of complex weighted networks. *Proc Natl Acad Sci U S A* 101, 3747-3752.

Deane, C. M., Salwinski, L., Xenarios, I., & Eisenberg, D. (2002). Protein interactions: two methods for assessment of the reliability of high throughput observations. *Mol Cell Proteomics* 1, 349-356.

Eisenberg, D., Marcotte, E. M., Xenarios, I., & Yeates, T. O. (2000). Protein function in the post-genomic era. *Nature* 405, 823-826.

Flake, G. W., Lawrence, C., Giles, C. L., & Coetzee, F. M. (2002). Self-organization and identification of Web communities. *IEEE Computer* 35, 66-71.

Gavin, A. C., Aloy, P., Grandi, P., Krause, R., Boesche, M., Marzioch, M., Rau, C., Jensen, L. J., Bastuck, S., Dumpelfeld, B., et al. (2006). Proteome survey reveals modularity of the yeast cell machinery. *Nature* 440, 631-636.

Gavin, A. C., Bosche, M., Krause, R., Grandi, P., Marzioch, M., Bauer, A., Schultz, J., Rick, J. M., Michon, A. M., Cruciat, C. M., et al. (2002). Functional organization of the yeast proteome by systematic analysis of protein complexes. *Nature* 415, 141-147.

Han, J. D., Bertin, N., Hao, T., Goldberg, D. S., Berriz, G. F., Zhang, L. V., Dupuy, D., Walhout, A. J., Cusick, M. E., Roth, F. P., & Vidal, M. (2004). Evidence for dynamically organized modularity in the yeast protein-protein interaction network. *Nature* 430, 88-93.

Hartuv, E., & Shamir, R. (2000). A clustering algorithm based on graph connectivity. *Information Processing Letters* 76, 175-181.

Hartwell, L. H., Hopfield, J. J., Leibler, S., & Murray, A. W. (1999). From molecular to modular cell biology. *Nature* 402, C47-52.

Hoglund, M., Frigyesi, A., & Mitelman, F. (2006). A gene fusion network in human neoplasia. *Oncogene* 25, 2674-2678.

Jansen, R., Lan, N., Qian, J., & Gerstein, M. (2002). Integration of genomic datasets to predict protein complexes in yeast. *J Struct Funct Genomics* 2, 71-81.

Jansen, R., Yu, H., Greenbaum, D., Kluger, Y., Krogan, N. J., Chung, S., Emili, A., Snyder, M., Greenblatt, J. F., & Gerstein, M. (2003). A Bayesian networks approach for predicting protein-protein interactions from genomic data. *Science* 302, 449-453.

Jeong, H., Mason, S. P., Barabasi, A. L., & Oltvai, Z. N. (2001). Lethality and centrality in protein networks. *Nature* 411, 41-42.

Lin, D. (1998). An information-theoretic definition of similarity. *The 15th International Conference on Mashine Learning* (Madison, WI).

Lord, P. W., Stevens, R. D., Brass, A., & Goble, C. A. (2003). Investigating semantic similarity measures across the Gene Ontology: the relationship between sequence and annotation. *Bioinformatics* 19, 1275-1283.

Lubovac, Z., Gamalielsson, J., & Olsson, B. (2006). Combining functional and topological properties to identify core modules in protein interaction networks. *Proteins* 64, 948-959.

Luo, F., & Scheuermann, R. H. (2006). Detecting Functional Modules from Protein Interaction Networks. *Proceeding of the First International Multi-Symposiums on Computer and Computational Scineces* (IMSCCS'06) (IEEE Computer Society).

Luo, F., Yang, Y., Chen, C. F., Chang, R., Zhou, J., & Scheuermann, R. H. (2007). Modular organization of protein interaction networks. *Bioinformatics* 23, 207-214.

Newman, M. E. J. (2001). The structure of scientific collaboration networks. *Proc Natl Acad Sci U S A* 98, 404-409.

Newman, M. E. J. (2003). Ego-centered networks and the ripple effect. *Socal networks* 25, 83-95.

Oltvai, Z. N., & Barabasi, A. L. (2002). Systems biology. Life's complexity pyramid. *Science* 298, 763-764.

Pereira-Leal, J. B., Enright, A. J., & Ouzounis, C. A. (2004). Detection of functional modules from protein interaction networks. *Proteins* 54, 49-57.

Petti, A. A., & Church, G. M. (2005). A network of transcriptionally coordinated functional modules in Saccharomyces cerevisiae. *Genome Res* 15, 1298-1306.

Poyatos, J. F., & Hurst, L. D. (2004). How biologically relevant are interaction-based modules in protein networks? *Genome Biol* 5, R93.

Przulj, N., Wigle, D. A., & Jurisica, I. (2004). Functional topology in a network of protein interactions. *Bioinformatics* 20, 340-348.

Qi, Y., & Ge, H. (2006). Modularity and dynamics of cellular networks. *PLoS Comput Biol* 2, e174.

Ravasz, E., Somera, A. L., Mongru, D. A., Oltvai, Z. N., & Barabasi, A. L. (2002). Hierarchical organization of modularity in metabolic networks. *Science* 297, 1551-1555.

Rives, A. W., & Galitski, T. (2003). Modular organization of cellular networks. *Proc Natl Acad Sci U S A* 100, 1128-1133.

Ruepp, A., Zollner, A., Maier, D., Albermann, K., Hani, J., Mokrejs, M., Tetko, I., Guldener, U., Mannhaupt, G., Munsterkotter, M., & Mewes, H. W. (2004). The FunCat, a functional annotation scheme for systematic classification of proteins from whole genomes. *Nucleic Acids Res* 32, 5539-5545.

Schlosser, G., & Wagner, G. P. (2004). Modularity in development and evolution: The University of Chicago Press).

Schwikowski, B., Uetz, P., & Fields, S. (2000). A network of protein-protein interactions in yeast. *Nat Biotechnol* 18, 1257-1261.

Segal, E., Shapira, M., Regev, A., Pe'er, D., Botstein, D., Koller, D., & Friedman, N. (2003). Module networks: identifying regulatory modules and their condition-specific regulators from gene expression data. *Nat Genet* 34, 166-176.

Spirin, V., & Mirny, L. A. (2003). Protein complexes and functional modules in molecular networks. *Proc Natl Acad Sci U S A* 100, 12123-12128.

Stelling, J., Klamt, S., Bettenbrock, K., Schuster, S., & Gilles, E. D. (2002). Metabolic network structure determines key aspects of functionality and regulation. *Nature* 420, 190-193.

Titz, B., Schlesner, M., & Uetz, P. (2004). What do we learn from high-throughput protein interaction data? *Expert Rev Proteomics* 1, 111-121.

Uhrig, J. F. (2006). Protein interaction networks in plants. Planta 224, 771-781.

Watts, D. J., & Strogatz, S. H. (1998). Collective dynamics of 'small-world' networks. *Nature* 393, 440-442.

von Mering, C., Krause, R., Snel, B., Cornell, M., Oliver, S. G., Fields, S., & Bork, P. (2002). Comparative assessment of large-scale data sets of protein-protein interactions. *Nature* 417, 399-403.

Xenarios, I., Rice, D. W., Salwinski, L., Baron, M. K., Marcotte, E. M., & Eisenberg, D. (2000). DIP: the database of interacting proteins. *Nucleic Acids Res* 28, 289-291.

Yook, S. H., Oltvai, Z. N., & Barabasi, A. L. (2004). Functional and topological characterization of protein interaction networks. *Proteomics* 4, 928-942.

5

Prediction of Combinatorial Protein-Protein Interaction from Expression Data Based on Conditional Probability

Takatoshi Fujiki, Etsuko Inoue,
Takuya Yoshihiro and Masaru Nakagawa
Wakayama University
Japan

1. Introduction

After the entire human DNA sequence was made public, many post-genome researchers began to investigate the systems of living creatures. Creatures consist of vast collections of proteins and their bodies are maintained by complex interactions among genes, proteins, and organic molecules. One major area of interest is how the characteristics of each creature are manifest and what kind of proteins, genes, and their interactions are related to them.

Much research to detect protein–protein interactions has been conducted. The most direct approach to tackle protein–protein interactions is to identify the evidence of the interactions through *in vitro* or *in vivo* experiments. Since several high-throughput experimental methods to detect physical interactions of proteins, such as yeast two-hybrid [1] and tandem affinity purification [2], have been developed, a significant number of protein interactions have been clarified that accelerated the exploration for protein functionality.

As vast amounts of genome sequences became available, computational approaches to infer protein–protein interactions became more focused. They typically assume some hypotheses of biological activity or property, and search biological databases with their own analytical methods for combinations of proteins to satisfy their hypotheses. Initially, many of these methods simply used gene or protein sequences, e.g., the method based on conservation of gene neighborhoods [3], the Rosetta Stone method [4][5], and the sequence-based co-evolution method [6]. Later, as various public databases became available, such as 3D-structures, domains, motifs, pathways, and phylogenetic profiles, various advanced methods to search for protein–protein interactions were developed. These methods and their results are available on the Web [7].

As one computational approach, gene or protein expression-based analysis is widely used to understand gene or protein interactions, which is the focus of this article. These methods were originally developed for microarray experiments that produced gene expression profiles, but they can apply to protein expression data as well. Because we can now obtain the expression profile of genes using high-throughput experiments such as microarray, protein chip, and 2D-electrophoresis, algorithms to derive interactions from expression data are increasingly valuable. As a basic analysis, the correlation coefficient of expression levels between two proteins is often used to measure the interaction level of protein pairs. (Note that, in this article, we call this type of interaction the *sole effect*, which refers to the effect on a protein from another single protein.) However, since protein interactions have more complex structures, more sophisticated analyses such as Bayesian networks [8] have been used to understand *combinatorial effects* among proteins. A Bayesian network provides the optimal network computed from a set of expression data, which shows the landscape of interaction effects among proteins. Although this network does not infer direct physical interactions, it helps us gain a better understanding of protein functions. However, since the process of Bayesian network analysis considers the sole effects and the combinatorial effects together, it cannot recognize the combinatorial effects alone.

In this article, we treat interactions among three proteins. We derive the combinatorial effect level, which emerges only when the three proteins are together, besides the sole effects that emerge between two proteins. The combinatorial effect level is estimated in a statistical manner, which will lead to a better understanding of protein interactions and a guide to deeper investigations.

The remainder of this paper is organized as follows. In Section 2, we describe related work to understand the current state of the art in this research area. In Section 3, we describe the model of protein–protein interactions used in our method, and present the method to retrieve the combinatorial effect of three proteins. In Section 4, we evaluate our method by applying it to real protein expression data, and finally in Section 5 present the conclusions.

2. Related work

In this section, we give a short introduction of the major approaches used to predict protein–protein interactions.

Many computational methods to predict protein–protein interactions have been proposed. They utilize various kinds of public data such as genome sequences, amino-acid sequences, pathways, domains, 3D-structures, motifs, and phylogenetic profiles, to identify a property of protein pairs in order to predict protein–protein interactions. One typical genome-sequence-based technique is based on conservation of gene neighbourhood [3]. This technique assumes that genes with similar functions or genes that are in the same pathways are transcribed together as a single unit known as an operon. Thus, finding two proteins that are neighbours in several genomes infers that they interact or have similar functions. Another typical sequence-based technique is called the Rosetta Stone method [4][5]. This method is based on the fact that several pairs of proteins interacting with each other have their homologs in other single proteins, called Rosetta Stone proteins. The phylogenetic profile method [6] uses a series of gene sequences in evolution and detects the set of genes that are simultaneously present or absent in the sequences. Since proteins in interaction tend to disappear simultaneously, finding the set of such genes predicts that the corresponding proteins interact. In addition, the in silico two-hybrid system [9] provides a fully alignment-based protein–protein interaction prediction. This technique tries to detect physical interaction of proteins within their 3D structures by means of correlation of sequences of sites among target proteins. Recently, docking analysis using 3D structures of proteins has progressed rapidly. The main difficulty in docking analysis is that there are many potential ways in which proteins can interact, and protein surfaces are flexible. Currently, one of the major approaches is a global search based on fast Fourier Transform [10]. Including the methods introduced in this brief discussion, there are a tremendous number of techniques to predict protein–protein interactions, and their algorithms and results are available in public databases. For more details, see [7][11].

Boolean networks [12] and Bayesian networks [8] are well known as computational methods to predict interactions from expression data. It is important to note that they treat gene interactions rather than protein interactions since most of them originally suppose microarray data as their source of analysis. However, they can also treat protein expression data.

A Boolean network [12] is a network that represents causal association and it is typically generated from a pattern of time-series expression data. In Boolean networks, a set of expression levels for a sample at time t is regarded as "state" at some time t, where each expression level is typically represented by "1 (expressed)" or "0 (not expressed)." To compute the network, the time-series state transition is analyzed to learn the functions to determine the state at time $t+1$ from the current state at time t. As a result, an expression level of a protein at time $t+1$ is determined depending on the expression level of several proteins at time t. This dependency indicates the protein–protein interaction, although it does not always indicate a direct interaction. There are several versions and extensions of Boolean networks. Akutsu et al. proposed a model and an algorithm of Boolean networks that is generated from non-time-series expression data [13]. Laubenbacher et al. proposed multistate Boolean networks [14]. However, these models cannot treat noise and, thus, often fail in computing networks. To overcome this problem, Shumulevich et al. proposed a model of probabilistic Boolean networks [15] that enables Boolean networks to apply to practical real expression data that includes noise.

A Bayesian network [8] is also a model of interactions often used in computational approaches that is typically built from expression data with discrete expression levels. Bayesian networks represent a joint distribution of random variables, and its direct edge between nodes represents causal association of those nodes. The learning process of a Bayesian network includes the optimization of network topology, where the evaluation of topologies is based on some information criterion, which is typically based on entropy. Note that it evaluates, for each node, the strength of the relationship between the node and its parents in the network, meaning that the sole effects and the combinatorial effects are evaluated together. Later, as an extension of the model, the Dynamic Bayesian network model was proposed [16], which handles time-series expression data. For details of this kind of network learning, there are several survey articles available, such as [17][18].

3. Method to retrieve combinatorial effects

3.1 Expression data used in our method

In this section, we explain the typical representation of protein expression data. Protein expression data represents the expression level of each protein i in sample j. Typically, the number of proteins in the data are several hundreds to thousands while the number of samples is usually several tens and at most hundreds.

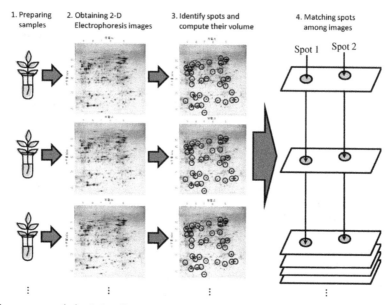

Fig. 1. The process of obtaining Proteome Expression Data.

Sample ID	Protain ID				
	A	B	C	D	...
1	0.000582	0.000107	0.000338	0.000451	...
2	0.000563		0.000475	0.000458	...
3	0.000495	0.000126	0.000433	0.000565	...
4	0.000553	0.000153	0.000382	0.000486	...
5	0.000536	0.000134	0.000536	0.000471	...
6	0.000601	0.000185	0.000457	0.000513	...
⋮	⋮	⋮	⋮	⋮	⋮

Fig. 2. The Data Format for Our Data Mining Process.

Protein expression data is obtained from several methods or devices such as protein arrays, 2D electrophoresis, and mass spectrometry. Among these, we now introduce a 2D electrophoresis-based method [19] as a typical way of generating protein expression data. The process of obtaining protein expression data is somewhat complicated compared to microarray data that measures gene expression levels (see Figure 1). First, we prepare target samples and obtain 2D electrophoresis images from each target sample through an experimental biological process. Second, we identify areas (in the rest of this article we call them *spots*) of separated proteins using image-processing software and measure the expression level of each spot. Third, we match the spots among different images such that the matched spots indicate the same protein. Finally, we normalize the values of expression levels using a normalization method as a preprocess to the data mining processes. As a result, we have a set of protein expression levels as shown in Figure 2, which shows the expression levels of each protein in each sample.

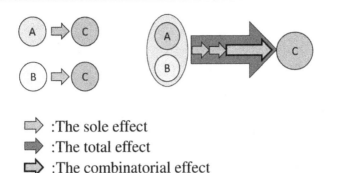

⇨ :The sole effect

⇛ :The total effect

⇨ :The combinatorial effect

Fig. 3. The Interaction Model to Predict.

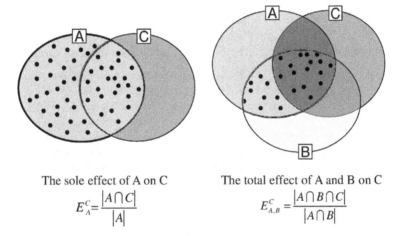

The sole effect of A on C

$$E_A^C = \frac{|A \cap C|}{|A|}$$

The total effect of A and B on C

$$E_{A,B}^C = \frac{|A \cap B \cap C|}{|A \cap B|}$$

Fig. 4. How to Measure Sole and Total Effect Level of Protein A and B on C.

3.2 Combinatorial protein-protein interaction model

The protein–protein interaction model we try to predict in this paper is shown in Figure 3. Three proteins, A, B, and C, are related to this model, where A and B individually effect the expression level of C, but if both A and B are expressed together, they have a far larger effect on the expression level of C. We call the effect from A to C (resp. B to C) the *sole effect*, and we call the whole effect from A and B on C the *total effect*. Note that the total effect consists of two sole effects and the *combinatorial effect* appears only if both A and B express. What we want to retrieve from expression data is the combinatorial effect of A and B on C.

To measure the combinatorial effect, we first estimate the amount of total effect of A and B on C. Then from the estimated total effect level, we subtract the two sole effects, i.e., the effect of A − C and B − C, to obtain the combinatorial effect level.

Note that the three proteins may interact directly or indirectly. We try to extract the three proteins that work in the same functional groups by identifying the behaviour of expression levels following our model of interaction.

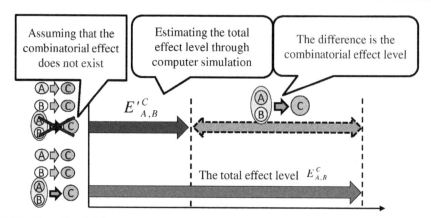

Fig. 5. Dividing Total Effect into Sole and Combinatorial Effect.

3.3 Estimating sole and total interaction levels based on conditional probability

We use conditional probability to retrieve this interaction from expression data. The probability of the sole interactions of $A - C$ and $B - C$ are measured by conditional probability, as shown in Figure 4. Namely, the sole interaction effect level of A on C is measured as the ratio of the number of samples in which the expression levels of both A and C are sufficiently high out of the number of samples in which the expression level of A is sufficiently high. The total interaction effect of A and B on C is also measured in a similar manner, i.e., the ratio of the number of samples in which the expression level of A, B, and C are all sufficiently high out of the number of samples in which the expression levels of both A and B are sufficiently high.

The definitions and formulation of our problems are as follows. We handle proteins i ($1 \leq i \leq I$) and samples j ($1 \leq j \leq J$), both of which are included in the input expression data. We also call the proteins A, B, C, ..., and so on. As a parameter, we define r ($0 < r < 1$) as the threshold of the ratio used to judge the expression, i.e., if the expression level of sample j for protein i is within the top r among all the expression levels of protein i, we call the protein i "expressed" in sample j. Let $|A|$ be the number of samples in which protein A is expressed, and similarly, let $|A \cap B|$ be the number of samples in which both protein A and B are expressed. Then, we define $E_A^C = \frac{|A \cap C|}{|A|}$ as the sole effect level of A on C. Similarly, the sole effect level of B on C is defined as $E_B^C = \frac{|B \cap C|}{|B|}$, and the total effect level of A and B on C is defined as $E_{A,B}^C = \frac{|A \cap B \cap C|}{|A \cap B|}$.

3.4 Retrieving combinatorial effect

What we want to estimate is the amount of the combinatorial interaction effect level, which can be estimated from the total interaction level (presented in the previous section) and the sole effect levels of $A - C$ and $B - C$ (see Figure 5). To estimate the combinatorial effect level for the combination of the three proteins A, B, and C, we split the total interaction effect into two parts, i.e., into two sole interaction effects and the combinatorial effect. Then, the difference between them is regarded as the combinatorial effect level that we wish to compute. To obtain the combinatorial effect level, we compute the statistical distribution of the total effect levels $E'^C_{A,B} = \frac{|A \cap B \cap C|}{|A \cap B|}$, which are computed through the simulation executed under the assumption that no combinatorial effect exists over A, B, and C. From the distribution of $E'^C_{A,B} = \frac{|A \cap B \cap C|}{|A \cap B|}$ and the total effect score $E^C_{A,B} = \frac{|A \cap B \cap C|}{|A \cap B|}$, which is the total effect level presented in the previous subsection, we can estimate the combinatorial effect level.

The computer simulation to compute the distribution of $E'^C_{A,B} = \frac{|A \cap B \cap C|}{|A \cap B|}$ is performed as follows. For the corresponding value of α and β, which are the sole effect values for the combination A − C and B − C, we first create distributions of A, B, and C randomly such that the sole effect levels of A − C and B − C are α and β, respectively. Since those distributions are created randomly, it is possible to assume that they do not include any combinatorial effect. Then we compute the total effect score of the combination A, B, and C. After a sufficient number of repetitions of this process, we obtain the distribution of $E'^C_{A,B}$ as the accumulation of the total effect scores. Note that we do not consider what kind of distribution A, B, and C follow in our method since we determine if the protein is expressed using the threshold r of the ranking in expression levels.

From this total effect distribution $E'^C_{A,B}$, we compute the combinatorial effect as a z-score in the distribution of $E'^C_{A,B}$. The z-score $z^C_{A,B}$ is defined as $z^C_{A,B} = \frac{(E^C_{A,B} - \mu)}{\sigma}$, where $E^C_{A,B}$ is the total effect level of A, B, and C obtained from the real data, and μ and σ are the average and the standard deviation of the distribution of $E'^C_{A,B}$ obtained from the computer simulation, respectively. Namely, the z-score is the difference between the average μ of the distribution of $E'^C_{A,B}$ and the real total effect level obtained from the real data, which is measured as the unit value σ. Intuitively, the z-score indicates the probability of the value $E^C_{A,B}$ assuming that the combinatorial effect does not exist, which implies the level of the combinatorial effect.

To compute the distribution of the total effect levels through the simulation, however, requires considerable computing time so it is desirable to precompute the distribution. Thus, we prepared a distribution table that shows the average and the standard deviation of the distribution for each value of α and β, as shown in Figure 6. Note that when we compute the distributions in Figure 6, we prepared the data of A, B, and C with 10,000 samples and we perform 5,000,000 trials for each pair of α and β. Because we computed the table for 20 values of α and β between 0 and 1, for obtaining the corresponding values of μ and σ we used the value in the table that is the closest to α and β of A, B, and C.

Now we summarize the proposed method. First, we enumerate every combination of the three proteins A, B, and C from the input data set. For each of the combinations, we compute the total effect level $E^C_{A,B}$ of A, B, and C. By referring to the precomputed distribution table, we find the distribution of $E'^C_{A,B}$ corresponding to the value α and β of A, B, and C. From the distribution of $E'^C_{A,B}$, and the total effect level $E^C_{A,B}$, we obtain the combinatorial effect level of A and B on C as the corresponding z-score. Finally, we create a ranking of all the combinations of the three proteins by ordering them by the z-score.

Average table

The sole effect of B on C

A on C \ B on C	0.00	0.05	0.10	0.15	0.20	0.25	0.30	0.35	0.40	0.45	0.50	0.55	0.60	0.65	0.70	0.75	0.80	0.85	0.90	0.95	1.00
0.00	0.0000	0.0000	0.0000	0.0000	0.0000	0.0000	0.0000	0.0000	0.0000	0.0000	0.0000	0.0000	0.0000	0.0000	0.0000	0.0000	0.0000	0.0000	0.0000	0.0000	0.0000
0.05	0.0000	0.0064	0.0135	0.0212	0.0298	0.0393	0.0500	0.0620	0.0757	0.0913	0.1094	0.1305	0.1556	0.1858	0.2228	0.2694	0.3296	0.4107	0.5255	0.7005	1.0000
0.10	0.0000	0.0135	0.0280	0.0437	0.0609	0.0795	0.1000	0.1225	0.1473	0.1750	0.2059	0.2406	0.2800	0.3251	0.3770	0.4377	0.5093	0.5953	0.7003	0.8315	1.0000
0.15	0.0000	0.0212	0.0437	0.0678	0.0933	0.1207	0.1500	0.1815	0.2154	0.2520	0.2917	0.3348	0.3819	0.4334	0.4901	0.5528	0.6224	0.7002	0.7877	0.8868	1.0000
0.20	0.0000	0.0298	0.0609	0.0933	0.1273	0.1628	0.2000	0.2390	0.2800	0.3231	0.3685	0.4163	0.4668	0.5201	0.5766	0.6365	0.7001	0.7679	0.8401	0.9173	1.0000
0.25	0.0000	0.0393	0.0795	0.1207	0.1628	0.2059	0.2500	0.2952	0.3415	0.3889	0.4376	0.4874	0.5385	0.5910	0.6449	0.7001	0.7569	0.8152	0.8751	0.9367	1.0000
0.30	0.0000	0.0500	0.1000	0.1500	0.2000	0.2500	0.3000	0.3500	0.4000	0.4500	0.5001	0.5501	0.6001	0.6501	0.7001	0.7501	0.8001	0.8501	0.9001	0.9500	1.0000
0.35	0.0000	0.0620	0.1225	0.1815	0.2390	0.2952	0.3500	0.4035	0.4559	0.5069	0.5569	0.6057	0.6534	0.7001	0.7457	0.7904	0.8341	0.8769	0.9188	0.9598	1.0000
0.40	0.0000	0.0757	0.1473	0.2154	0.2800	0.3415	0.4000	0.4559	0.5091	0.5601	0.6088	0.6554	0.7001	0.7429	0.7841	0.8236	0.8616	0.8982	0.9334	0.9673	1.0000
0.45	0.0000	0.0913	0.1750	0.2520	0.3231	0.3889	0.4500	0.5069	0.5601	0.6097	0.6563	0.7001	0.7412	0.7801	0.8167	0.8514	0.8843	0.9154	0.9450	0.9732	1.0000
0.50	0.0000	0.1094	0.2059	0.2917	0.3685	0.4376	0.5001	0.5569	0.6088	0.6563	0.7000	0.7405	0.7779	0.8126	0.8449	0.8750	0.9033	0.9297	0.9546	0.9780	1.0000
0.55	0.0000	0.1305	0.2406	0.3348	0.4163	0.4874	0.5501	0.6057	0.6554	0.7001	0.7405	0.7770	0.8106	0.8412	0.8694	0.8954	0.9194	0.9417	0.9625	0.9819	1.0000
0.60	0.0000	0.1556	0.2800	0.3819	0.4668	0.5385	0.6001	0.6534	0.7001	0.7412	0.7779	0.8106	0.8399	0.8667	0.8909	0.9131	0.9334	0.9520	0.9692	0.9852	1.0000
0.65	0.0000	0.1858	0.3251	0.4334	0.5201	0.5910	0.6501	0.7001	0.7429	0.7801	0.8126	0.8412	0.8667	0.8894	0.9100	0.9286	0.9455	0.9609	0.9750	0.9880	1.0000
0.70	0.0000	0.2228	0.3770	0.4901	0.5766	0.6449	0.7001	0.7457	0.7841	0.8167	0.8449	0.8694	0.8909	0.9100	0.9269	0.9423	0.9561	0.9686	0.9800	0.9904	1.0000
0.75	0.0000	0.2694	0.4377	0.5528	0.6365	0.7001	0.7501	0.7904	0.8236	0.8514	0.8750	0.8954	0.9131	0.9286	0.9423	0.9545	0.9655	0.9754	0.9844	0.9925	1.0000
0.80	0.0000	0.3296	0.5093	0.6224	0.7001	0.7569	0.8001	0.8341	0.8616	0.8843	0.9033	0.9194	0.9334	0.9455	0.9561	0.9655	0.9738	0.9815	0.9882	0.9944	1.0000
0.85	0.0000	0.4107	0.5953	0.7002	0.7679	0.8152	0.8501	0.8769	0.8982	0.9154	0.9297	0.9417	0.9520	0.9609	0.9686	0.9754	0.9815	0.9868	0.9917	0.9960	1.0000
0.90	0.0000	0.5255	0.7003	0.7877	0.8401	0.8751	0.9001	0.9188	0.9334	0.9450	0.9546	0.9625	0.9692	0.9750	0.9800	0.9844	0.9882	0.9917	0.9947	0.9975	1.0000
0.95	0.0000	0.7005	0.8315	0.8868	0.9173	0.9367	0.9500	0.9598	0.9673	0.9732	0.9780	0.9819	0.9852	0.9880	0.9904	0.9925	0.9944	0.9960	0.9975	0.9988	1.0000
1.00	0.0000	1.0000	1.0000	1.0000	1.0000	1.0000	1.0000	1.0000	1.0000	1.0000	1.0000	1.0000	1.0000	1.0000	1.0000	1.0000	1.0000	1.0000	1.0000	1.0000	1.0000

(Row axis: The sole effect of A on C)

Standard deviation table

The sole effect of B on C

A on C \ B on C	0.00	0.05	0.10	0.15	0.20	0.25	0.30	0.35	0.40	0.45	0.50	0.55	0.60	0.65	0.70	0.75	0.80	0.85	0.90	0.95	1.00
0.00	0.0000	0.0000	0.0000	0.0000	0.0000	0.0000	0.0000	0.0000	0.0000	0.0000	0.0000	0.0000	0.0000	0.0000	0.0000	0.0000	0.0000	0.0000	0.0000	0.0000	0.0000
0.05	0.0000	0.0024	0.0032	0.0040	0.0046	0.0053	0.0059	0.0065	0.0070	0.0076	0.0082	0.0088	0.0095	0.0103	0.0111	0.0122	0.0137	0.0156	0.0183	0.0209	0.0000
0.10	0.0000	0.0032	0.0048	0.0055	0.0064	0.0071	0.0078	0.0084	0.0090	0.0096	0.0101	0.0107	0.0112	0.0118	0.0125	0.0133	0.0141	0.0150	0.0155	0.0144	0.0000
0.15	0.0000	0.0040	0.0055	0.0069	0.0076	0.0084	0.0090	0.0096	0.0101	0.0105	0.0109	0.0114	0.0117	0.0121	0.0125	0.0129	0.0132	0.0132	0.0127	0.0108	0.0000
0.20	0.0000	0.0046	0.0064	0.0076	0.0087	0.0092	0.0098	0.0103	0.0107	0.0110	0.0113	0.0115	0.0117	0.0119	0.0120	0.0121	0.0120	0.0116	0.0106	0.0085	0.0000
0.25	0.0000	0.0053	0.0071	0.0084	0.0092	0.0099	0.0103	0.0107	0.0110	0.0112	0.0113	0.0114	0.0115	0.0115	0.0114	0.0112	0.0109	0.0102	0.0091	0.0070	0.0000
0.30	0.0000	0.0059	0.0078	0.0090	0.0098	0.0103	0.0107	0.0109	0.0111	0.0112	0.0112	0.0112	0.0111	0.0109	0.0107	0.0103	0.0098	0.0090	0.0078	0.0059	0.0000
0.35	0.0000	0.0065	0.0084	0.0096	0.0103	0.0107	0.0109	0.0111	0.0111	0.0111	0.0110	0.0108	0.0106	0.0104	0.0100	0.0095	0.0089	0.0080	0.0068	0.0050	0.0000
0.40	0.0000	0.0070	0.0090	0.0101	0.0107	0.0110	0.0111	0.0111	0.0113	0.0109	0.0107	0.0104	0.0101	0.0098	0.0093	0.0087	0.0080	0.0071	0.0060	0.0044	0.0000
0.45	0.0000	0.0076	0.0096	0.0105	0.0110	0.0112	0.0112	0.0111	0.0109	0.0112	0.0103	0.0100	0.0096	0.0092	0.0086	0.0080	0.0073	0.0064	0.0053	0.0038	0.0000
0.50	0.0000	0.0082	0.0101	0.0109	0.0113	0.0113	0.0112	0.0110	0.0107	0.0103	0.0109	0.0095	0.0091	0.0085	0.0080	0.0073	0.0066	0.0057	0.0047	0.0033	0.0000
0.55	0.0000	0.0088	0.0107	0.0114	0.0115	0.0114	0.0112	0.0108	0.0104	0.0100	0.0095	0.0095	0.0085	0.0079	0.0073	0.0067	0.0059	0.0051	0.0042	0.0029	0.0000
0.60	0.0000	0.0095	0.0112	0.0117	0.0117	0.0115	0.0111	0.0106	0.0101	0.0096	0.0091	0.0085	0.0096	0.0073	0.0067	0.0061	0.0054	0.0046	0.0037	0.0026	0.0000
0.65	0.0000	0.0103	0.0118	0.0121	0.0119	0.0115	0.0109	0.0104	0.0098	0.0092	0.0085	0.0079	0.0073	0.0085	0.0061	0.0055	0.0048	0.0041	0.0033	0.0023	0.0000
0.70	0.0000	0.0111	0.0125	0.0125	0.0120	0.0114	0.0107	0.0100	0.0093	0.0086	0.0080	0.0073	0.0067	0.0061	0.0075	0.0049	0.0043	0.0036	0.0029	0.0020	0.0000
0.75	0.0000	0.0122	0.0133	0.0129	0.0121	0.0112	0.0103	0.0095	0.0087	0.0080	0.0073	0.0067	0.0061	0.0055	0.0049	0.0062	0.0038	0.0032	0.0025	0.0017	0.0000
0.80	0.0000	0.0137	0.0141	0.0132	0.0120	0.0109	0.0098	0.0089	0.0080	0.0073	0.0066	0.0059	0.0054	0.0048	0.0043	0.0038	0.0050	0.0027	0.0021	0.0015	0.0000
0.85	0.0000	0.0156	0.0150	0.0132	0.0116	0.0102	0.0090	0.0080	0.0071	0.0064	0.0057	0.0051	0.0046	0.0041	0.0036	0.0032	0.0027	0.0039	0.0018	0.0012	0.0000
0.90	0.0000	0.0183	0.0155	0.0127	0.0106	0.0091	0.0078	0.0068	0.0060	0.0053	0.0047	0.0042	0.0037	0.0033	0.0029	0.0025	0.0021	0.0018	0.0025	0.0010	0.0000
0.95	0.0000	0.0209	0.0144	0.0108	0.0085	0.0070	0.0059	0.0050	0.0044	0.0038	0.0033	0.0029	0.0026	0.0023	0.0020	0.0017	0.0015	0.0012	0.0010	0.0013	0.0000
1.00	0.0000	0.0000	0.0000	0.0000	0.0000	0.0000	0.0000	0.0000	0.0000	0.0000	0.0000	0.0000	0.0000	0.0000	0.0000	0.0000	0.0000	0.0000	0.0000	0.0000	0.0000

(Row axis: The sole effect of A on C)

Fig. 6. The Distribution Table of $E'^{C}_{A,B}$ Created through Simulation.

4. Evaluation

4.1 Property of expression data used in our method

In this section, we explain the preprocess applied to the expression data, and also describe the basic property of the data. The expression data used in this experiment originated from the sample of fat near the kidney of black cattle. We performed 2D electrophoresis on each sample and measured the volume of each separated spot that corresponds to each protein. For details of the protocol of the experiment, see [19].

We preprocessed the expression data to improve the reliability of the expression data. Our preprocess consists of the following three steps. First, we removed from the data the

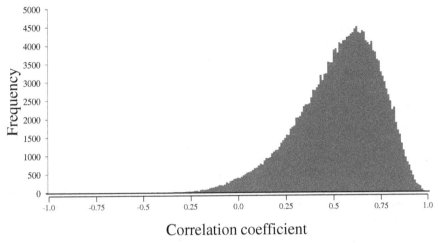

Correlation coefficient

Fig. 7. The histogram of correlation coefficient between proteins.

samples and the proteins that included more than 10% of null expression levels. This was done because samples or proteins with so many null values significantly reduce the reliability of the expression data. Next, we normalized the expression data with the global scaling method [20], where for every sample a scale factor is applied such that the total sum of the protein expression levels in the sample is 1. Finally, we removed the samples with high repetition error. Note that, in fact, in this data set, we performed 2D electrophoresis twice for each sample to confirm the accuracy of each electrophoresis experiment. To maintain the reliability of the data, we removed the sample in which more than 30% of the spots have a high repetation error or null value. Specifically, we consider a spot to have high repetition error if the larger expression level is larger than 1.3 times the value of the smaller expression level. Otherwise, the average of the two expression levels is used for each sample-protein pair. As a result, the expression data used for our evaluation consist of 124 samples and 670 proteins.

In order to indicate a characteristic of this data, we investigated the correlation between proteins. See Figure 7 for the results of calculating correlation coefficients for all pairs of the proteins. Note that the number of pairs is $_{670}C_2$ in total. Figure 7 is the histogram where the horizontal axis shows the correlation coefficient separated into classes with 0.05 intervals and the vertical axis shows the frequency of each class. From this result, we can see that most of the correlation coefficients take positive values, and many of them take relatively large values.

4.2 Evaluation experiment of retrieving combinatorial effect

4.2.1 Methods

We performed the experiment to evaluate the performance of the proposed method by applying it to the expression data described in Section 4.1. As a parameter of the experiment, we used the values of 50% and 30% as the threshold r to define the phenomenon that a protein is expressed.

To maintain statistical reliability, we excluded from the analysis the combinations of three proteins where the number of samples was insufficient. Namely, we ignored the combinations of the three proteins if $|A \cap B|$, which is the denominator in the total effect level $E_{A,B}^C$, was less than 35 in case of r is 50%, and less than 20 in case r is 30%. Similarly, we also removed the combinations if $|A \cap B \cap C|$ was less than 18 in case of r is 50%, and less than 10 in case r is 30%. Furthermore, for the computation, we only used the samples in which all the expression levels of the three proteins are not null.

4.2.2 Results

In this section, we describe the results of the evaluation experiments. Figure 8 shows the histogram of the case of $r = 50\%$, where the horizontal axis indicates the z-scores separated into classes with 0.5 intervals, and the vertical axis indicates the number of combinations in each class. Figure 9 shows the ranking of the top 30 combinations of proteins in terms of z-score. This table includes the columns of the spot numbers of proteins A, B, C, z-score of the combinations, E_A^C and E_B^C (the sole effect levels), $E_{A,B}^C$ (the total effect level), $|A \cap B|$ and $|A \cap B \cap C|$ (the number of samples contained in each phenomenon).

Under the significance level of 1%, we extracted 462,706 combinations in which a strong combinatorial effect is inferred. Here, we caluculate the corresponding p-value to the significance level of 1% using the formula of the Bonferroni correction presented in [21], i.e., p-value $= 1 - e^{\frac{log(1-\gamma)}{n}}$, where n is the number of combinations of three proteins and γ is the significance level. This suggests that if p-value $= 1 - e^{\frac{log(1-0.01)}{149,708,820}} = 6.713 \times 10^{-11}$ or less, the combinatorial effect exists. When the p-value is 6.713×10^{-11}, then the corresponding z-score is 6.423. This is computed as the point in the normal distribution where the probability that the value will become more than the point is p-value $= 6.713 \times 10^{-11}$. Figure 8 shows only the part where the z-score is larger than 6.423. Note that the probability of a z-score larger than 6.423 is only 6.713×10^{-11} if we assume that there is no combinatorial effect. This and the results of Figure 8 imply that our expression data includes many combinations in which the combinatorial effect exists.

Figure 9 shows that most of the sole effects of the shown combinations occur between 0.4 and 0.45, and the total effects occur between 0.45 and 0.55. Moreover, in most of the combinations, $|A \cap B|$ takes values close to 35, which is the threshold value to judge statistical reliability. This implies that combinations of lower $|A \cap B|$ tend to have larger z-scores. Although it is not shown in Figure 9, the combinations of lower ranks have larger values of $|A \cap B|$.

Figures 10 and 11 show the results with $r = 30\%$. Compared to Figure 8, z-scores tend to have lower values. In addition, the number of combinations with z-scores larger than 6.423 decreases to 167,320. Here, 6.423 is the corresponding p-value with the significance level of 1%. In Figure 11, all of the total effects take a value of 1.0 and all of $|A \cap B|$ take a value of 20, which is the threshold value to judge statistical reliability. Furthermore, about 97.8% of the total effects take 1.0 in the retrieved 167,320 combinations. This means that in most of retrieved combinations, protein C is expressed in all the samples in which both proteins A and B are expressed. This appears to be an unusual tendency. Since in the case of 30% the number of samples in the phenomenon "express" is smaller than in the case of 50%, it is possible that the number of samples is not sufficient to ensure a reliable statistical analysis. One of our future projects will be to clarify why this result appears in the case of $r = 30\%$.

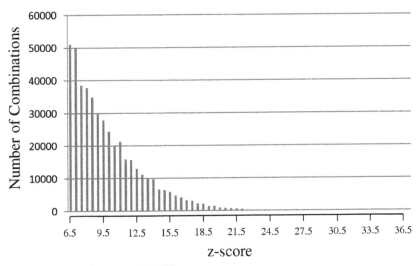

Fig. 8. The histogram of z-score (r=50%).

rank	A (spot No.)	B (spot No.)	C (spot No.)	z-score	E_A^c	E_B^c	$E_{A,B}^c$	$|A \cap B|$	$|A \cap B \cap C|$
1	5052	6080	5895	37.3456	0.4583	0.3958	0.5429	35	19
2	3554	5639	5895	36.2082	0.4600	0.4000	0.5429	35	19
3	2742	3554	5895	34.4911	0.4490	0.4490	0.5714	35	20
4	4015	5735	3100	33.7957	0.4348	0.4348	0.5405	37	20
5	5812	5866	1767	33.7458	0.4468	0.4255	0.5429	35	19
6	4798	6080	4849	33.5581	0.4000	0.4000	0.4737	38	18
7	5052	5731	5895	33.4141	0.4468	0.3830	0.5000	36	18
8	5739	6043	4838	33.2666	0.4043	0.4255	0.5000	38	19
9	5812	5866	5895	32.7405	0.4490	0.4286	0.5429	35	19
10	5052	5730	5895	32.6462	0.4375	0.3958	0.5000	36	18
11	3861	6111	5649	32.6423	0.3958	0.3958	0.4615	39	18
12	2318	5940	1765	32.5554	0.4130	0.4348	0.5135	37	19
13	926	5739	5895	32.3921	0.4800	0.4000	0.5429	35	19
14	168	6162	5695	31.9159	0.4667	0.4444	0.5714	35	20
15	5738	6043	3657	31.8151	0.4222	0.4444	0.5278	36	19
16	5639	6242	5895	31.3446	0.3600	0.4400	0.4615	39	18
17	5612	5732	5895	31.3436	0.4375	0.4167	0.5135	37	19
18	6043	6080	4849	31.2987	0.4348	0.3913	0.4865	37	18
19	4015	5735	4838	31.2948	0.4130	0.4565	0.5278	36	19
20	5735	6043	4838	31.2367	0.4348	0.4348	0.5278	36	19
21	4201	5808	3646	31.1739	0.4468	0.4681	0.5714	35	20
22	5726	6242	5895	30.9849	0.4082	0.4490	0.5143	35	18
23	5940	6080	1767	30.7235	0.4255	0.4468	0.5278	36	19
24	2318	4134	5895	30.6533	0.4082	0.5102	0.5714	35	20
25	5734	5866	3467	30.5461	0.4255	0.4043	0.4865	37	18
26	3880	6162	1763	30.5325	0.4444	0.4444	0.5429	35	19
27	5620	5639	5895	30.4974	0.4800	0.3800	0.5135	37	19
28	3554	5621	5895	30.4920	0.4490	0.4694	0.5714	35	20
29	4015	5849	3100	30.4629	0.4222	0.4222	0.5000	36	18
30	5622	5731	5895	30.3865	0.4800	0.4200	0.5526	38	21

Fig. 9. The Top 30 Combinations in z-score (r=50%).

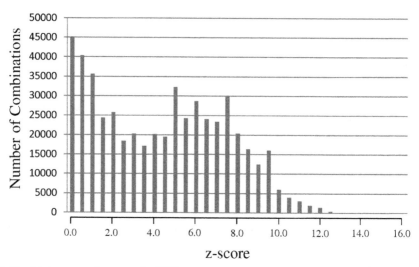

Fig. 10. The histogram of z-score (r=30%).

rank	A (spot No.)	B (spot No.)	C (spot No.)	z-score	E_A^c	E_B^c	$E_{A,B}^c$	|A∩B|	|A∩B∩C|
1	932	4257	4284	16.1614	0.6061	0.6970	1.0000	20	20
2	932	4284	4257	16.1614	0.6061	0.6970	1.0000	20	20
3	934	5140	828	15.9453	0.6176	0.6765	1.0000	20	20
4	932	6240	4134	15.5417	0.7059	0.6176	1.0000	20	20
5	2319	4056	6039	15.5417	0.6176	0.7059	1.0000	20	20
6	934	4284	4257	15.0960	0.6364	0.6970	1.0000	20	20
7	975	4134	4284	15.0960	0.6364	0.6970	1.0000	20	20
8	3998	4795	5045	15.0960	0.6970	0.6364	1.0000	20	20
9	4479	5724	4009	15.0960	0.6970	0.6364	1.0000	20	20
10	5045	5715	5194	15.0960	0.6970	0.6364	1.0000	20	20
11	5573	5954	5218	15.0960	0.6970	0.6364	1.0000	20	20
12	5615	6240	5965	15.0960	0.6970	0.6364	1.0000	20	20
13	2318	4013	5943	15.0958	0.6061	0.7273	1.0000	20	20
14	2318	5943	4013	15.0958	0.6061	0.7273	1.0000	20	20
15	932	4755	5954	14.9425	0.6286	0.7143	1.0000	20	20
16	3972	4755	4476	14.8927	0.7188	0.6250	1.0000	20	20
17	4476	4755	3972	14.8927	0.7188	0.6250	1.0000	20	20
18	4134	6240	5724	14.8927	0.7188	0.6250	1.0000	20	20
19	5724	6240	4134	14.8927	0.7188	0.6250	1.0000	20	20
20	5731	6065	6158	14.8927	0.6250	0.7188	1.0000	20	20
21	5731	6158	6065	14.8927	0.6250	0.7188	1.0000	20	20
22	5733	6065	6158	14.8927	0.6250	0.7188	1.0000	20	20
23	5733	6158	6065	14.8927	0.6250	0.7188	1.0000	20	20
24	934	6240	4134	14.6466	0.7059	0.6471	1.0000	20	20
25	2319	6158	5207	14.6466	0.6471	0.7059	1.0000	20	20
26	5622	5639	5955	14.6466	0.7059	0.6471	1.0000	20	20
27	1762	6034	5965	14.5887	0.7097	0.6452	1.0000	20	20
28	5965	6034	1762	14.5887	0.7097	0.6452	1.0000	20	20
29	1764	3626	5965	14.5887	0.7097	0.6452	1.0000	20	20
30	3626	5965	1764	14.5887	0.6452	0.7097	1.0000	20	20

Fig. 11. The Top 30 Combinations in z-score (r=30%).

4.3 Evaluation experiment of exchangeable proteins

4.3.1 Procedure to exchange proteins

In this section, for the combinations that have high z-scores, we investigate the z-scores when we exchange protein A with protein D in the case where D has a high correlation coefficient with A. Figure 9 shows that many high z-score combinations include C as the common protein, although A and B are also found as common proteins. Since our method defines the samples with the top r expression levels as expressed, having similar z-scores is intuitively inferred if we exchange A with D when D has a high correlation coefficient with A. We believe this is because there are many pairs of proteins in our data set that have a high correlation coefficient allowing us to retrieve so many combinations with a high combinatorial effect. In order to confirm this, we performed an experiment where we exchanged proteins.

The experiment is as follows. First, we create the list of proteins for D that have correlation coefficients against A that are larger than a certain threshold value. Next, we exchange A with D, and calculate the z-score $z_{D,B}^{C}$ for all combinations of proteins D, B, and C.

4.3.2 Result of exchanging protein

Figure 12 shows the value of the z-scores $z_{D,B}^{C}$ when A and D are exchanged in the highest z-score combination of A, B, and C in the case $r = 50\%$, where A is exchanged with D if D has the correlation coefficient with A larger than 0.8. This table includes the columns of the spot numbers of proteins A, B, C, protein D exchanged with A, correl(A,D) (the correlation coefficient of A and D), E_{D}^{C} (the sole effect level when A and D are exchanged), E_{B}^{C} (the sole effect level of before exchanging), $E_{D,B}^{C}$ (the total effect level), |D ∩ B| and |D ∩ B ∩ C| (the number of samples contained in each phenomenon). In addition, this table is sorted in descending order of z-score.

Figure 12 shows that the lowest z-score as a result of exchanging is 5.503. Note that there are only three combinations that have a z-score less than 6.423, by which the combinatorial effect is inferred under the significance level of 1%. This means that the z-score tends to be high when two proteins with a strong correlation are exchanged. Accordingly, one of the reasons that so many combinations that have a combinatorial effect are retrieved in our data seems to be that our data includes so many pairs of proteins in which the correlation coefficient is high.

5. Conclusion

In this paper, we proposed a method to retrieve the combinatorial protein–protein (or gene-gene) interactions from expression data using statistics of conditional probability. We suppose a model of protein–protein interactions in which the expression level of C takes a large value only if proteins A and B are expressed together. This is the first study to estimate the combinatorial effect level apart from the sole effect. In this study we described our method to treat protein interactions, but note that our method is also applicable to gene expression data generated from microarray experiments.

We evaluated our method using real expression data obtained from a 2D electrophoresis-based experiment. We performed two evaluation experiments with two different parameters, i.e., $r = 50\%$ and $r = 30\%$. As a result, the real expression data used in our experiment

A (spot No.)	B (spot No.)	C (spot No.)	D (spot No.)	correl(A,D)	z_score	E_D^c	E_B^c	$E_{D,B}^c$	$\|D \cap B\|$	$\|D \cap B \cap C\|$
			5142	0.8222	30.7904	0.5306	0.4286	0.6129	31	19
			6019	0.9351	27.2615	0.4898	0.3878	0.5143	35	18
			4275	0.8750	26.9008	0.5600	0.4000	0.5938	32	19
			2312	0.8205	26.3425	0.5000	0.4565	0.5882	34	20
			6043	0.8442	26.2674	0.4600	0.4200	0.5128	39	20
			926	0.8302	26.1577	0.4902	0.4314	0.5526	38	21
			4001	0.8393	25.2836	0.5600	0.4200	0.6061	33	20
			4269	0.8817	25.0023	0.5882	0.4118	0.6250	32	20
			5706	0.9268	24.8728	0.5319	0.3617	0.5161	31	16
			5281	0.8255	24.7993	0.5000	0.3913	0.5152	33	17
			4225	0.8406	24.7963	0.5417	0.3542	0.5172	29	15
			5298	0.8360	24.7883	0.4783	0.4130	0.5161	31	16
			4243	0.8686	24.6493	0.5102	0.3673	0.5000	34	17
			5612	0.8929	24.4295	0.4706	0.4314	0.5250	40	21
			4256	0.8447	24.2406	0.6939	0.4082	0.7308	26	19
			6020	0.9019	24.0511	0.5000	0.3800	0.5000	34	17
			5703	0.9148	24.0087	0.4800	0.4400	0.5405	37	20
			5961	0.8195	23.9841	0.5490	0.4314	0.6000	35	21
5052	6080	5895	2595	0.8112	23.9176	0.5400	0.4000	0.5588	34	19
			⋮	⋮	⋮	⋮	⋮	⋮	⋮	⋮
			4257	0.8637	14.3240	0.6800	0.4200	0.6774	31	21
			914	0.8112	14.2823	0.5686	0.4314	0.5714	35	20
			4185	0.8303	14.0312	0.6531	0.4286	0.6552	29	19
			4710	0.8103	13.5711	0.5200	0.4400	0.5278	36	19
			5978	0.8501	13.5634	0.7000	0.4400	0.7143	28	20
			5207	0.8104	13.4411	0.5294	0.4314	0.5278	36	19
			2589	0.8287	13.1487	0.5294	0.4314	0.5263	38	20
			921	0.8219	12.5048	0.6000	0.4000	0.5625	32	18
			6057	0.8128	12.4964	0.5652	0.4348	0.5625	32	18
			6012	0.8380	12.4960	0.5625	0.4375	0.5625	32	18
			6181	0.8033	11.6221	0.5714	0.4490	0.5789	38	22
			5060	0.8653	11.1105	0.5800	0.4200	0.5556	36	20
			942	0.8278	10.6944	0.7000	0.4400	0.7000	30	21
			5193	0.8156	8.1077	0.5800	0.4200	0.5405	37	20
			6276	0.8043	8.0482	0.6739	0.4348	0.6538	26	17
			5968	0.8026	7.7789	0.6939	0.4490	0.6875	32	22
			5615	0.8195	6.9890	0.6000	0.4400	0.5758	33	19
			975	0.8314	6.2982	0.7200	0.4400	0.7000	30	21
			4261	0.8536	5.7678	0.6600	0.4200	0.6129	31	19
			978	0.8142	5.5027	0.7000	0.4200	0.6552	29	19

Fig. 12. The ranking of z-score about exchangeable proteins (r=50%).

included a considerable number of combinations in which combinatorial effect is inferred. However, the results are quite different between the two parameters of r that we used in our expeirment. This may be because the number of samples is not sufficient for statistical analysis, and we hope to clarify the validity of our method in detail in our future work. Further, we confirmed that we can exchange protein of A with D when D has strong correlation with A, and we found that the combinatorial effect is still strong even when A is exchanged with D.

In the future, we would like to perform more experiments to further validate our proposed method. In addition, we would like to develop an algorithm for the analytical computation of the statistical distribution under the assumption of no combinatorial effect, i.e., we would like to compute the distribution shown in Figure 6 without simulation. If such fast computation is possible, it enables us to easily vary the threshold r, and it also enables us to compute a more accurate analysis. Finally, we also would like to find the known interactions in our results verify the value of this data-mining method.

6. Acknowledgment

This work was partly supported by the Program for Promotion of Basic and Applied Researches for Innovations in Bio-oriented Industry.

7. References

[1] Fields, S., & Song, O. (1989). A novel genetic system to detect protein-protein interactions, *Nature*, Vol. 340, pp. 245-246.

[2] Rigaut, G., Shevchenko, A., Rutz, B., Wilm, M., Mann, M. & Séraphin, B. (1999). A generic protein purification method for protein complex characterization and proteome exploration, *Nature Biotechnology*, Vol. 17, pp. 1030 - 1032.

[3] Overbeek, R., Fonstein, M., D'Souza, M., Pusch, G.D. & Maltsev, N. (1999). The use of gene clusters to infer functional coupling, *Proc. Natl Acad Sci U S A*, Vol. 96, No. 6, pp. 2896-2901.

[4] Enright, A.J., Iliopoulos, I., Kyrpides, N.C. & Ouzounis, C.A. (1999). Protein interaction maps for complete genomes based on gene fusion events, *Nature*, Vol. 402, No. 6757, pp. 86-90.

[5] Marcotte, E.M., Pellegrini, M., Ng, H.L., Rice, D.W., Yeates, T.O.& Eisenberg, D. (1999). Detecting protein function and protein-protein interactions from genome sequences, *Science*, Vol. 285, No. 5428, pp. 751-753.

[6] Pellegrini, M., Marcotte, E.M., Thompson, M.J., Eisenberg, D. & Yeates, T.O. (1999). Assigning protein functions by comparative genome analysis: protein phylogenetic profiles, *Proc. Natl Acad Sci U S A*, Vol. 96, No. 8, pp. 4285-4288.

[7] Tuncbag, N., Kar, G., Keskin, O., Gursoy, A. & Nussinov, R. (2009). A Survey of Available Tools and Web Servers for Analysis of Protein-Protein Interactions and Interfaces, *Briefings in Bioinformatics*, Vol. 10, No.3, pp.217-232.

[8] Friedman, N., Linial, M., Nachman, I. & Pe'er, D. (2000). Using Bayesian Networks to Analyze Expression Data, *Journal of Computational Biology*, Vol. 7, No. 3/4, pp. 601-620.

[9] Pazos, F. & Valencia, A. (2002). In silico two-hybrid system for the selection of physically interacting protein pairs. *Proteins: Structure, Function and Genetics*, Vol. 47, No. 2, pp. 219-227.

[10] Comeau, S.R., Gatchell, D.W., Vajda, S. & Camacho, C.J. (2004). ClusPro: A Fully Automated Algorithm for Protein-protein Docking, *Nucleic Acids Research*, Vol. 32(Web server issue), pp. W96-99.

[11] Jothi, R. & Przytycka, T.M. (2008). Computational approaches to predict protein-protein and domain-domain interactions, In: *Bioinformatics Algorithms: Techniques and Applications*, Mondoiu, I.I. and Zelikovsky, A. of Editors, pp. 465-492, Wiley Press, ISBN 978-047-0097-73-1.

[12] Liang, S., Fuhrman, S. & Somogyi, R. (1998). REVEAL, a General Reverse Engineering Algorithm for Inference of Genetic Network Architectures, *Proc. Pacific Symposium on Biocomputing '98*, pp. 18-29.

[13] Akutsu, T., Kuhara, S., Maruyama, O. & Miyano, S. (1998). A system for identifying genetic networks from gene expression patterns produced by gene disruptions and overexpressions, *Genome Informatics*, Vol. 9, pp. 151-160.

[14] Laubenbacher, R. & Stigler, B. (2004). A Computational Algebra Approach to the Reverse Engineering of Gene Regulatory Network, *Journal of Theoretical Biology*, Vol. 229, No. 4, pp. 523-537.

[15] Shmulevich, I., Dougherty, E.R., Kim, S. & Zhang, W. (2002). Probabilistic Boolean Networks: A Rule-based Uncertainty Model for Gene Regulatory Networks, *Bioinformatics*, Vol. 18, No. 2, pp. 261-274

[16] Husmeier, D. (2003). Sensitivity & Specificity of Inferring Genetic Regulatory Interactions From Microarray Experiments with Dynamic Bayesian Networks, *Bioinformatics*, Vol. 19, No. 17, pp. 2271-2282.

[17] Huang, Y., Tienda-Luna, I.M. & Wang, Y. (2009). A Survey of Statistical Models for Reverse Engineering Gene Regulatory Networks, *IEEE Signal Process Mag*, Vol. 26, No. 1, pp. 76-97.

[18] Sima, C., Hua, J. & Jung, S. (2009). Inference of Gene Regulatory Networks Using Time-Series Data: A Survey, *Current Genomics*, Vol. 10, No. 6, pp. 416-429.

[19] Nagai, K., Yoshihiro, T., Inoue, E., Ikegami, H., Sono, Y., Kawaji, H., Kobayashi, N., Matsuhashi, T., Ohtani, T., Morimoto, K., Nakagawa, M., Iritani, A. & Matsumoto, K. (2008). Developing an Integrated Database System for the Large-scale Proteomic Analysis of Japanese Black Cattle, *Animal Science Journal*, Vol. 79, No. 4. (in Japanese)

[20] Lu, C. (2004). Improving the Scaling Normalization for High-density Oligonucleotide GeneChip Expression microarrays, *BMC Bioinformatics*, Vol. 5, pp. 103.

[21] Spelman, R.J., Coppieters, W., Karim, L., van-Arendonk, J.A.M., & Bovenhuis, H. (1996). Quantitative Trait Loci Analysis for Five Milk Production Traits on Chromosome Six in the Dutch Holstein-Friesian Population, *Genetics*, Vol. 144, No. 4, pp. 1799-1808.

6

Mining Protein Interaction Groups

Lusheng Wang
Department of Computer Science
City University of Hong Kong
Hong Kong

1. Introduction

Proteins with interactions carry out most biological functions within living cells such as gene expression, enzymatic reactions, signal transduction, inter-cellular communications and immunoreactions. As the interactions are mediated by short sequence of residues among the long stretches of interacting sequences, these interacting residues or so-called interaction (binding) sites are at the central spot of proteome research. Although many imaging wet-lab techniques like X-ray crystallography, nuclear magnetic resonance spectroscopy, electron microscopy and mass spectrometry have been developed to determine protein interaction sites, the solved amount of protein interaction sites constitute only a tiny proportion among the whole population due to high cost and low throughput. Computational methods are still considered as the major approaches for the deep understanding of protein binding sites, especially for their subtle 3-dimensional structure properties that are not accessible by experimental methods.

The classical graph concept—maximal biclique subgraph (also known as maximal complete bipartite subgraph)—has been emerged recently for bioinformatics research closely related to topological structures of protein interaction networks and biomolecular binding sites. For example, Thomas *et al.* introduced complementary domains in (Thomas et al., 2003), and they showed that the complementary domains can form near complete bipartite subgraphs in PPI networks. A lock-and-key model has been proposed by Morrison *et al.* which is also based on the concept of maximal complete bipartite subgraphs (Morrison et al., 2006). Very recently, Andreopoulos *et al.* used clusters in PPI networks for identifying locally significant protein mediators (Andreopoulos et al., 2007). Their idea is to cluster common-friend proteins, which are in fact complete-bipartite proteins, based on their similarity to their direct neighborhoods in PPI networks. Other computational methods studying bipartite structures of PPI networks include (Bu et al., 2003; Hishigaki et al., 2001) which are focused on protein function prediction.

To identify motif pairs at protein interaction sites, Li *et al.* introduced a novel method with the core idea related to the concept of complete bipartite subgraphs from PPI networks (Li et al., 2006). The first step of the algorithm in (Li et al., 2006) finds large subnetworks with all-versus-all interactions (complete bipartite subgraphs) between a pair of protein groups. As the proteins within these protein groups have similar protein interactions and may share the same interaction sites, the second step of Li's algorithm is to compute conserved motifs (possible interaction sites) by multiple sequence alignments within each protein group. Thus, those conserved motifs can be paired with motifs identified from other protein groups to model protein interaction sites. One of the novel aspects of the algorithm in (Li et al., 2006) is that it combines two types of data: the PPI data and the associated sequence data for modeling binding motif pairs.

Each protein in the above PPI networks is represented by a vertex and every interaction between two proteins is represented by an edge. Discovering complete bipartite subgraphs in PPI networks can thus be formulated as the following biclique problem: Given a graph, the biclique problem is to find a subgraph which is bipartite and complete. The objective is to maximize the number of vertices or edges in the bipartite complete subgraph. We note that the maximum vertex biclique problem is polynomial time solvable (Yannakakis, 1981). This problem is also equivalent to the maximum independent set problem on bipartite graphs which is known to be solvable by a minimum cut algorithm. However, the maximum vertex balanced biclique problem is NP-hard (Garey & Johnson, 1979). The maximum edge biclique problem is proved to be NP-hard as well (Peeters, 2003).

In this paper, we consider incompleteness of biological data, as the interaction data of PPI networks is usually not fully available. On the other hand, within an interacting protein group pair, some proteins in one group may only interact with a proportion of the proteins in the other group. Therefore, many subgraphs formed by interacting protein group pairs are not perfect bicliques. They are more often near complete bipartite subgraphs. Therefore, methods of finding bicliques may miss many useful interacting protein group pairs. To deal with this problem, we use quasi-bicliques instead of bicliques to find interacting protein group pairs. With the quasi-biclique, even though some interactions are missing in a protein interaction subnetwork, we can still find the two interacting protein groups. In this paper, we introduce and investigate the maximum vertex quasi-biclique problem. We show that the problem is NP-hard. We also propose approximation and heuristic algorithms for finding large quasi-bicliques in PPI networks. The applications for finding protein-protein binding sites are illustrated.

2. Bicliques and quasi-bicliques

Let $\mathcal{G} = (\mathcal{V}, \mathcal{E})$ be an undirected graph, where each vertex represents a protein and there is an edge connecting two vertices if the two proteins have an interaction. Since $\mathcal{G}$ is an undirected graph, any edge $(u, v) \in \mathcal{E}$ implies $(v, u) \in \mathcal{E}$. For a selected edge (u, v) in $\mathcal{G}$, in order to find the two groups of proteins having the similar pairs of binding sites, we translate the graph $\mathcal{G} = (\mathcal{V}, \mathcal{E})$ into a bipartite graph. Let $X = \{x | (x, v) \in \mathcal{E}\}$, $Y_1 = \{y | (u, y) \in \mathcal{E} \& y \notin X\}$ and $Y_2 = \{w | (u, w) \in \mathcal{E} \& w \in X\}$. For a vertex $w \in Y_2$, w is incident to both u and v in $\mathcal{G}$. Thus both X and Y_2 contain w. We keep w in X and replace w in Y_2 with a new virtual vertex $\overline{w}$. After replacing all vertices w in Y_2 with $\overline{w}$, we get a new vertex set $\overline{Y_2}$. Let $Y = Y_1 \cup \overline{Y_2}$ and $E = \{(x, y) | (x, y) \in \mathcal{E} \& x \in X \& y \in Y_1\} \cup \{(x, \overline{w}) | (x, w) \in \mathcal{E} \& x \in X \& \overline{w} \in \overline{Y_2}\}$. In this way, we have a bipartite graph $G = (X \cup Y, E)$. A biclique in G corresponds to two subsets of vertices, say, subset A and subset B, in $\mathcal{G}$. In $\mathcal{G}$, every vertex in A is adjacent to all the vertices in B, and every vertex in B is adjacent to all the vertices in A. Moreover, $A \cap B$ may not be empty. In this case, for any vertex $w \in A \cap B$, $(w, w) \in \mathcal{E}$. This is the case, where the protein has a self-loop. Self-loops are very common in practice. When a self-loop appears, one protein molecule interacts with another identical protein molecule. For example, two identical protein subunits can assemble together to form a homodimeric protein.

In the following, we focus on the bipartite graph $G = (X \cup Y, E)$. For a vertex $x \in X$ and a vertex set $Y' \subseteq Y$, the degree of x in Y' is the number of vertices in Y' that are adjacent to x, denoted by $d(x, Y') = |\{y|y \in Y'\&(x,y) \in E\}|$. Similarly, for a vertex $y \in Y$ and $X' \subseteq X$, we use $d(y, X')$ to denote $|\{x|x \in X'\&(x,y) \in E\}|$. Now, we are ready to define the δ-quasi-biclique.

Definition 1. *For a bipartite graph $G = (X \cup Y, E)$ and a parameter $0 < \delta \leq \frac{1}{2}$, G is called a δ-quasi-biclique if for each $x \in X, d(x, Y) \geq (1 - \delta)|Y|$ and for each $y \in Y, d(y, X) \geq (1 - \delta)|X|$.*

Similarly, a δ-quasi-biclique in G corresponds to two subsets of vertices, say, subset A and subset B, in $\mathcal{G}$. In $\mathcal{G}$, every vertex in A is adjacent to at least $(1 - \delta)|B|$ vertices in B, and every vertex in B is adjacent to at least $(1 - \delta)|A|$ vertices in A. Moreover, according to the translation and the definition, $A \cap B$ may not be empty. Again, if a protein appears in both sides of a δ-quasi-biclique and there is an edge between the two corresponding vertices, the protein has a self-loop. In our experiments, we observe that about 22% of the δ-quasi-bicliques produced by our program contain self-loop proteins.

In many applications, due to various reasons, some edges in a clique/biclique may be missing and a clique/biclique becomes a quasi-clique/quasi-biclique. Thus, finding quasi-cliques/quasi-bicliques is more important in practice. Here we show that large quasi-bicliques may not contain any large bicliques.

Theorem 1. *Let $G = (X \cup Y, E)$ be a random graph with $|X| = |Y| = n$, where for each pair of vertices $x \in X$ and $y \in Y$, (x, y) is chosen, randomly and independently, to be an edge in E with probability $\frac{2}{3}$. When $n \rightarrow \infty$, with high probability, G is a $\frac{1}{2}$-quasi-biclique, and G does not contain any biclique $G' = (X' \cup Y', E')$ with $|X'| \geq 2\log n$ and $|Y'| \geq 2\log n$.*

In the biological context, Theorem 1 indicates that it is possible that some large interacting protein groups cannot be obtained by simply finding a maximal biclique if a few (interaction) edges are missing. As large interacting protein groups are more useful, according to this theorem, we have to develop new computational algorithms to extract from PPI networks large interacting protein groups which form quasi-bicliques.

In terms of false positive edges, both quasi-biclique and biclique can handle spurious edges very well. If very few spurious edges are added, in most cases, an irrelative protein will not be included in the quasi-bicliques or biclique unless $(1 - \delta)|A|$ spurious edges are simultaneously added to the protein that has no interaction with any of the proteins in A, where A is one of the two interaction groups.

The maximum vertex quasi-biclique problem is defined as follows.

Definition 2. *Given a bipartite graph $G = (X \cup Y, E)$ and $0 < \delta \leq \frac{1}{2}$, the maximum vertex δ-quasi-biclique problem is to find $X' \subseteq X$ and $Y' \subseteq Y$ such that the $X' \cup Y'$ induced subgraph is a δ-quasi-biclique and $|X'| + |Y'|$ is maximized.*

The maximum vertex biclique problem, where $\delta = 0$, can be solved in polynomial time (Yannakakis, 1981). Here we show that the maximum vertex δ-quasi-biclique problem when $\delta > 0$ is NP-hard. The reduction is from $X3C$ (Exact Cover by 3-Sets), which is known to be NP-hard (Karp, 1972).

Theorem 2. *For any constant integers $p > 0$ and $q > 0$ such that $0 < \frac{p}{q} \leq \frac{1}{2}$, the maximum vertex $\frac{p}{q}$-quasi-biclique problem is NP-hard.*

3. A polynomial time approximation scheme

The following lemma that is originally from (Li et al., 2002) will be repeatedly used in our proofs.

Lemma 1. *Let $X_1, X_2, \ldots, X_n$ be n independent random 0-1 variables, where X_i takes 1 with probability p_i, $0 < p_i < 1$. Let $X = \sum_{i=1}^{n} X_i$, and $\mu = E[X]$. Then for any $0 < \epsilon \leq 1$,*

$$\mathbf{Pr}(X > \mu + \epsilon n) < exp(-\frac{1}{3}n\epsilon^2),$$

$$\mathbf{Pr}(X < \mu - \epsilon n) \leq exp(-\frac{1}{2}n\epsilon^2).$$

The Main Ideas and Techniques: The problem can be formulated as a quadratic programming problem. We use a random sampling technique and a randomized rounding method to get a good approximate solution for the quadratic programming problem under the conditions that $|X_{opt}| = \Omega(|X|)$ and $|Y_{opt}| = \Omega(|Y|)$. The random sampling technique involves to randomly select $r1 = \Omega(\log |X_{opt}|)$ vertices from X_{opt} when X_{opt} is not known. This can be done when $|X_{opt}| = \Omega(|X|)$ and $|Y_{opt}| = \Omega(|Y|)$.

In order to make sure that $|X_{opt}| = \Omega(|X|)$ and $|Y_{opt}| = \Omega(|Y|)$, we design a combinatorial approach to find a subset $X' \subseteq X$ and a subset $Y' \subseteq Y$ such that $|X'| = \Omega(|X_{opt}| + |Y_{opt}|)$, $|X' \cap X_{opt}| \geq (1-\epsilon)|X_{opt}|$, $|Y'| = \Omega(|X_{opt}| + |Y_{opt}|)$ and $|Y' \cap Y_{opt}| \geq (1-\epsilon)|Y_{opt}|$. See Lemma 2. Thus, we can work on a bipartite graph induced by X' and Y'. Without loss of generality, we can assume that $|Y_{opt}| \geq |X_{opt}|$. Now, two subcases arise: Case 1: $|X_{opt}| \leq \epsilon |Y_{opt}|$, and Case 2: $|X_{opt}| > \epsilon |Y_{opt}|$. For case 1, we can use linear programming approach and a brute-force approach to solve the problem. For case 2, we can use the quadratic programming approach to solve the problem.

Let $G = (X \cup Y, E)$ be the input bipartite graph. Let $X_{opt} \subseteq X$ and $Y_{opt} \subseteq Y$ be the optimal biclique for the maximum quasi-biclique problem. Without loss of generality, we can assume that

Assumption 1: $|Y_{opt}| \geq |X_{opt}|$.

The basic idea of our algorithm is to (1) formulate the problem into a quadratic programming problem and (2) use a random sampling approach to approximately solve the problem. In order to make the random sampling approach work, we have to make sure that

$$|X_{opt}| = \Omega(|X|) \tag{1}$$

and

$$|Y_{opt}| = \Omega(|Y|). \tag{2}$$

However, for any input bipartite graph $G = (X \cup Y, E)$, there is no guarantee that (1) and (2) hold. Here we propose a method to find a subset X' of X and Y' of Y such that for any $t > 0$, $|X_{opt}| = \Omega(|X'|)$, $|X_{opt} \cap X'| \geq \frac{t-1}{t}|X_{opt}|$, $|Y_{opt}| = \Omega(|Y'|)$, and $|Y_{opt} \cap Y'| \geq \frac{t-1}{t}|Y_{opt}|$. If we can obtain this kind of X' and Y', then we can work on the induced bipartite graph $G' = (X' \cup Y', E')$, where $E' = \{(u, v) | u \in X', v \in Y' \text{ and } (u, v) \in E\}$. Obviously, any good approximate solution of G' is also a good approximate solution of G.

Let x_i be a vertex in the bipartite graph $G = (X \cup Y, E)$. Define $D(x_i, Y)$ to be the set of vertices in Y that are incident to x_i. The following lemma tells us how to obtain X' and Y'.

Lemma 2. For any $t > 0$, there exist k vertices $x_1, x_2, \ldots, x_k$ in X for $k = \lceil \delta t \rceil$ such that $|\bigcup_{i=1}^{k} D(x_i, Y)| \leq k(|Y_{opt}| + |X_{opt}|)$ and $|Y_{opt} \cap \bigcup_{i=1}^{k} D(x_i, Y)| \geq \frac{t-1}{t}|Y_{opt}|$. Similarly, there exists k vertices $y_1, y_2, \ldots, y_k$ in Y for $k = \lceil \delta t - 1 \rceil$ such that $|\bigcup_{i=1}^{k} D(y_i, X)| \leq k(|Y_{opt}| + |X_{opt}|)$ and $|X_{opt} \cap \bigcup_{i=1}^{k} D(y_i, X)| \geq \frac{t-1}{t}|X_{opt}|$.

Though we do not know which k vertices in X we should choose, we can try all possible size k subsets of X in $O(|X|^k)$ time for constant k. The value of k is $\lceil \delta t \rceil$ and is determined by t later. Thus, from now on, we assume that the k vertices $x_1, x_2, \ldots, x_k$ are known. Let $X' = \bigcup_{i=1}^{k} D(y_i, X)$ and $Y' = \bigcup_{i=1}^{k} D(x_i, Y)$. We will focus on finding a quasi-biclique in the sub-graph $G' = (X' \cup Y', E')$ of G induced by X' and Y'.

Let $X'_{opt} \subseteq X'$ and $Y'_{opt} \subseteq Y'$ be a quasi-$(\delta + \frac{1}{t})$-biclique with maximum number of vertices in G'. From Lemma 2, $|X'_{opt}| + |Y'_{opt}| \geq (1 - \frac{1}{t})(|X_{opt}| + |Y_{opt}|)$ since $X' \cap X_{opt}$ and $Y' \cap Y_{opt}$ also form a quasi-$\delta + \frac{1}{t}$-biclique of size $(1 - \frac{1}{t})(|X_{opt}| + |Y_{opt}|)$. From now on, we will try to find a good approximate solution for X'_{opt} and Y'_{opt}.

If $|X'_{opt}|$ and $|Y'_{opt}|$ are approximately the same, then we have $|X'_{opt}| = \Omega(|X'|)$ and $|Y'_{opt}| = \Omega(|Y'|)$. That is, (1) and (2) hold for graph G'. Therefore, we can use quadratic programming approach to solve the problem. Nevertheless, there is no guarantee that $|X'_{opt}|$ and $|Y'_{opt}|$ are approximately the same. For any $\epsilon > 0$, we consider two cases.

Case 1: $|X'_{opt}| < \epsilon|Y'_{opt}|$. In this case, the number of vertices in Y'_{opt} will dominate the size of the whole quasi-biclique. If we select a vertex $x \in X'_{opt}$, then x and $D(x, Y')$ form a biclique of size at least $1 + (1 - \delta)|d(x, Y')| \geq 1 + (1 - \delta)|Y'_{opt}|$. When the value of δ is big with respect to ϵ, we do not have the desired quasi-biclique. If we try to add more vertices from Y', we have to guarantee that for every selected vertex y in Y', y is incident to at least $(1 - \delta)|X'|$ selected vertices in X'. This is impossible if x is the only selected vertex from X'. Therefore, we have to consider to add more vertices from both X' and Y'. It is clear that the task here is non-trivial.

In the following lemma, we will show that there exists a subset of r vertices (for some constant r) $X_r \subseteq X'$ and a subset $Y''_{opt} \subseteq Y'_{opt}$ such that X_r and Y''_{opt} form a quasi-$(\delta + \epsilon'')$-biclique with $|Y''_{opt}| \geq (1 - \epsilon'')|Y_{opt}|$ for some $\epsilon'' > 0$. Here r and ϵ'' are closely related.

Lemma 3. Let $\frac{1}{t} = \epsilon'$. There exists a subset X'_r of X'_{opt} containing $r = \frac{2}{\epsilon'^2} \log(\frac{1}{\epsilon'})$ elements and a subset Y''_{opt} of Y'_{opt} with $|Y''_{opt}| \geq (1 - \frac{r(r-1)}{2|X_{opt}|} - 2\epsilon')|Y'_{opt}|$ such that X'_r and Y''_{opt} form a quasi-$(\delta + \frac{r(r-1)}{2|X'_{opt}|} + 2\epsilon')$-biclique.

Based on Lemma 3, we can design an algorithm that finds a quasi-$(\delta + 4\epsilon')$-biclique with size at least $(1 - 4\epsilon' - \epsilon)(|X'_{opt}| + |Y'_{opt}|)$. Let $G' = (X' \cup Y', E')$ be the sub-graph obtained from Lemma 2. For any $\epsilon' > 0$, define $r = \frac{2}{\epsilon'^2} \log(\frac{1}{\epsilon'})$.

Case 1.1. $|X'_{opt}| \geq \frac{r(r-1)}{\epsilon'}$: When $|X'_{opt}| \geq \frac{r(r-1)}{\epsilon'}$, $\frac{r(r-1)}{2|X'_{opt}|} \leq \epsilon'$. Thus, there exist a quasi-$(\delta + 3\epsilon')$-biclique $X_r \subset X'$ and Y''_{opt} as described in Lemma 3.

We select r vertices from X'. For each subset $X_r \subseteq X'$ of r vertices $\{v_1, v_2, \ldots, v_r\}$, we define the following integer linear programming. Let $c_{i,j}$ be a constant, where $c_{i,j} = 1$ if $(v_i, u_j) \in E'$; and $c_{i,j} = 0$ if $(v_i, u_j) \notin E'$. Let y_i be a 0/1 variable, where $y_i = 1$ indicates that the vertex u_i in Y' is selected in the quasi-biclique and $y_i = 0$ otherwise.

$$y_i(\sum_{j=1}^{r} c_{i,j}) \geq (1 - \delta - \frac{1}{t} - \epsilon')r \tag{3}$$

$$\sum_{i=1}^{|Y'|} y_i c_{i,j} \geq (1 - \delta - 3\epsilon')|Y'_{opt}| \text{ for } j = 1, 2, \ldots, r, \tag{4}$$

Here we do not know $|Y'_{opt}|$. However, we can guess the value of $|Y'_{opt}|$ by trying $|Y'_{opt}| = 1, 2, \ldots, |Y'|$. The integer programming problem formulated by (3) and (4) has no objective function and we just want a feasible solution to fit (3) and (4). The integer programming problem is hard to solve. However, we can obtain a fractional solution $\bar{y}_i$ for (3) and (4) with $0 \leq \bar{y}_i \leq 1$ in polynomial time. After obtaining the fractional solution $\bar{y}_i$, we randomly set y_i to be 1 with probability $\bar{y}_i$.

Lemma 4. *Assume that $\frac{1}{2}(1 - \delta - 3\epsilon')|Y'_{opt}|\epsilon'^2 \geq 2\log r$ and $\frac{1}{t} = \epsilon'$. With probability at least $1 - \frac{1}{r}$, we can get a pair of subsets $X_A \subseteq X'$ and $Y_A \subseteq Y'$ (an integer solution) by randomized rounding according to the probability $\bar{y}_i$ such that X_A and Y_A form a quasi-$(\delta + 4\epsilon')$-biclique with $|X_A| + |Y_A| \geq (1 - \delta - 4\epsilon')|Y'_{opt}|$.*

A standard method in (Li et al., 2002) can give a de-randomized algorithm.

When $\frac{1}{2}(1 - \delta - 3\epsilon')|Y'_{opt}|)\epsilon'^2 < 2\log r$, we can enumerate all possible subsets of size $(1 - \delta - 3\epsilon')|Y'_{opt}|$ in Y' in polynomial time to get the desired solution.

Case 1.2. $|X'_{opt}| < \frac{r(r-1)}{\epsilon'}$: In this case, X'_{opt} and Y'_{opt} form the desired quasi-δ-biclique. Instead of selecting r vertices in X', we select $|X'_{opt}|$ vertices in X'. Though we do not know the value of $|x'_{opt}|$, we can guess the value for $|x'_{opt}| = 1, 2, \ldots, \frac{r(r-1)}{\epsilon'}$. We also solve the integer linear programming (3) and (4) in the same way as in Case 1.1. The algorithm for Case 1 is given in Fig. 1.

Theorem 3. *Assume $|X'_{opt}| \leq \epsilon|Y'_{opt}|$. We set $\frac{1}{t} = \epsilon'$ in the algorithm. With probability at least $1 - \frac{1}{r}$, Algorithm 1 finds a quasi-$(\delta + 4\epsilon')$-biclique $X_A \subseteq X$ and $Y_A \subseteq Y$ with $|X_A| + |Y_A| \geq (1 - \delta - 4\epsilon')(|X_{opt}| + |Y_{opt}|)(1 - \epsilon')/(1 + \epsilon)$ in time $O((|X||Y|)^{\lceil \delta t \rceil}[|X||Y||Y'|^{\frac{4\log r}{\epsilon'^2}} + |X'|^{\frac{r(r-1)}{\epsilon'}}\frac{r(r-1)}{\epsilon'}(|X| + |Y|)^3)])$.*

Algorithm 1: Algorithm for Solving Case 1: $\|X'_{opt}\| \le \epsilon \|Y'_{opt}\|$.
Input: a bipartite graph $G = (X \cup Y, E)$, a real number $0 \le \delta \le 0.5$, a number $t > 0$, a number $\epsilon > 0$, and a number $\epsilon' > 0$.
0.   Let $k = \lceil \delta t \rceil$.
1.   **for** any $v_1, v_2, \ldots, v_k \in X$ and any $u_1, u_2, \ldots, u_k \in Y$ **do**
2.   Set $X' = \cup_{i=1}^k D(v_i, Y)$ and $Y' = \cup_{i=1}^k D(u_i, X)$.
3     $r == \frac{2}{\epsilon'^2} \log(\frac{1}{\epsilon'})$
4   Guess $\|X'_{opt}\|$ and $\|Y'_{opt}\|$ assuming $\|X'_{opt}\| \le \epsilon \|Y'_{opt}\|$.
5   **if** $\frac{1}{2}(1 - \delta - 3\epsilon')\|Y'_{opt}\|\epsilon'^2 < 2\log r$ **then** enumerate all possible subsets of size $(1 - \delta - 3\epsilon')\|Y'_{opt}\|$ in Y' in polynomial time to get the desired solution.
6   **if** $\frac{1}{2}(1 - \delta - 3\epsilon')\|Y'_{opt}\|)\epsilon'^2 > 2\log r)$ **then**
7   **for** $i = r, r+1, \ldots \frac{r(r-1)}{\epsilon'}$ **do**
8   **for** every i-elements subset $X_i = \{x_1, x_2, \ldots, x_i\}$ **do**
9   give a fractinal solution $\bar{y}_i$ for (3) and (4).
10   randomly set $y_i = 1$ with probability $\bar{y}_i$.
8.   Output a $\delta + \frac{1}{t} + 4\epsilon'$ quasi-biclique with the biggest $\|X_A\| + \|Y_A\|$.

Fig. 1. The algorithm for solving Case 1.

Case 2: $\|X'_{opt}\| \ge \epsilon \|Y'_{opt}\|$. In this case, we have $\|X'_{opt}\| = \Omega(\|X'\|)$ and $\|Y'_{opt}\| = \Omega(\|Y'\|)$. We will use a quadratic programming approach to solve the problem. We can formulate the quasi-biclique problem for the bipartite graph $G' = (X' \cup Y', E')$ into the following quadratic programming problem.

Quadratic programming formulation:

Let x_i and y_j be 0/1 variables, where $x_i = 1$ indicates that vertex v_i in X' is in the quasi-biclique and $y_j = 1$ indicates that vertex u_j in Y' is in the quasi-biclique. Define $e_{i,j} = 1$ if $(v_i, u_j) \in E'$ and $e_{i,j} = 0$ otherwise. Let c_1 and c_2 be two integers representing the sizes of X'_{opt} and Y'_{opt}, respectively. We can guess the values of c_1 and c_2 in polynomial time though we do not know c_1 and c_2. We have the following inequalities:

$$y_i \left(\sum_{j=1}^{|X'|} e_{i,j} x_j \right) \ge (1 - \delta - \frac{1}{t}) y_i c_1 \text{ for } i = 1, 2, \ldots, |Y'| \tag{5}$$

$$x_i \left(\sum_{j=1}^{|Y'|} e_{i,j} y_j \right) \ge (1 - \delta - \frac{1}{t}) x_i c_2 \text{ for } i = 1, 2, \ldots, |X'| \tag{6}$$

$$\sum_{i=1}^{|Y'|} y_i = c_1, \tag{7}$$

$$\sum_{i=1}^{|X'|} x_i = c_2. \tag{8}$$

(5) and (6) indicate that $x_i > 0$ and $y_i > 0$ imply that $\sum_{j=1}^{|X'|} e_{i,j} x_j \ge (1 - \delta - \frac{1}{t}) c_1$ and $\sum_{j=1}^{|Y'|} e_{i,j} y_j \ge (1 - \delta - \frac{1}{t}) c_2$, respectively.

Let $\hat{x}_i$ and $\hat{y}_j$ be the 0/1 integer solution for the quadratic programming problem (5)-(8). Let $\hat{r}_i = \sum_{j=1}^{|X'|} e_{i,j}\hat{x}_j$ and $\hat{s}_i = \sum_{j=1}^{|Y'|} e_{i,j}\hat{y}_j$. To deal with the quadratic programming problem, the key idea here is to estimate the values of $\hat{r}_i$ and $\hat{s}_j$. If we know the values of $\hat{r}_i$ and $\hat{y}_j$, then (5) and (6) become

$$y_i\hat{r}_i \geq y_i c_1\left(1 - \delta - \frac{1}{t}\right) \text{ for } i = 1, 2, \ldots, |Y'| \tag{9}$$

$$x_i\hat{s}_i \geq x_i c_2\left(1 - \delta - \frac{1}{t}\right) \text{ for } i = 1, 2, \ldots, |X'|, \tag{10}$$

where $\hat{r}_i$ and $\hat{s}_i$ in (9) and (10) are constants and the quadratic inequalities become linear inequalities.

Estimating $\hat{r}_i$ and $\hat{s}_i$.

The approach for giving a good estimation of $\hat{r}_i$ and $\hat{s}_i$ is to randomly and independently select a subset $B_{X'}$ of $O(\log(|X'_{opt}|))$ vertices and a subset $B_{Y'}$ of $O(\log(|Y'_{opt}|))$ vertices in X'_{opt} and Y'_{opt}, respectively. Let $c_1 = |X'_{opt}|$ and $c_2 = |Y'_{opt}|$. We do not know c_1 and c_2, but we can guess them in $O(|X'| \times |Y'|)$ time. Then we can use $\frac{c_1}{k}\sum_{v_j \in B_{X'}} e_{i,j}$ and $\frac{c_2}{k}\sum_{u_j \in B_{Y'}} e_{i,j}$ to estimate $\hat{r}_i$ and $\hat{s}_i$, respectively. Since we do not know X'_{opt} and Y'_{opt}, it is not easy to randomly and independently select vertices from X'_{opt} and Y'_{opt}. We develop a method to randomly select $p \times \log|Y'|$ vertices in Y'_{opt} from Y' when Y'_{opt} is not known. Here p is a constant to be determined later.

Finding $p\log|Y'|$ vertices in Y'_{opt} when Y'_{opt} is not known

Let $|Y'| = c|Y'_{opt}|$. The idea here is to randomly and independently select a subset B of $(c + 1) \times p \times \log|Y'|$ vertices from Y' and enumerate all size $p \times \log|Y'|$ subsets of B in time $C_{p(c+1)\log|Y'|}^{p\log|Y'|} \leq O(|Y'|^{p(c+1)})$. We can show that with high probability, we can get a set of $p\log|Y'|$ vertices randomly and independently selected from Y'_{opt}.

Lemma 5. *With probability at least* $1 - |Y'|^{-\frac{p}{2c^2(c+1)}}$, B *contains a size* $p\log|Y'|$ *subset of* Y'_{opt}.

Proof. Let us consider the probability that B contains less than $p\log|Y'|$ vertices in Y'_{opt}. Let b be the expected number of vertices in B that are also in Y'_{opt}. Recall that $|Y'| = c|Y'_{opt}|$. If we randomly select a vertex in Y', the probability that the vertex is in Y'_{opt} is $\frac{1}{c}$. Let μ be the expected number of vertices in B that are in Y'_{opt}. We have $\mu = \frac{|B|}{c} = \frac{1}{c}\lceil(c+1)p\log|Y'|\rceil$. Let $X_1, X_2, \ldots, X_{|B|}$ be $|B|$ independent random 0/1 variables, where $X_i = 1$ with probability $\frac{1}{c}$ indicating that the selected vertex is in Y_{opt}. Thus,

$$b = \sum_{i=1}^{|B|} X_i \tag{11}$$

and

$$\mu = E(\sum_{i=1}^{|B|} X_i) = \frac{1}{c} \lceil (c+1)p \log |Y'| \rceil. \tag{12}$$

Since we selected $(c+1)p \log |Y'|$ vertices,

$$|B| = \lceil (c+1)(p \log |Y'|) \rceil. \tag{13}$$

Based on Lemma 1, we have

$$\mathbf{Pr}(b < p \log |Y'|) \leq \mathbf{Pr}(b < (\frac{1}{c} - \frac{1}{c(c+1)}) \lceil (c+1)(p \log |Y'|) \rceil)$$

$$= \mathbf{Pr}(\sum_{i=1}^{|B|} X_i < \mu - \frac{1}{c(c+1)}|B|) \quad \text{(From (11), (12) and (13))}$$

$$\leq exp(-\frac{1}{2c^2(c+1)^2}|B|)$$

$$\leq exp(-\frac{1}{2c^2(c+1)^2}(c+1)(p \log |Y'|))$$

$$= exp(-\frac{p \log |Y'|}{2c^2(c+1)}) = |Y'|^{-\frac{p}{2c^2(c+1)}}.$$

Therefore, with probability at most $|Y'|^{-\frac{p}{2c^2(c+1)}}$, B does not contain any size $p \log |Y'|$ subset of Y'_{opt}. This completes the proof. □

Let $B_{X'}$ and $B_{Y'}$ be the sets of randomly and independently selected vertices in X'_{opt} and Y'_{opt}. Let $|B_{X'}| = p_1 \log |X'|$ and $|B_{Y'}| = p_2 \log |Y'|$. We define $\bar{r}_i = \sum_{v_j \in B_{X'}} e_{i,j}$ and $\bar{s}_i = \sum_{u_j \in B_{Y'}} e_{i,j}$. The following lemma shows that $\frac{c_1}{|B_{X'}|}\bar{r}_i$ and $\frac{c_2}{|B_{Y'}|}\bar{s}_i$ are good approximations of $\hat{r}_i$ and $\hat{s}_i$.

Lemma 6. *With probability at least* $1 - 2|Y'||X'|^{-\frac{\epsilon^2}{3}p_1} - 2|X'||Y'|^{-\frac{\epsilon^2}{3}p_2}$, *for any* $i = 1, 2, \ldots, |X'|$ *and* $j = 1, 2, \ldots, |Y'|$,

$$(1-\epsilon)\hat{r}_i \leq \frac{c_1}{|B_{X'}|}\bar{r}_i \leq (1+\epsilon)\hat{r}_i$$

and

$$(1-\epsilon)\hat{s}_j \leq \frac{c_2}{|B_{Y'}|}\bar{s}_j \leq (1+\epsilon)\hat{s}_j.$$

Now, we set $r_i = \frac{c_1}{|B_{X'}|}\bar{r}_i$ and $s_i = \frac{c_2}{|B_{Y'}|}\bar{s}_j$. We consider the following linear programming problem.

$$y_i r_i \geq y_i c_1 (1 - \epsilon)(1 - \delta) \text{ for } i = 1, 2, \ldots, m, \tag{14}$$

$$x_i s_i \geq x_i c_2 (1 - \epsilon)(1 - \delta) \text{ for } i = 1, 2, \ldots, m, \tag{15}$$

$$\sum_{i=1}^{|Y'|} y_i = c_1, \tag{16}$$

$$\sum_{i=1}^{|X'|} x_i = c_2 \tag{17}$$

$$\sum_{j=1}^{|X'|} e_{i,j} x_j \geq \frac{r_i}{1 + \epsilon} \tag{18}$$

$$\sum_{j=1}^{|Y'|} e_{i,j} y_j \geq \frac{s_i}{1 + \epsilon}. \tag{19}$$

The term $(1 - \epsilon)$ in (14) and (15) ensures that the quadratic programming problem has a solution when the estimated values of r_i and s_i are smaller than $\hat{r}_i$ and $\hat{s}_i$. Similarly, the term $(1 + \epsilon)$ in (18) and (19) ensures that the quadratic programming problem has a solution when the estimated values of r_i and s_i are bigger than $\hat{r}_i$ and $\hat{s}_i$.

Randomized rounding

Let x_i' and y_j' be a fractional solution for (14) -(19). In order to get a 0/1 solution, we randomly set x_i and y_j to be 1 using the fractional solution as the probability. That is, we randomly set x_i and y_j to be 1's with probability x_i' and y_j', respectively. (Otherwise, x_i and y_j will be 0.)

Lemma 7. *With probability* $1 - 2exp(-\frac{1}{3}|X'|\epsilon^2) - 2exp(-\frac{1}{3}|Y'|\epsilon^2) - |Y'|exp(-\frac{1}{2}|X'|\epsilon^2) - |X'|exp(-\frac{1}{2}|Y'|\epsilon^2)$, *we can find a subset* $\hat{X} \subseteq X'$ *and a subset* $\hat{Y} \subseteq Y'$ *with* $(1 - \epsilon)c_1 \leq |\hat{X}| \leq (1 + \epsilon)c_1$ *and* $(1 - \epsilon)c_2 \leq |\hat{Y}| \leq (1 + \epsilon)c_2$ *such that for any* $x \in \hat{X}$, $d(x, Y') \geq (1 - \delta - 4\epsilon)|\hat{Y}|$ *and for any* $y \in \hat{Y}$, $d(y, X) \geq (1 - \delta - 4\epsilon)|\hat{X}|$.

The complete algorithm for Case 2 is given in Fig. 2. Let $k = \lceil \delta t \rceil$ as defined in Lemma 2. Here c_x, c_y are set to be $k(1 + \frac{1}{\epsilon})$ and $2k$, respectively. $p_1 = p_2 = \frac{5}{\epsilon^2}$.

Theorem 4. *With probability at least* $1 - o(1)$, *Algorithm 2 finds a quasi-*$(\delta + 4\epsilon + \frac{1}{t})$*-biclique of size* $(1 - \frac{1}{t} - \epsilon)(|X_{opt}| + |Y_{opt}|)$ *in* $O((k \times \frac{1}{\epsilon^2}|X||Y|)^{\lceil \delta t \rceil}(|X|^{\frac{5}{\epsilon^2}k(1 + \frac{1}{\epsilon})} + |Y|^{\frac{5}{\epsilon^2}2k})(|X| + |Y|^3))$ *time.*

We can derandomize the algorithm to get a polynomial time deterministic algorithm. Step 3 can be derandomized by using the standard method. For instance, instead of randomly and independently choosing $p_1 \log(|X'|)$ and $p_2 \log(|Y'|)$ vertices from X' and Y', we can pick the vertices encountered on a random walk of the same length on a constant degree expander. Obviously, the number of such random walks on a constant degree expander is polynomial. Thus, by enumerating all random walks of length $p_1 \log(|X'|)$ and $p_2 \log(|Y'|)$, we have a polynomial time deterministic algorithm.

Algorithm 2: Algorithm for Soving Case 2: $|X'_{opt}| > \epsilon|Y'_{opt}|$.

Input: a bipartite graph $G = (X \cup Y, E)$, a real number $0 \le \delta \le 0.5$, a number
         $t > 0$ and a number $\epsilon > 0$.

0. Let $k = \lceil \delta t \rceil$, $p_1 = p_2 = \frac{5}{\epsilon^2}$, $c_x = k(1 + \frac{1}{\epsilon})$ and $c_y = 2k$.

1. **for** any $v_1, v_2, \dots, v_k \in X$ and any $u_1, u_2, \dots, u_k \in Y$ **do**

2.   Set $X' = \cup_{i=1}^{k} D(v_i, Y)$ and $Y' = \cup_{i=1}^{k} D(u_i, X)$.

3   Randomly and independently select a set $S_{X'}$ of $(c_x + 1)p_1 \log |X'|$
     vertices in X' and a set $S_{Y'}$ of $(c_y + 1)p_2 \log |Y'|$ vertices in Y'.

4   **for** any size $p_1 \log |X'|$ subset $B_{X'}$ of $S_{X'}$ and size $p_2 \log |X'|$ subset $B_{Y'}$
    of $S_{Y'}$ **do**

        (a) $\bar{r}_i = \frac{c_1}{|B_{X'}|} \sum_{v_i \in |B|_{X'}} e_{i,j}$

        (b) $\bar{s}_i = \frac{c_2}{|B_{Y'}|} \sum_{u_i \in |B|_{Y'}} e_{i,j}$

        (c) Get a fractional solution x'_i and y'_i for $x_i \in X'$ and $y_i \in Y'$ of
           (11)-(16)

        (d) do randomrized rouding according to x'_i and y'_i

        (e) $X_A = \{v_i | x_i = 1\}$ and $Y_A = \{u_i | y_i = 1\}$

5. Output a $\delta + \frac{1}{t} + 4\epsilon$ quasi-biclique with the biggest $|X_A| + |Y_A|$.

Fig. 2. The algorithm for Case 2.

Step 4 (d) can be derandomized by using Raghavan's conditional probabilities method (Raghavan, 1988). From Case 1 and Case 2, we can immediately obtain the following theorem.

Theorem 5. *There exists a polynomial time approximation scheme that outputs a quasi-biclique* $X_A \subseteq X$ *and* $Y_A \subseteq Y$ *with* $|X_A| + |Y_A| \ge (1 - \epsilon)(|X_{opt}| + |Y_{opt}|)$ *such that any vertex* $x \in X_A$ *is incident to at least* $(1 - \delta - \epsilon)|Y_A|$ *vertices in* Y_A *and any vertex* $y \in Y_A$ *is incident to at least* $(1 - \delta - \epsilon)|X_A|$ *vertices in* X_A *for any* $\epsilon > 0$, *where* X_{opt} *and* Y_{opt} *form the optimal solution.*

4. The heuristic algorithm

In practice, we need to find large quasi-bicliques in PPI networks. Here, we propose a heuristic algorithm to find large quasi-bicliques. Consider a PPI network $\mathcal{G} = (\mathcal{V}, \mathcal{E})$. Our heuristic algorithm has two steps. First, we construct a bipartite graph from the graph $\mathcal{G}$ based on a pair of interacting proteins (u, v). Using the method described at the beginning of Section 2, we can get a bipartite graph $G = (X \cup Y, E)$. Second, we find quasi-bicliques in G. The bipartite graph G contains all proteins that have interactions with u or v. So we can find large quasi-bicliques containing u and v in the bipartite graph.

In the algorithm for finding quasi-bicliques in G, we have two parameters δ and τ, which control the quality and sizes of the quasi-bicliques. We use a greedy method to get the seeds for finding large quasi-bicliques in G. At the beginning, we set $X' = \phi$ and $Y' = Y$. In each step, we find a vertex with the maximum degree in $X - X'$. The vertex is added into the biclique vertex set X', and we eliminate all vertices y in Y' such that $d(y, X') < (1 - \delta)|X'|$. We will continue this process until the size of Y' is less than τ. At each step, we get a seed for finding large quasi-bicliques.

The seeds may miss some possible vertices in the quasi-bicliques. We can extend the seeds to find larger quasi-bicliques. Let $X'' = X'$ and $Y'' = Y'$ be a pair of seed vertex sets. In the first step, we can find a vertex x in $X - X''$ with the largest degree $d(x, Y'')$ in $X - X''$. If

$d(x, Y'') \geq (1 - \delta)|Y''|$, we add the vertex x to X''. In the second step, we can find a vertex y in $Y - Y''$ with the largest $d(y, X'')$ in $Y - Y''$. If $d(y, X'') \geq (1 - \delta)|X''|$, we add the vertex y to Y''. We repeat the above two steps until no vertex can be added. The whole algorithm is shown in Fig. 3. We can also exchange the two vertex sets X and Y to find more quasi-bicliques using the algorithm.

Let n be the number of vertices in the bipartite graph G. In the greedy algorithm, the time complexity of Steps $3 - 5$ and Step 10 is $O(n)$, and the time complexity of Steps $6 - 9$ is $O(n^2)$. So the time complexity of Steps $3 - 10$ is dominated by $O(n^2)$. Since Steps $3 - 10$ is repeated $O(n)$ times, the time complexity of the whole algorithm is $O(n^3)$.

The Greedy Algorithm
Input   A bipartite graph $(X \cup Y, E)$ and two parameters δ and τ.
Output   A set of δ-quasi-bicliques $(X' \cup Y', E')$ with $|X'| \geq \tau$ and $|Y'| \geq \tau$.
1.   Let $X' = \phi$ and $Y' = Y$.
2.   **while** $|Y'| \geq \tau$ and $X' \neq X$ **do**
3.   Find the vertex $x \in X - X'$ with the maximum degree $d(x, Y')$.
4.   Add x into X', $X' = X' \cup \{x\}$, and delete from Y' all vertices $y \in Y'$ such that $d(y, X') < (1 - \delta)|X'|$.
5. $X'' = X'$ and $Y'' = Y'$.
6.   **repeat**
7.   Find the vertex $x \in X - X''$ with the maximum degree $d(x, Y'')$. If $d(x, Y'') \geq (1 - \delta)|Y''|$, add x to X'', $X'' = X'' \cup \{x\}$.
8.   Find the vertex $y \in Y - Y''$ with the maximum degree $d(y, X'')$. If $d(y, X'') \geq (1 - \delta)|X''|$, add y to Y'', $Y'' = Y'' \cup \{y\}$.
9.   **until** no vertex is added in the steps 7 and 8.
10.   **if** $|X''| \geq \tau$, $|Y''| \geq \tau$, for each $x \in X''$, $d(x, Y'') \geq (1 - \delta)|Y''|$, for each $y \in Y''$, $d(y, X'') \geq (1 - \delta)|X''|$, output $(X'' \cup Y'')$ as a quasi-biclique.

Fig. 3. The greedy algorithm.

5. Finding motifs from the multiple sequence alignment of computed δ-bicliques.

We implemented the heuristic algorithm described in the last section in JAVA. The software is called PPIExtend. In the implementation, we added a new parameter α to speed up the algorithm. In Step 3, instead of selecting one vertex with the best degree, we can select the best α vertices in $X - X'$ and add all the α vertices into X' in Step 4. As shown in the last step of the algorithm, some vertices in X'' may be adjacent to less than $(1 - \delta)|Y''|$ vertices in Y'', but the average degree of the vertices in X'' is no less than $(1 - \delta)|Y''|$. Similarly, some vertices in Y'' may be adjacent to less than $(1 - \delta)|X''|$ vertices in $|X''|$, but the average degree of the vertices in Y'' is no less than $(1 - \delta)|X''|$. In our experiments, these quasi-bicliques are still output to get more useful quasi-bicliques.

Our algorithm for PPIExtend consists of two steps: (i) find interacting protein group pairs (quasi-bicliques) using the greedy algorithm, (ii) find conserved motifs from multiple sequence alignments for each of the protein groups. (We use the existing multiple sequence alignment software PROTOMAT (Pietrokovski, 1996).)

The motifs found by PROTOMAT can be viewed as a *block*, that is a conserved region in a multiple sequence alignment of the proteins in a group. For each biclique X and Y obtained by the greedy algorithm, we use S_X and S_Y to denote the sets of motifs obtained by the multiple sequence alignments of protein sequences in X and Y, respectively. Any pair of motifs (m_1, m_2) with $m_1 \in S_X$ and $m_2 \in S_Y$ is a candidate protein-protein interaction motif pair. Thus, our algorithm can also output lots of motif pairs as candidate protein-protein interaction motif pairs.

We look at the numbers of motifs found by the programs PPIExtend and FPClose* that are also in the two block databases, BLOCKS (Pietrokovski, 1996) and PRINTS (Attwood & Beck, 1994). The LAMA program (Pietrokovski, 1996) is used to find the local optimal alignment of two blocks (the motif output by PPIExtend/FPClose* and a block in the databases), where the Z-score is computed to measure the alignments. The default threshold of Z-score was used in the experiments. The results are reported in Table 1. From this table, we can see that our method has more mappings to BLOCKS and PRINTS than FPClose* (Li et al., 2006; Grahne & Zhu, 2003).

	BLOCKS		PRINTS		BOTH	
	blocks	domains	blocks	domains	blocks	domains
FPClose*	6408/24294	3128/4944	2174/11170	1093/1850	24.1%	62.1%
PPIExtend	9325/29767	4191/6149	2423/11435	1160/1900	28.5%	66.4%

Table 1. The mappings between the motifs and the two databases: BLOCKS and PRINTS. FPClose* uses BLOCKS 14.0 and PRINTS 37.0. Our PPIExtend method uses BLOCKS 14.3 and PRINTS 38.0. Each entry a/b means the motifs are mapped to a blocks(domains) in all b blocks(domains) in the databases.

	BLOCKS	PRINTS	Pfam	iPfam
Version	14.3	38.0	20.0	20.0
Number of domains	6149	1900	8296	2883
Number of entries	29767	11435	8296	3019

Table 2. Databases used in the experiments.

We look at the numbers of motif pairs found by the two programs PPIExtend and FPClose* that can be mapped into domain-domain interaction pairs in the domain-domain interaction database iPfam (Finn et al., 2005). The versions of the databases are shown in Table 2. The iPfam database is built on top of the Pfam database (Sonnhammer et al., 1997) which stores the information of protein domain-domain interactions. To examine whether the motif pairs found by PPIExtend and FPClose* can match some pairs of interacting domains in iPfam, we map our motif pairs to domain pairs in iPfam through the integrated protein family database InterPro (Apweiler et al., 2001) which integrates a number of databases. In fact, we strictly follow the procedure as suggested in (Li et al., 2006). (1) We map our motifs to domains (protein groups) in the database BLOCKS or PRINTS; (2) we map a protein group of BLOCKS to a protein group of InterPro based on the one-to-one mapping between an entry of BLOCKS and an entry of InterPro; (Note that both PRINTS and Pfam are member databases of InterPro, and the mapping between PRINTS and Pfam is clear.) (3) we use existing cross-links between protein groups of InterPro and domains of Pfam to determine the crosslinks between the motifs found by PPIExtend/FPClose* and Pfam domains. In this way, we can map our motif pairs into domain pairs with Pfam domain entries. Note that the mapping between motif pairs and domain pairs is not one-to-one.

We observed that the motif pairs found by PPIExtend can map to 81 distinct domain pairs in *i*Pfam. However, only 18 domain pairs were reported in (Li et al., 2006). This is a significant improvement and the main reason is the use of quasi-bicliques. In the 81 domain pairs, 48 pairs are domain-domain interactions on one protein (self-loops) and 33 pairs are domain-domain interactions on different proteins. Although the self-loops form a large portion, we still find many other domain-domain interactions that are not self-loops.

6. Protein interaction sites: a case study

In this section, we present detailed information about binding motif pairs that can be mapped to interacting domain pairs. The first motif pair is derived from a protein group pair in which the left protein group contains 7 proteins and the right protein group contains 10 proteins. There are 66 interactions between the two groups of proteins. Using the hypergeometric probability model, the p-value of the protein group pair is less than 1.57×10^{-191}. PROTOMAT finds two left blocks and two right blocks in this protein group pair. The second left block contains 20 positions and the first right block contains 12 positions. By the mapping method, the positions $1 - 19$ of the second left block can be aligned with the positions $9 - 27$ of block IPB001425B in BLOCKS, and the positions $4 - 12$ of the first right block can be aligned with the positions $1 - 9$ of block IPB003660A in BLOCKS. Block IPB001425B is in the Bac_rhodopsin domain, and block IPB003660A is in the HAMP domain. See Table 3 for more details. Our binding motif pair can map into the domain pair (PF00672, PF01036) in *i*Pfam. *i*Pfam shows that the HAMP domain interacts with the Bac_rhodopsin domain in protein complexes such as lh2s. 1h2s is the complex of *Natronobacterium pharaonis* sensory rho-dopsin II (sRII) with receptor-binding domain of HtrII. The X-ray structure of 1h2s was obtained at 1.93 Å resolution (Gordeliy et al., 2002) and it provided an atomic picture of the first step of the signal transduction. The interactions in the sRII-HtrII complex have been intensively investigated to find the signal relay mechanism from the receptor to the transducer (Bergo et al., 2005; Inoue et al., 2007; Sudo et al., 2007). The 3D structure of the interactions is shown in Fig. 4(a) and 4(b), which are generated by Protein Explorer (Martz, 2002). The shortest residue-residue distance between the two motifs in a pair is also interesting. In protein complex 1h2s, there are two chains: chain A (1h2s_A) and chain B (1h2s_B). The left motif is located at positions $168 - 186$ of 1h2s_A, and the right motif is located at positions $61 - 69$ of 1h2s_B (Table 3). We downloaded the coordinate information of 1h2s from http://www.ebi.ac.uk/msd-srv/msdlite/atlas/summary/1h2s.html, and computed the residue-residue distances between the two motifs. The shortest residue-residue distance is 4.07 Å between atom 1346 of residue 177 in 1h2s_A and atom 2018 of residue 69 in protein 1h2s_B (Fig. 4(b)). The average shortest residue-residue distance is 9.17Å. From these

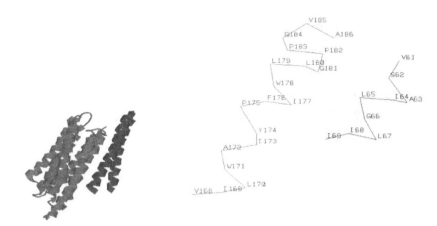

(a) The 3D structure of 1h2s (asymmetric unit).

(b) The backbone structure of the two motifs in 1h2s.

Fig. 4. (a) The 3D structure (best viewed in color) of the interactions between the Bac_rhodopsin domain and the HAMP domain in 1h2s. The left part is chain A and contains the Bac_rhodopsin domain. The right part is chain B and contains the HAMP domain. (b) The backbone structure of the interactions between segment [168V,186A] in 1h2s_A and segment [61V,69I] in 1h2s_B.

calculation and information, we may conclude that the positions 1 − 19 of the second left block and the positions 4 − 12 of the first right block are possibly interaction sites.

7. Prediction of binding sites

After obtaining candidate domains (conserved regions) in multiple sequence alignment, we can further verify if a pairs of predicted domains really interact with each other by using some tools for protein binding site prediction. Here we briefly introduce a method originally in (Guo & Wang, 2011). This method assumes that the 3D structures of the two given proteins are known.

Given two complete protein structures, the task is to find the binding sites between the two proteins. The method contains three steps. Firstly, we do local sequence alignment at the atom level to get the alignments of conserved regions. Those alignments of conserved regions may contain some gaps. Secondly, among the conserved regions obtained in Step 1, we use the 3D structure information to identify the surface segments. Finally, for any pair of the surface segments identified in Step 2, we compute a rigid transformation to compare the similarity of the two substructures in 3D space and output the qualified pairs as binding sites. When computing the rigid transformations, we treat each protein as a molecule with some volume and introduce a method to ensure that the two whole protein 3D structures have no overlap under such a rigid transformation in 3D space. The software package is available at http://sites.google.com/site/guofeics/bsfinder.

```
AC   118493xB;
     distance from previous block=(4,396)
DE   none BL   IIK motif=[6,0,17] motomat=[1,1,-10]
     width=20 seqs=7
DIP:8095N   ( 206)  VIGILIISYTKATCDMLAGK
DIP:4973N   ( 536)  MILILIAQFWVAIAPIGEGK
DIP:5150N   ( 417)  LIKDEINNDKKDNADDKYIK
DIP:5371N   ( 384)  IILALIVTILWFMLRGNTAK
DIP:676N    ( 402)  VIVAWIFFVVSFVTTSSVGK
...

pdb 1h2s_A  ( 168)  VILWAIYPFIWLLGPPGVA
Bac_rhodopsin:      VVLWLAYPVVWLLGPEGIG

AC   r18493xA;
     distance from previous block=(7,177)
DE   none BL   LLL motif=[6,0,17] motomat=[1,1,-10]
     width=12 seqs=8
DIP:7371N   (  10)  LALIILYLSIPL
DIP:8128N   (  35)  LSLRFLALIFDL
DIP:4176N   ( 106)  LVLTSLSLTLLL
DIP:7280N   (  11)  LSLFLPPVAVFL
DIP:5331N   ( 178)  LSFFVLCGLARL
...

pdb 1h2s_B  (  61)      VSAILGLII
HAMP:                   IALLLALLL
```

Table 3. Left block l18493xB aligning with the Bac_rhodopsin domain and right block r18493xA aligning with the HAMP domain. For brevity, only 5 sequences in each of the two blocks are shown. In line Bac_rhodopsin and line HAMP, each letter is the amino acid with the highest frequency in the corresponding column in the multiple alignment. Pdb 1h2s_A and pdb 1h2s_B are chain A and chain B in protein complex 1h2s, respectively.

8. Conclusion

We have proposed algorithms for finding the maximum vertex quasi-biclique problem. We illustrate the applications of the proposed algorithms for finding protein-protein binding sites. The general approach contains three steps: (1) find quasi-bicliques from PPI networks; (2) do multiple sequence alignment for each of the groups in the quasi-biclique and identify possible domains on the protein sequences. (3) use other methods, e.g., the one in (Guo & Wang, 2011), to further confirm the binding sites.

9. References

Yannakakis, M. (1981). Node deletion problems on bipartite graphs. *SIAM Journal on Computing*, Vol. 10, 1981, 310–327.

Thomas, A.; Cannings, R.; Monk, N. A. M. & Cannings, C. (2003). On the structure of protein-protein interaction networks. *Biochemical Society Transanctions*, Vol. 31, Dec 2003, 1491–1496.

Morrison, J. L.; Breitling, R.; Higham, D. J. & Gilbert, D. R. (2006). A lock-and-key model for protein-protein interactions. *Bioinformatics*, Vol. 22, No. 16, Aug 2006, 2012–2019.

Andreopoulos, B.; An, A.; Wang, X.; Faloutsos, M. & Schroeder, M. (2007). Clustering by common friends finds locally significant proteins mediating modules. *Bioinformatics*, Vol. 23, No. 9, 2007, 1124–1131.

Bu, D.; Zhao, Y.; Cai, L.; Xue, H.; Zhu, X.; Lu, H.; Zhang, J.; Sun, S.; Ling, L.; Zhang, N.; Li, G. & Chen, R. (2003). Topological structure analysis of the protein-protein interaction network in budding yeast. *Nucleic Acids Research*, Vol. 31, No. 9, May 2003, 2443–2450.

Hishigaki, H.; Nakai, K.; Ono, T.; Tanigami, A. & Takagi, T. (2001). Assessment of prediction accuracy of protein function from protein–protein interaction data. *Yeast*, Vol. 18, No. 6, Apr 2001, 523–531.

Li, H.; Li, J.; Wong, L. (2006). Discovering motif pairs at interaction sites from protein sequences on a proteome-wide scale. *Bioinformatics*, Vol. 22, No. 8, 2006, 989–996.

Garey, M. R. & Johnson, D.S. (1979). *Computers and Intractability, A guide to the theory of NP-completeness*. Freeman, San Francisco, 1979.

Peeters, R. (2003). The maximum edge biclique problem is NP-complete. *Discrete Applied Mathematics*, Vol. 131, No. 3, 2003, 651–654.

Karp, R. M. (1972). Reducibility among combinatorial problems. *Complexity of Computer Computations (R. E. Miller and J. W. Thatcher, eds.)*, 1972, 85–103.

Peleg, D.; Schechtman, G. & Wool, A. (1993). Approximating bounded 0-1 integer linear programs, *Proceedings of the 2nd Symposium on Theory of Computing and Systems*. IEEE Computer Society, 1993.

Pietrokovski, S. (1996). Searching databases of conserved sequence regions by aligning protein multiple-alignments. *Nucleic Acids Research*, Vol. 24, 1996, 3836–3845.

Attwood, T. K. & Beck, M. E. (1994). PRINTS-a protein motif fingerprint database. *Protein Engineering, Design and Selection*, Vol. 7, 1994, 841–848.

Finn, R. D.; Marshall, M.; & Bateman,A. (2005). iPfam: visualization of protein-protein interactions in PDB at domain and amino acid resolutions. *Bioinformatics*, Vol. 21, No. 3, Feb 2005, 410–412.

Grahne, G. & Zhu, J. (2003). Efficiently using prefix-trees in mining frequent itemsets. *Proceedings of the Workshop on Frequent Itemset Mining Implementataions (FIMI)*, 2003.

Sonnhammer, E. L.; Eddy, S. R. & Durbin, R. (1997). Pfam: A comprehensive database of protein domain families based on seed alignments. *Proteins: Structure, Function and Genetics*, Vol. 28, 1997, 405–420.

Apweiler, R.; Attwood, T. K.; Bairoch, A.; Bateman, A.; Birney, E.; Biswas, M.; Bucher, P.; Cerutti, L.; Corpet, F.; Croning, M. D.; Durbin, R.; Falquet, L.; Fleischmann, W.; Gouzy, J.; Hermjakob, H.; Hulo, N.; Jonassen, I.; Kahn, D.; Kanapin, A.; Karavidopoulou, Y.; Lopez, R.; Marx, B.; Mulder, N. J.; Oinn, T. M.; Pagni, M.; Servant, F.; Sigrist, C. J. & Zdobnov, E. M. (2001). The InterPro database, an integrated documentation resource for protein families, domains and functional sites. *Nucleic Acids Research*, Vol. 29, No. 1, Jan 2001, 37–40.

Gordeliy, V. I.; Labahn, J.; Moukhametzianov, R.; Efremov, R.; Granzin, J.; Schlesinger, R.; Büldt, G.; Savopol, T.; Scheidig, A. J.; Klare, J. P. & Engelhard, M. (2002). Molecular basis of transmembrane signalling by sensory rhodopsin II-transducer complex. *Nature*, Vol. 419, No. 6906, Oct 2002, 484–487.

Bergo, V. B.; Spudich, E. N.; Rothschild, K. J. & Spudich, J. L. (2005). Photoactivation perturbs the membrane-embedded contacts between sensory rhodopsin II and its transducer. *Journal of Biological Chemistry*, Vol. 280, No. 31, Aug 2005, 28 365–28 369.

Inoue, K.; Sasaki, J.; Spudich, J. L. & Terazima, M. (2007). Laser-induced transient grating analysis of dynamics of interaction between sensory rhodopsin II D75N and the HtrII transducer. *Biophysical Journal*, Vol. 92, No. 6, Mar 2007, 2028–2040.

Sudo, Y.; Furutani, Y.; Spudich, J. L. & Kandori, H. (2007). Early photocycle structural changes in a bacteriorhodopsin mutant engineered to transmit photosensory signals. *Journal of Biological Chemistry*, Vol. 282, No. 21, May 2007, 15 550–15 558.

Martz, E. (2002). Protein Explorer: easy yet powerful macromolecular visualization. *Trends in Biochemical Sciences*, Vol. 27, No. 3, 2002, 107–109.

Sivaraman, J.; Li, Y.; Banks, J.; Cane, D. E.; Matte, A. & Cygler, M. (2003). Crystal structure of Escherichia coli PdxA, an enzyme involved in the pyridoxal phosphate biosynthesis pathway. *Journal of Biological Chemistry*, Vol. 278, No. 44, Oct 2003, 43 682–43 690.

Sakai, A.; Kita, M. & Tani, Y. (2004). Recent progress of vitamin B6 biosynthesis. *Journal of Nutritional Science and Vitaminology (Tokyo)*, Vol. 50, No. 2, Apr 2004, 69–77.

Fitzpatrick, T. B.; Amrhein, N.; Kappes, B.; Macheroux, P.; Tews, I. & Raschle, T. (2007). Two independent routes of de novo vitamin B6 biosynthesis: not that different after all. *Biochemistry Journal*, Vol. 407, No. 1, Oct 2007, 1–13.

Guo, F. & Wang, L. (2011). Computing the Protein Binding Sites. *ISBRA*, 2011, 25-36.

Liu, X.; Li, J. & Wang, L. (2010). Modeling Protein Interacting Groups by Quasi-Bicliques: Complexity, Algorithm, and Application. *IEEE/ACM Trans. Comput. Biology Bioinform.*, Vol. 7, No. 2, 2010, 354-364.

Wang, L. (2011). Near Optimal Solutions for Maximum Quasi-bicliques, *Journal of Combinatorial Optimization*, on-line available at DOI 10.1007/s10878-011-9392-4.

Li, M.; Ma, B. & Wang, L. (2002). On the closest string and substring problems. *Journal of the ACM*, Vol. 49, No. 2, 2010, 157-171.

Raghavan, P. (1988). Probabilistic construction of deterministic algorithms: Approximate packing integer programs. *JCSS*, Vol. 37, No. 2, 2010, 130-143.

7

Inferring Protein-Protein Interactions (PPIs) Based on Computational Methods

Shuichi Hirose

Nagase & Co. Ltd. Research & Development Center, 2-2-3 Murotani,
Nishi-ku, Kobe, Hyogo,
Computational Biology Research Center, Advanced Industrial Science and
Technology, 2-4-7 Aomi, Koto-ku, Tokyo,
Japan

1. Introduction

Proteins are involved in many essential cellular processes, such as metabolism and signalling. They function by interacting with other molecules within the cell. Thus, protein interaction is one of the important keys to understand protein functions. As a consequence of the development of high-throughput experimental methods for detecting protein interactions, large volumes of data are now available. Although the data are valuable, there are limitations to their application. Therefore, computational methods are helpful tools for predicting protein interactions. With the increase in genome sequence data, the importance of computational methods in this field is growing more and more.

Another important factor to understand protein function is flexibility, because a protein molecule is not a rigid body. Flexible regions are often necessary for proteins to perform their functions, e.g. by enabling their flexible conformations to interact with other molecules and proteins. Therefore, it is important to understand the relationship between protein flexibility and protein interactions. In accordance with the increasing numbers of available protein structures, several databases that deal with protein flexibility have been built. Computational methods for analyzing protein motion are also being developed, for applications to PPI (protein-protein interaction) data.

The aim of this chapter is to provide a review on PPI prediction by computational techniques. In the first half of this chapter, the concepts and applications of several methods for inferring PPIs are introduced. They use genomic information based on evolutionary events. In the second half, the databases and prediction methods that deal with protein flexibility are introduced, and the possibility of inferring PPIs from protein flexibility will be discussed.

2. Computational methods to infer PPIs

The prediction of PPIs can be regarded as a binary classification problem, whereby the aim is to identify pairs of proteins as either interacting or non-interacting. PPIs can be divided into three types (Brown *et al.*, 2010). The first is direct protein interactions, which involve direct physical contacts between proteins. The second is indirect functional association. In this case, the interacting protein pair does not have a direct physical contact, but it indirectly interacts, such as in the formation of a complex. The third is a member of biological pathway. In this case, the protein pairs do not form a complex, but their interactions occur in a logical order (for instance, in a signalling pathway).

Proteins exert their biological functions by participating in a PPI network. Protein interactions, as well as their biochemical functions, work as a type of selective pressure during evolution. Therefore, they influence the genome structure associated with the protein interactions. Conversely, analyzing the changes in the patterns on the genome makes it possible to infer PPIs. Several evolutionary events function as factors that impact the evolution of PPIs, including horizontal gene transfer, operon structure, co-evolution, co-expression, and lineage-specific gene loss. Several methods that infer PPIs using these various types of evolutionary information have been developed (Valencia and Pazos, 2002; Skrabanek et al., 2008). In this section, the principles and applications of PPI prediction methods are introduced. The different PPI prediction methods (Shoemaker and Panchenko, 2007) are listed in Table 1. First, the genomic inference methods (Rosetta stone, conservation of gene neighborhood, and phylogenetic profile), which predict functional association using genomic context, are presented. Next, the methods (mirror tree, in silico two-hybrid system) based on co-evolution are introduced, which are applicable to domain interactions as well as protein interactions. Finally, the sequence signature and machine learning based-methods are presented.

Method	Interaction Type	Interaction
Rosetta stone	Indirect functional association	Protein
Conservation of gene neighborhood	Indirect functional association	Protein
Phylogenetic profiles	Indirect functional association	Protein/domain
Mirror tree	Indirect functional association	Protein/domain
In silico two-hybrid system	Direct physical interaction	Protein/domain
Sequence signature	Direct physical interaction	Protein/domain
Supervised classification	Direct physical interaction	Protein/domain

Table 1. Summary of PPI prediction methods.

The second column represents the interaction type predicted by the method. The third column shows whether the method is designed to predict protein or domain interaction.

2.1 Rosetta stone

The Rosetta stone approach infers PPIs by comparing different genomes. It is often observed that two proteins that interact in genome i have a homologous protein that is fused into one protein in genome j (Fig. 1). These two gene products are functionally related in many cases (Enright et al., 1999). The fused protein is called a 'Rosetta stone protein', since it serves as a key for unlocking the functional relationship between two genes that are encoded independently in the genome. The Rosetta stone approach estimates functionally related protein pairs based on such a concept. The benefit of this approach is applicable to all genomes, including those of Eukaryote. Not surprisingly, the inference of a protein interaction is restricted to the case where the gene fusion can be detected.

Hence, the approach searches for the proteins that are conserved between different organisms. The following two points must be considered, in order to obtain higher prediction accuracy. First, the proteins that interact with many other proteins, such as the HRG domain, and the CBS domain, which binds to DNA, should be removed. Second, the analysis is focused on the case where the pairs of genes that are fused together are orthologous. As an extension of the Rosetta stone approach, it can predict a functionally related gene cluster by combining several results. Four proteins, A, B, C, and D, are considered to be functionally related if the Rosetta stone proteins of the A-B, B-C, and C-D pairs are found.

Applying the Rosetta stone approach to many genomes revealed 6,809 potentially interacting pairs in *Escherichia coli* and 45,502 pairs in yeast (Marcotte *et al.*, 1999). The two proteins in each pair have significant sequence similarity to a single fused protein in another genome. Some proteins interact with several other proteins, and these connections apparently represent functional interactions, such as complexes or pathways.

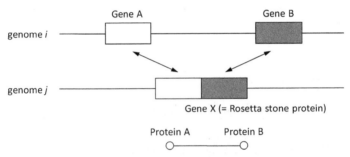

Gene X is the Rosetta stone protein, indicating that protein A and protein B are functionally related.

Fig. 1. The concept of the Rosetta stone approach.

2.2 Conservation of gene neighborhood

The genome comparison among Bacteria or Archaea indicated that the gene order and the operon structure are not conserved on the genome. This is because they have changed with evolutionary events, such as recombination, gene disruption, gene formation, and horizontal gene transfer. These phenomena suggest that the gene order is basically not subjected to selective pressure. However, the gene order or gene clusters on a genome are conserved if the gene products physically interact with each other, such as by complex formation, or if the proteins are transcribed as a single unit (Dandekar *et al.*, 1998). Briefly, it is often observed that the genes encoding proteins that either form a complex via physical interaction or work together in the same pathway are encoded in the same operon in different genomes. Thus, the gene order is conserved among different genomes, although the operon structure is fundamentally unstable during evolution (Fig. 2(A)). The conservation of the gene neighborhood approach infers proteins that are involved in the same biological process, using genome information. Many of the functionally related genes predicted by this approach encode proteins that either interact with each other directly, participate in the formation of the same complex, or work in the same metabolic pathway.

The conserved clusters of genes in an operon are detected by various concepts, such as Run or BBH (bidirectional best hit) (Overbeek *et al.*, 1999) (Fig. 2(B)). A set of genes is called a "run" if they all occur on the same strand and the gaps between adjacent genes are 300 bases or less. Any pair of genes occurring within a single run is called "close". If gene Xi in genome i is closest to Xj in genome j and Xj is closest to Xi, then Xi and Xj are called BBH. Genes (Xi, Yi) from genome i and genes (Xj, Yj) from genome j form a PCBBH (pair of close bidirectional best hit) if two pairs of BBHs are considered. The conservation of gene neighborhood approach uses such virtual operons and orthologs to infer PPIs. That is, two orthologous groups are considered to have a connection if they co-occur in the same potential operon two or more times. The advantage of this approach is that the conservation of gene order or gene co-occurrence in the Run is stricter than the Rosetta stone and phylogenetic profile approaches, and it can cover a wider range of genes. However, the application of this approach is limited to Bacteria or Archaea that have operon structures.

Snel *et al.* reported 3,033 orthologous groups with 8,178 pairwise significant associations, by comparing 38 genomes (Snel *et al.*, 2002). Among them, 88% of the 516 small, disjointed clusters, containing 2.7 orthologous groups on average, have a more homogeneous functional composition, in terms of the COG functional category. They are regarded as functional modules.

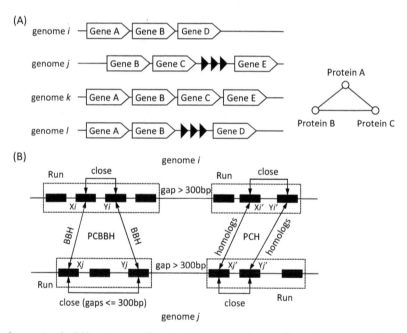

Different boxes signify different genes. The triangles represent genes that lack a conserved gene order. Protein A, protein B, and protein C, which line up in the same order among different organisms, are considered to be functionally associated.

Fig. 2. Illustration of (A) the concept of the conservation of gene neighborhood approach and (B) the definitions of BBH, PCBBH, and PCH (pairs of close homologs).

2.3 Phylogenetic profile

This approach is based on a concept derived from the lineage-specific gene loss. The genes encoding proteins that interact with each other co-occur in different genomes. If one gene is absent in a genome, and then the other gene that interacts with it also is lost. On the basis of this hypothesis, the phylogenetic profile approach infers PPIs from genome comparisons. The phylogenetic profile approach is based on the co-occurrence of gene pairs, while the conservation of gene neighborhood approach is based on the gene order or co-occurrence of genes. The advantage is that it is applicable to Eukaryote, since it is not necessary to consider operon structure. In addition, this approach is different from the prediction method based on the operon, in that the rate of predicted genes that belong to the same biological process is higher. The approach has two disadvantages. The first point is that the analysis targets are limited to the organisms with completely sequenced genomes, because whether a certain gene is actually encoded in the genome must be known. The second is that this approach is not applicable for the proteins encoded in all organisms that are analysis subjects.

The functional relationship between two genes is detected by comparing their phylogenetic profiles (Fig. 3) (Pellegrini *et al.*, 1999). A phylogenetic profile is constructed for each protein, as a vector of N elements, where N is the number of genomes. Each position of the profile represents whether the protein that is homologous to the target protein is absent (signified by 0) or present (1) in each genome. Consequently, the phylogenetic distribution is shown by a long binary number along with each genome. A functionally related protein pair is detected by searching for the same phylogenetic distribution patterns. This method is applicable to domains as well as proteins (Pagel *et al.*, 2004).

Pellegrini *et al.* applied a phylogenetic profile approach to the *Escherichia coli* genome and 16 other fully sequenced genomes, in order to predict the functions of uncharacterized proteins. When the function of a protein is assumed to be the same as that of its neighbors in the phylogenetic-profile space, 18% of the neighbor keywords overlapped the known keywords of the query protein. This indicates that the phylogenetic profile approach has the ability to assign functions to uncharacterized proteins.

	genome *i*	genome *j*	genome *k*	genome *l*	genome *m*
gene A	1	0	1	0	1
gene B	1	1	0	1	1
gene C	1	0	1	0	1
gene D	1	0	1	1	1
gene E	0	1	1	0	1

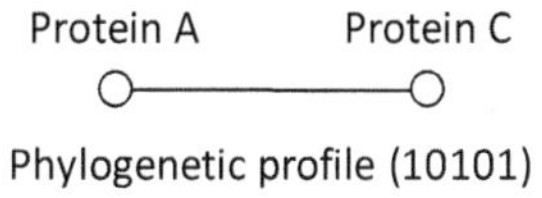

Protein A   Protein C

O————————O

Phylogenetic profile (10101)

Protein A and protein C are considered to interact with each other, since they have the same profile (10101).

Fig. 3. An example of the phylogenetic profile approach.

2.4 Mirror tree

Pairs of physically contacting proteins co-evolve, such as insulin and its receptors (Fryxell, 1996). Co-evolution refers to the phenomenon in which the evolution found in one protein has a considerable effect on the evolution of its partner protein, in order to maintain the protein interaction. Therefore, the amino acid substitutions are expected to occur at the same time in the interacting proteins. As a result, the two phylogenetic trees drawn for the interacting proteins show a greater degree of similarity than those drawn for proteins without interactions (Goh et al., 2000). The mirror tree approach infers two protein/domain interaction pairs, using the similarity between the phylogenetic trees as an indicator. The advantage of this approach is that it can be applied to an organism whose genome has not been completely sequenced. Conversely, the approach is not applicable to a gene that shows a species-specific loss. In addition, the applications of this approach are limited to the cases where high-quality and complete multiple sequence alignments, including sequences from the common organisms, can be obtained.

The similarity between two proteins/domains can be quantified as follows (Fig. 4) (Pazos and Valencia, 2001). First, for two proteins or domains, the multiple sequence alignments are built using orthologous proteins that are collected from N organisms. Next, the distant matrices are constructed from the genetic distances among all sequences, based on the multiple sequence alignment. The correlation coefficient between the two distance matrices is calculated. The value can be considered as an indicator that shows the intensity of co-evolution. Hence, if the value is close to one, it is judged that the two phylogenetic trees , and the two proteins are considered to interact. The mirror tree approach does not depend on the method used to construct the phylogenetic tree, since it does not compare them directly.

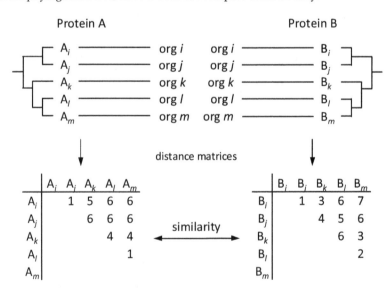

The trees have the same number of leaves and the same organisms in the leaves.

Fig. 4. The flow of the mirror tree approach.

The mirror tree method was applied to six protein families of ligand-receptor pairs, to predict the interaction partners (Goh and Cohen, 2002). Consequently, 79% of all known binding partners on average were detected. In addition, potentially new binding partner in the syntaxin/Unc-18 protein and TGF-β/TGF-β receptor families were found among previously characterized proteins.

2.5 *In silico* two-hybrid system

The *in silico* two-hybrid system infers physical contact sites by computing the correlation coefficient of amino acid variation between two sites, using the multiple sequence alignments for protein pairs (Göbel *et al.*, 1994). That is, in the residue pairs that are in physical contact or related functionally, the amino acids tend to change at the same time. This type of correlated mutation is called co-variation. The similarity of the variation patterns is thought to be related to compensatory mutation. The *in cilico* two-hybrid system infers PPIs by expanding this concept. This system can detect an interaction accompanied by physical contact, and estimates the protein binding sites as well as interacting protein pairs. Meanwhile, the main limitation of this system is the requirement of high quality alignments that include a wide range of common organisms encoding the two proteins, in the same manner as the mirror tree approach.

The *in silico* two-hybrid system quantifies the degree of co-variation between pairs of residues (Fig. 5) (Pazos and Valencia, 2002). First, a multiple sequence alignments are built

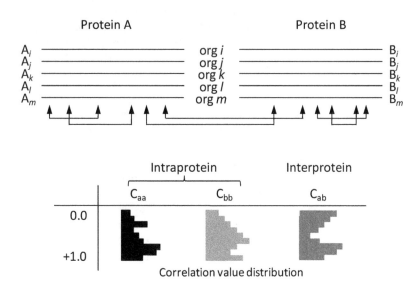

On the top, the alignments are built for two different proteins (protein A and protein B), including the corresponding sequences from different organisms (org $i, j, k...$). On the bottom, the distributions of the correlated coefficients for the pair of residues internal to the two proteins (C_{aa} and C_{bb}) and for the pair of residues from each of the two proteins (C_{ab}) are represented.

Fig. 5. A schematic representation of the *in silico* two-hybrid system.

using orthologs derived from the common organism for two proteins. Next, the correlation coefficients between all combinations of sites in a protein are computed, and the frequency distribution of the values that are computed between two sites is investigated. Similarly, the correlation coefficients between all combinations of two sites from different proteins are calculated, and the frequency distribution is computed. Finally, the interaction index score is computed by using the three frequency distributions of correlation coefficients. If the value is close to one, then the two proteins are considered to interact.

Pazos *et al.* applied this system to four test sets: 1) 14 two-domain proteins with a tight intradomain interaction, from the PDB, 2) 53 proteins including 31 known interactions, 3) 195 pairs with 15 possible interactions, derived from 749 predicted interactions, and 4) 321 pairs, 17 of which are known to interact, from the SPIN database. As a result, it discriminated between true and false interactions in a significant number of cases.

2.6 Sequence signature

The sequence signature approach, which predicts interacting proteins based on domain information, has developed separately from the methods using genome comparison or protein sequence analysis. This approach utilizes sequence and/or structure motifs in order to discriminate interacting proteins. In this approach, the characteristic pairs of sequence signatures are prepared from a database including experimentally determined interacting proteins, where one protein contains one sequence-signature and its interacting partner contains the other sequence-signature. The pairs that occur with high frequency are termed "correlated sequence-signatures", and they can be used for the prediction of putative interacting partners. The prediction result provides the pairs of protein/domain groups that include the correlated sequence-signature, while the other methods described above predict one-on-one protein pairs. Combining this approach with other techniques can yield higher performance.

In this approach, the sequence-signature of the signature combinations must be constructed to identify the correlated sequence-signatures (Fig. 6) (Sprinzak and Margalit, 2001). First, the experimentally determined interacting protein pairs are collected. Then, the sequence-signatures, defined by a motif database such as InterPro, are identified for each sequence. Each entry (a,b) in the table shows the number of protein pairs, composed of one protein containing signature a and its partner containing signature b. Next, the occurrence frequencies of the sequence-signature are converted into the log-odds. The sequence-signature with a positive log-odds value is considered to be observed more frequently in the interactive pairs. Therefore, they are regarding as having a correlated sequence-signature. Finally, this approach searches for the protein or domain pairs that contain the correlated sequence-signature.

An example of applying the Myb domain and the Bromodomain that are correlated sequence-signature to the yeast *S. cerevisiae* is shown (Sprinzak and Margalit, 2001). There are 19 and 10 protein sequences containing the Myb domain and the Bromodomain, respectively. Therefore, in this case, 190 protein interaction pairs are predicted, out of which five interactions were already known.

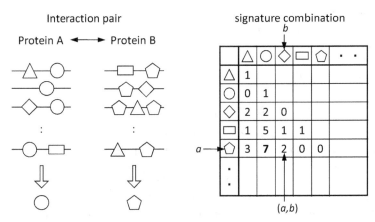

In the left panel, each row contains the sequences of the pair of protein A and protein B. Each sequence has a sequence-signature, illustrated by shapes. In the right panel, a contingency table of the signature combination is described, where each entry (a,b) in the table shows the number of protein pairs. For example, the sequence-signature pair represented by a square and a pentagon appears in two pairs of interacting proteins. The most abundant pair of sequence-signatures is indicated by bold type.

Fig. 6. A scheme for detecting correlated sequence-signatures in interacting proteins.

2.7 Supervised classification

The PPI prediction can be defined as a binary classification problem. Therefore, a statistical model or machine learning method can be applied to the problem of determining whether a pair of proteins is interacting or non-interacting. The K-Nearest Neighbor (KNN) (Qi et al., 2006), Naïve Bayesian (NB) (Jansen et al., 2003; Lu et al., 2005), support vector machines (SVM) (Lo et al., 2005), Artificial Neural Networks (ANN) (Ma et al., 2007), and Random Forest (RF); (Chen and Liu, 2005; Qi et al., 2005) methods were previously applied to this problem. The advantage of these methods is to use data that integrated different datasets. Datasets that do not directly measure PPI, such as sequence and structure information, can be used to infer PPIs. Conversely, the weak point is that the predictive performance varies widely, depending on the quality of the dataset and the selection of statistical methods.

In a statistical model, protein pairs are expressed by N dimensional vectors, where N is the number of features. For example, gene co-expression, GO biological process similarity, MIPS functional similarity, and essentiality are used as features in Jansen's work (Jansen et al., 2003). In addition, sequence information, such as homology and domain data, is used. Two points must be considered when the prediction model is built. The first is that it is necessary to pay attention to the quality of the experimental data used for training and evaluating statistical model, since the performance of the prediction model strongly depends on them. A high-throughput experimental method, such as Yeast Two-Hybrid (Y2H), Mass Spectrometry and Tandem Affinity Purification (MS TAP), and gene co-expression, can detect proteomic-wide PPIs, yielding vast amounts of protein interaction data within the cell. However, these data are often noisy, incomplete, and low-reproducible, since they contain contradictory values. The second is that the selection of an appropriate classification technique is an important task.

The statistical model was developed to infer PPIs in the human and yeast genomes (Lee *et al.*, 2004; Rhodes *et al.*, 2005). Qi *et al.* applied six different classifiers (RF, KNN, NB, Decision Tree, Logistic Regression, and SVM) to predict PPIs, and among them, the RF classifier exhibited the highest performance (Qi *et al.*, 2006). In addition, gene expression is the most important feature for prediction.

3. Computational methods to infer protein flexibility

A protein molecule is not a rigid body. The scale of protein motions is very broad: motions range from local fluctuations, such as those seen in loop regions, to global ones involving changes in the relative positions of rigid domains. Protein motion is often necessary for proteins to perform their specific biological functions. For example, a protein possesses certain conformations in order to interact with its partner protein in many cases. Therefore, structural flexibility is an important feature to consider for understanding protein functions.

Experimental methods that analyze protein dynamics have been developed. Nuclear magnetic resonance (NMR) is a powerful experimental technique (Williams, 1989). NOEs and relaxation experiments provide information related to picosecond-microsecond-scale motions of the backbone atoms (Chill *et al.*, 2004; Gitti *et al.*, 2005). Also, model-free analysis enables quantitative determination of the fluctuations and slow conformational changes of the backbone amide vectors (Lipari and Szabo, 1982a; Lipari and Szabo, 1982b). Although NMR provides a detailed view of protein dynamics, it is time-consuming and suffers from size limitations.

In contrast, computational methods are useful to calculate the dynamics of proteins for which structures are available. They are divided into two types of method. One method compares the structures of a protein crystallized under different conditions or different conformers obtained by NMR. The structural differences indicate flexible regions (Shatsky *et al.*, 2002; Ye and Godzik, 2004). Another computational method is to simulate protein dynamics by methods such as Normal Model Analysis (NMA) and Molecular Dynamics (MD). With the increasing number of available protein structures and the development of high-performance computers, databases that treat protein dynamics have been developed (Table 2). Some databases are introduced below.

ProMode

ProMode is a database including NMA results from analyses performed with a full-atom model for many proteins. It displays realistic three-dimensional motions at an atomic level, using a free plug-in, Charm. In addition, the dynamic domains and their mutual screw motions defined from NMA results are displayed.

MolMovDB

The database of macromolecular movements (MolMovDB) is a collection of quantitative data for flexibility and a number of graphical representations. The motions are generated from alignments of pairs of structures from the Protein Data Bank (PDB). The motions are divided into various classes (e.g. 'hinged domain' or 'allosteric'), according to the type of conformational change.

DynDom database

DynDom, a domain motion analysis program, analyzes the conformational change in terms of dynamic domains, interdomain screw axes, and interdomain bending regions, by comparing two structures when at least two X-ray conformers are available. The DynDom database displays details on the conformational changes obtained from the DynDom analysis results.

iGNM

The database contains visual and quantitative information on the collective modes predicted by the Gaussian Network Model (GNM) for the structure in the PDB. The output includes the equilibrium fluctuations of residues and comparisons with X-ray crystallographic B-factors, the sizes of residue motions in different collective modes, the cross-correlations between the residue fluctuations or domain motions, and other useful information.

Database name	HTTP address	Description	Reference
ProMode	http://cube.socs.waseda.ac.jp/pages/jsp/index.jsp	Large-scale collection of animations of the normal mode vibrating proteins with the full-atom models.	Wako *et al.*, 2004
MolMovDB	http://www.molmovdb.org/	Visualization and classification of molecular motions according to their size and mechanism.	Echols *et al.*, 2003
DynDom database	http://fizz.cmp.uea.ac.uk/dyndom/	Collection of domains, hinge axes and hinge bending residues in proteins.	Lee *et al.*, 2003
iGNM	http://ignm.ccbb.pitt.edu/	Static and animated images for describing the conformational mobility of proteins by computing the GNM dynamics.	Yang *et al.*, 2005

Table 2. List of databases that deal with conformational changes.

4. PPI prediction from protein flexibility

A Structural flexibility is an important characteristic of protein that is frequently related to their functions, as reviewed in section 3. Flexible regions are often necessary for proteins to bind a ligand or another protein. When we focus on the motion of a protein backbone segment, the movement can be classified conceptually into two forms: internal motion and external motion (Nishikawa and Go, 1987). An internal motion is the deformation of the segment itself, while an external motion involves only rotational and translational motions, as a rigid body. The segment fluctuates as a rigid body by changes in the dihedral angles of the flanking residues. For this reason, internal and external motions are considered to be fundamentally different.

This section introduces a means for the calculation of internal and external motions in a protein, by the constriction of statistical models, called "FlexRetriver", and its application to PPI data (Hirose *et al.*, 2010).

4.1 Development of a method for predicting internal and external motions

This subsection introduces the RF-based method for predicting the internal and external motions defined by the NMA from the sequence information.

4.1.1 Calculation of internal and external motions

Using FEDER/2 (Wako *et al.*, 2004), the NMA was performed for the energy-minimized conformation, with the PDB data as the starting conformation. In the NMA, the mean-square displacement of atom a, $<D_a^2>$, in the thermal fluctuations is given as the sum of contributions from individual modes

$$<D_a^2> = \sum_{k=1}^{N} D_{ak}^2 \,,$$

where D_{ak} is the displacement vector of atom a in the k-th normal mode, and N is the number of dihedral angles used as independent variables, i.e., the number of normal modes.

In this study, two conformations for a nine-residue segment in each normal mode are considered. The displacement vector of atom a by this purely translational and rotational motion is designated as D_{ak}^e, and the residual one is designated as D_{ak}^i. Then, D_{ak} is decomposed as

$$D_{ak} = D_{ak}^e + D_{ak}^i \,.$$

The superscripts e and i respectively stand for external and internal. The mean square deviation of atom a is given as

$$<D_a^2> = \sum_k \left|D_{ak}^e\right|^2 + \sum_k \left|D_{ak}^i\right|^2 + \sum_k 2 D_{ak}^e \cdot D_{ak}^i$$

$$= <\left|D_a^e\right|^2> + <\left|D_a^i\right|^2> + 2 <\left|D_a^e \cdot D_a^i\right|> \,.$$

The third term on the right-hand side of this equation is usually much smaller than the first two terms. Therefore, the mean-square deviation of atom a is decomposed approximately into external (first term) and internal (second term) ones. In this case, we are interested in the main-chain fluctuation; for simplicity, only the C_α atom in this decomposition is considered. This means that we selected data for the C_α atoms from the results obtained using NMA with a full-atom model.

4.1.2 Dataset

The dataset was created by selecting protein chains from ProMode, as follows. Proteins with a root mean square deviation (RMSD) of more than 2Å between the energy-minimized structure and the PDB structure were excluded. Protein chains with redundant SCOP IDs were excluded, multi-domain proteins defined by SCOP were then removed. Next, some proteins were discarded so that the maximum pairwise sequence identity was limited to 25%. The resulting dataset comprised 481 chains (87,236 residues).

We calculated the internal and external motions using NMA with a full-atom model for all proteins in the dataset. Raw NMA values were normalized to correct for the variability among proteins in the dataset.

4.1.3 Structure-specific protein mobility propensity

The protein mobility propensities of amino acids are associated with their secondary structures and accessible surface areas (ASAs). The protein mobility propensity was divided into three types of protein mobility: the high and low groups comprised amino acids with normalized NMA scores higher than 1 and lower than -1, respectively, while the normal group comprised amino acids with normalized NMA scores between 1 and -1. The structure-specific protein mobility propensity ($SpecProg(n,s,g)$) was calculated as

$$SpecProp(n,s,g) = log_2 \, freq(n,s,g)/freq(n,s) \, ,$$

where $freq(n,s,g)$ and $freq(n,s)$ respectively represent the relative frequencies of amino acid n in the g protein mobility group of the s state dataset and in the s dataset. The s state indicates a secondary structure or ASA.

The results of the structure-specific protein mobility propensity are shown in Fig. 7. For most amino acids, the protein mobility propensity pattern (in the high, normal, and low groups in the same type of secondary structure) depended on the type of secondary structure (Fig. 7(A)). For example, for both motions, the high mobility propensity of proline (Pro) was low in alpha helices and beta sheets, but high in other structures. This might be because Pro is a secondary structure breaker and its amide nitrogen cannot form a hydrogen bond. On the other hand, the low mobility propensity of hydrophobic amino acids tended to be high in

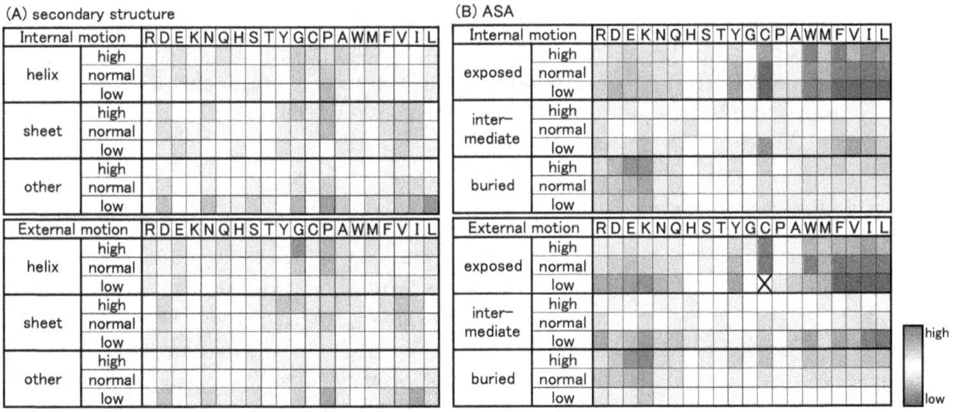

The upper and lower tables represent the results of internal and external motions, respectively. The terms high, normal, and low stand for the protein mobility of the high, normal and low states, respectively. The protein mobility propensity is colored with a gradient from negative (blue) to positive (red). The other in secondary structure is a region without helix and sheet. The cross mark signifies that no data exist.

Fig. 7. The protein mobility propensity associated with (A) secondary structures and (B) ASA.

alpha helices and beta sheets, but low in other structures. The distribution of high mobility propensity in other structures is similar to that of the propensity in hinge regions (Flores *et al.*, 2007). Similarly, the protein mobility propensity pattern changes, depending on the ASA (Fig. 7(B)). The high mobility propensity became higher with increasing ASA, as seen for hydrophilic amino acids. In contrast, high mobility propensity of hydrophobic amino acids was lower with increasing ASA. The external motion might be more strongly influenced by the ASA, as compared to the internal motion. Altogether, these results strongly suggest that the secondary structure and the ASA influence the degrees of the internal and external motions.

4.1.4 Construction of a prediction method

A method for predicting internal and external motions was built by applying RF, which is a type of supervised classification algorithm. The sequence in the sliding window (with sizes of 11 and 17 residues for internal and external motions, respectively) was encoded by using paired amino acid information, corresponding to the variable. The variables were obtained by adding two features, which are derived from the amino acid pairs of the central amino acid with the other amino acids in the window. In total, 18 features were defined, and they were divided into four groups, designated as physicochemical, mobility, secondary structure (predicted by psipred (Jones, 1999) or PHD (Rost, 1996)), and ASA (predicted by sable (Adamczak *et al.*, 2005) or RVPnet (Ahmad *et al.*, 2003)). The profile-based predictors (psipred and sable) have higher prediction accuracy than the amino-acid propensity-based predictors (PHD and RVPnet). The value of a feature of an amino acid was set to one if the amino acid satisfied a feature's definition, and to zero otherwise.

The RF algorithm was used to build a prediction model for classifying amino acids into the three classes: flexible, intermediate, and rigid. Three RF prediction models were trained for the three categories of window location: the center of a secondary structure (CS), the remote area from a secondary structure (RS), and the periphery of a secondary structure (PS). The RF prediction models classified the windows into the three classes, and their prediction results were attributed to the central residue in the window. The results of the classification obtained from the RF were then converted into a score.

4.1.5 Prediction performance

The prediction results were assessed on a residue basis, by which the predicted score in the sequence was compared to the normalized NMA score. The prediction performance was evaluated by using three criteria: the mean absolute error (MAE), correlated coefficient (CC), and Receiver Operating Characteristic (ROC) curves. The MAE was defined as the absolute difference between two values. The MAE value approaches 0 as the prediction improves. The CC was also computed between two values. The CC ranges from -1 to 1, and a large, positive value represents a better prediction. The ROC curve was obtained by plotting the false positive rate against the true positive rate. A larger area under the ROC curve (AUC) indicates a more robust algorithm.

The prediction performance of FlexRetriever was compared with those of three published methods (PROFbval (Schlessinger *et al.*, 2006), POODLE-S (Shimizu *et al.*, 2007), FlexPred

(Kuznetsov and McDuffie, 2008)), and the naïve model. The naïve model is based on the simple idea that protein motion tends to be large in a coil or loop region and small in a secondary structure. The FlexRetriver, which implemented psipred and sable, yielded the lowest MAE and the highest CC among all prediction methods for both motions (Table 3). However, it is noteworthy that the distribution of CC varied widely. Additionally, in AUC, FlexRetriever exhibited the best performance among all methods.

(A) Internal motion			
Method	MAE	CC	AUC
FlexRetriever (PHD & RVPnet)	0.621	0.482	0.765
FlexRetriever (psipred & sable)	0.605	0.525	0.786
Naïve model (PHD)	0.988	0.248	0.653
Naïve model (psipred)	0.952	0.293	0.672
PROFbval	0.743	0.367	0.693
POODLE-S	-	-	0.730
FlexPred	-	-	0.741
(B) External motion			
Method	MAE	CC	AUC
FlexRetriever (PHD & RVPnet)	0.571	0.541	0.777
FlexRetriever (psipred & sable)	0.542	0.597	0.806
Naïve model (PHD)	0.970	0.262	0.661
Naïve model (psipred)	0.929	0.320	0.681
PROFbval	0.608	0.547	0.784
POODLE-S	-	-	0.783
FlexPred	-	-	0.777

The CC and MAE were estimated by performing a five-fold cross validation test. The highest scores in each criterion are underlined. PHD and psipred in parentheses signify the secondary structure predictor. Similarly, RVPnet and sable represent the ASA predictor. "-": scores could not be calculated.

Table 3. Comparison of prediction performance.

4.2 Applying FlexRetriever to PPI data

In this study, we utilized the set of 20 proteins that undergo large conformational changes upon association (> 2Å C_α RMSD) created by Dobbins et al., with which they demonstrated the relationship between normal mode fluctuations and conformational change (Dobbins et al., 2008). They regarded protein motions as being associated with their functions, because they are observed along with the PPI. We compared the internal motion with the observed conformational change region, because it was defined as the deformation of a segment itself. To begin with, we present three typical results, in which the observed conformational change regions are located in a binding site, a hinge region, and other regions. We will then discuss the overall results.

i.   Ecotin

Ecotin, a homodimeric protein, is an inhibitor of a group of homologous serine proteases, such as trypsin, chymotrypsin, and elastase. One dimeric inhibitor binds to a protease molecule. From a comparison of two structures determined with different crystalline environments, an inherent flexible loop was identified in the binding site with trypsin. It was necessary for its inhibitory function (Shin *et al.*, 1996). FlexRetriever predicted high internal motion for the corresponding loop (Fig. 8(A)).

ii.   Fab fragment

The fragment antigen binding (Fab fragment) region is the site where an antibody binds to antigens. It is a heterodimer of the heavy and light chains in each of the two composed domains. The hinge region between the two domains changed its conformation when Fab bound to hemagglutinin derived from a flu virus (Fleury *et al.*, 1998). FlexRetriever predicted high internal motion at the hinge region in each chain (Fig. 8(B)).

iii.   Erythropoietin

Erythropoietin (EPO) is a hormone produced primarily in the kidneys. It has a four-helical bundle topology with two long loops, and binds to the extracellular domain of the EPO receptor. The CD loop, which is located in a region remote from the binding site, changed its conformation (Cheetham *et al.*, 1998). FlexRetriever predicted high internal motion for the corresponding loop (Fig. 8(C)).

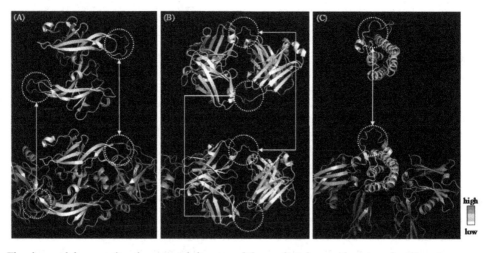

The observed degrees of conformational change and the predicted scores for internal motion are mapped, respectively, with a gradient from zero (white) to a high score (dark red) onto their structures in the upper and lower sections. The regions enclosed with a yellow dotted line are the regions with observed conformational changes. The free-state and complex-state structures are displayed, respectively, in the upper and lower sections.

Fig. 8. Example of the relationship between the predicted internal motions and the observed conformational changes of (A) ectin, (B) Fab fragment, and (C) erythropoietin.

Overall results

When FlexRetriever was applied to a set of 20 proteins, three or more consecutive residues with high internal scores were regarded as candidates for the regions undergoing conformational changes. From a comparison between the observed conformational change regions with the predicted high internal motion regions, at least one overlap was found in 85% of the proteins studied. If the analysis object was limited to the 16 proteins that interact with only one partner, then the overlap was observed in 15 proteins (94% of the proteins studied). These observations suggest that FlexRetriever is a sensitive method for the detection of protein motions related to PPIs, including binding sites.

4.3 Web server

The presented method is implemented in the FlexRetriever server, which has been designed with a user-friendly interface to provide easily interpretable prediction results.

The server accepts the submission of a single amino acid sequence with less than 1,000 amino acids in the FASTA format (Fig. 9(A)). The user is asked to choose a calculation mode.

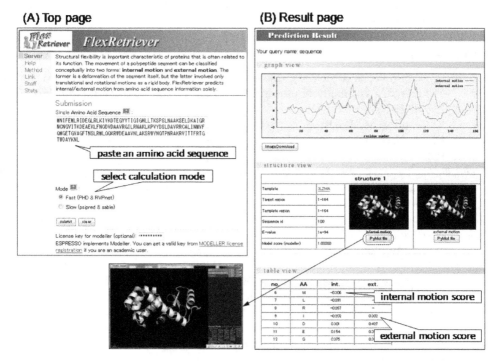

On the result page, a graph is displayed on the top, and two structures on which the scores of the internal and external motions are mapped are shown in the middle. They can be downloaded as a PyMol file. The table with the raw scores is displayed below the structures.

Fig. 9. Images of FlexRetriever's (A) top page and (B) result page.

The calculation time of the fast mode, which uses PHD for secondary structure prediction and RVPnet for ASA prediction, is shorter than that of the slow mode, but its performance is poorer.

The results page is divided into three sections (Fig. 9(B)). The first section (graph view) provides the graph which contains the prediction results of both motions. The second section (structure view) presents the degrees of internal and external motions on the three-dimensional structure. The third section (table view) lists the amino acids and the raw scores of their internal and external motions.

FlexRetriever is available at http://mbs.cbrc.jp/FlexRetriever and is free.

5. Conclusion

This chapter provides an overview of the computational methods to infer pairs of interacting proteins and to study the relevance of protein flexibility. Genomic information and experimental data are now readily available, and thus computational methods will become more important tools in the field of analyzing or inferring PPIs. In addition, as a novel attempt to predict PPIs, we have presented an efficient algorithm for predicting flexible regions in proteins, and shown its application to PPIs. The tool is expected to be useful for inferring motions associated with PPIs.

6. References

Adamczak, R.; Porollo, A. & Meller, J. (2005) Combining prediction of secondary structure and solvent accessibility in proteins. *Proteins*, Vol.59, No.3, (May 15), pp. 467-475, ISSN 0887-3585

Ahmad, S.; Gromiha, M.M & Sarai, A. (2003) Real value prediction of solvent accessibility from amino acid sequence. *Proteins*, Vol.50, No.4, (Mar 1), pp. 629-635, ISSN 0887-3585

Brown, F.; Zheng, H.; Wang, H.Y. & Azuaje, F. (2010) From experimental approaches to computational techniques: A review on the prediction of protein-protein interaction. *Advances in Artificial Intelligence*, Vol.2010, ISSN 16877470

Chill, J.H.; Quadt, S.R. & Anglister, J. (2004) NMR backbone dynamics of the human type I interferon binding subunit, a representative cytokine receptor. *Biochemistry*, Vol.43, No.31, (Aug 10), pp. 10127-10137, ISSN 0006-2960

Cheetham, J.C.; Smith, D.M.; Aoki, K.H.; Stevenson, J.L.; Hoeffel, T.J.; Syed, R.S.; Egrid, J. & Harvey, T.S. (1998) NMR structure of human erythropoietin and a comparison with its receptor bound conformation. *Nat Struct Biol*, Vol.5, No.10, (Oct 1998), pp. 861-868, ISSN 1072-8368

Chen, X.W. & Liu, M. (2005) Prediction of protein-protein interactions using random decision forest framework. *Bioinformatics*, Vol.21, No.24, (Dec 15), pp. 4394-4400, ISSN 0973-2063

Dandekar, T.; Snel, B.; Huynen, M. & Bork, P. (1998) Conservation of gene order: a fingerprint of proteins that physically interact. *Trends Biochem Sci*, Vol.23, No.9, (Sep 1998), pp. 324-328, ISSN 0968-0004

Dobbins, S.E.; Lesk, V.I. & Sternberg, M.J. (2008) Insights into protein flexibility: The relationship between normal modes and conformational change upon protein-protein docking. *Proc Natl Acad Sci U S A*, Vol.105, No.30, (July 29), pp. 10390-10395, ISSN 0027-8424

Echols, N.; Milburn, D. & Gerstein, M. (2003) MolMovDB: analysis and visualization of conformational change and structural flexibility. *Nucleic Acids Res*, Vol.31, No.1, (Jan 1), pp. 478-482, ISSN 0305-1048

Enright, A.J.; Iliopoulos, I.; Kyrpides N.C. & Ouzounis, C.A. (1999) Protein interaction maps for complete genomes based on gene fusion events. *Nature*, Vol.402, No.6757, (Nov 4), pp. 86-90, ISSN 0028-0836

Fleury, D.; Wharton, S.A.; Skehel, J.J.; Knossow, M. & Bizebard, T. (1998) Antigen distortion allows influenza virus to escape neutralization. *Nat Struct Biol*, Vol.5, No.2, (Feb 1998), pp. 119-123, ISSN 1072-8368

Flores, S.C.; Lu, L.J.; Yang, J.; Carriero, N. & Gerstein, M.B. (2007) Hinge Atlas: relating protein sequence to site of structural flexibility. *BMC Bioinformatics*, Vol.8, No.167, (May 22), ISSN 1471-2105

Fryxell, K.J. (1996) The coevolution of gene family trees. *Trends Genet*, Vol.12, No.9, (Sep 1996), pp. 364-369, ISSN 0168-9525

Gitti, R.K.; Wright, N.T.; Margolis, J.W.; Varney, K.M.; Weber, D.J. & Margolis, F.L. (2005) Backbone dynamics of the olfactory marker protein as studied by 15N NMR relaxation measurements. *Biochemistry*, Vol.44, No.28, (Jul 19), pp. 9673-9679, ISSN 0006-2960

Göbel, U.; Sander, C.; Schneider, R. & Valencis, A. (1994) Correlated mutations and residue contacts in proteins. *Proteins*, Vol.18, No.4, (Apr 1994), pp. 309-317, ISSN 0887-3585

Goh, C.S.; Bogan, A.A.; Joachimiak, M.; Walther, D. & Cohen, F.E. (2000) Co-evolution of proteins with their interaction partners. *J Mol Biol*, Vol.299, No.2, (Jun 2), pp. 283-293, ISSN 0022-2836

Goh, C.S. & Cohen, F.E. (2002) Co-evolutional analysis reveals insights into protein-protein interactions. *J Mol Biol*, Vol.324, No.1, (Nov 15), pp. 177-192, ISSN 0022-2836

Hirose, S.; Yokota, K.; Kuroda, Y.; Wako, H.; Endo, S.; Kanai, S. & Noguchi, T. (2010) Prediction of protein motions from amino acid sequence and its application to protein-protein interaction. *BMC Struct Biol*, Vol.10, No.20, (Jul 13), ISSN 1472-6807

Jansen, R.; Yu, H.; Greenbaum, D.; Kluger, Y.; Krogan, N.J.; Chung, S.; Emili, A.; Snyder, M.; Greenblatt, J.F. & Gerstein, M. (2003) A Baysian networks approach for predicting protein-protein interactions from genomic data. *Science*, Vol. 302, No.5644, (Oct 17), pp. 449-453, ISSN 0036-8075

Jones, D.T. (1999) Protein secondary structure prediction based on position-specific scoring matrices. *J Mol Biol*, Vol.292, No.2, (Sep 17), pp. 195-202, ISSN 0022-2836

Kuznetsov, I.B. & McDuffie M. (2008) FlexPred: a web-server for predicting residue positions involved in conformational switches in proteins. *Bioinformation*, Vol.3, No.3, (Nov 5), pp. 134-136, ISSN 0973-2063

Lee, I.; Date, S.V.; Adai, A.T. & Marcotte E.M. (2004) A probabilistic functional network of yeast genes. *Science*, Vol.306, No.5701, (Nov 26), pp. 1555-1558, ISSN 0036-8075

Lee, R.A.; Razaz, M. & Hayward, S. (2003) The DynDom database of protein domain motions. *Bioinformatics*, Vol.19, No.10, (Jul 1), pp.1290-1291, ISSN 1460-2059

Lipari, G. & Szabo, A. (1982a) Model-free approach to the interpretation of nuclear magnetic resonance relaxation in macromolecules. 1. Theory and range of validity. *J Am Chem Soc*, Vol.104, No.17, (August 1982), pp. 4546-4559, ISSN 0002-7863

Lipari, G. & Szabo, A. (1982b) Model-free approach to the interpretation of nuclear magnetic resonance relaxation in macromolecules. 2. Analysis of experimental results. *J Am Chem Soc*, Vol.104, No.17, (August 1982), pp. 4559-4570, ISSN 0002-7863

Lo, S.L.; Cai, C.Z.; Chen, Y.Z. & Chung, M.C. (2005) Effect of training datasets on support vector machine prediction of protein-protein interactions. *Proteomics*, Vol.5, No.4, (Mar 2005), pp. 876-884, ISSN 1615-9861

Lu, L.J.; Xia, Y.; Paccanaro, A.; Yu, H. & Gerstein, H. (2005) Assessing the limits of genomic data integration for predicting protein networks. *Genome Res*, Vol.15, No.7, (Jul 2005), pp. 945-953, ISSN 1088-9051

Ma, Z.; Zhou, C.; Lu, L.; Lu, L.; Ma, Y.; Sun, P. & Cui, Y. Predicting protein-protein interactions based on BP neural network, *Proceedings of the IEEE International Conference on Bioinformatics and Biomedicine Workshops*, pp. 3-7, ISBN 978-1-4244-1604-2, Fremont, CA, Nov 2-4, 2007

Marcotte, E.M.; Pellegrini, M.; Ng, H.L.; Rice D.W.; Yeates, T.O. & Eisenberg, D. (1999) Detecting protein function and protein-protein interactions from genome sequence. *Science*, Vol.285, No.5428, (Jul 30), pp. 751-753, ISSN 0036-8075

Nishikawa, T. & Go, N. (1987) Normal modes of vibration in bovine pancreatic trypsin inhibitor and its mechanical property. *Proteins*, Vol.2, No.4, (1987), pp. 308-329, ISSN 0887-3585

Overbeek, R.; Fonstein, M.; D'Souza, M.; Pusch, G.D. & Maltsev, N. (1999) The use of gene clusters to infer functional coupling. *Proc Natl Acad Sci U S A*, Vol.96, No.6, (Mar 16), pp. 2896-2901, ISSN 0027-8424

Pagel, P.; Wong, P. & Frishman, D. (2004) A domain interaction map based on phylogenetic profiling. *J Mol Biol*, Vol.344, No.5, (Dec 10), pp. 1331-1346, ISSN 0022-2836

Pazos, F. & Valencia, A. (2001) Similarity of phylogenetic trees as indicator of protein-protein interaction. *Protein Engineering*, Vol.14, No.9, (Sep 14), pp. 609-614, ISSN 0269-2139

Pazos, F. & Valencia, A. (2002) *In silico* two-hybrid system for the selection of physically interacting protein pairs. *Proteins*, Vol.47, No.2, (May 1), pp. 219-227, ISSN 0887-3585

Pellegrini, M.; Marcotte, E.M.; Thompson, M.J.; Eisenberg, D. & Yeates, T.O. (1999) Assigning protein functions by comparative genome analysis: protein phylogenetic profiles. *Proc Natl Acad Sci U S A*, Vol.96, No.8, (Apr 13), pp. 4285-4288, ISSN 0027-8424

Rost, B. (1996) PHD: predicting one-dimensional protein structure by profile-based neural networks. *Methods Enzymol*, Vol.266, (1996), pp. 525-539, ISSN 0076-6879

Qi, Y.; Klein-Seetharaman, J & Bar-Joseph, Z. (2005) Random forest similarity for protein-protein interaction prediction from multiple sources. *Pac Symp Biocomput*, pp. 531-542, ISSN 1793-5091

Qi, Y.; Bar-Joseph, Z. & Klein-Seetharaman, J. (2006) Evaluation of different biological data and computational classification methods for use in protein interaction prediction. *Proteins*, Vol.63, No.3, (May 15), pp. 490-500, ISSN 0887-3585

Rhodes, D.R.; Tomlins, S.A.; Varambally, S.; Mahavisno, V.; Barrette, T.; Kalyana-Sundaram, S.; Ghosh, D.; Pandey, A. & Chinnalyan, A.M. (2005) Probabilistic model of the human protein-protein interaction network. *Nat Biotechnol*, Vol.23, No.8, (Aug 2005), pp. 951-959, ISSN 1087-0156

Schlessinger, A.; Yachday, G. & Rost, B. (2006) PROFbval: predict flexible and rigid residues in proteins. *Bioinformatics*, Vol.22, No.7, (Apr 1), pp. 891-893, ISSN 1460-2059

Shatsky, M.; Nussinov, R. & Wolfson, H.J. (2002) Flexible protein alignment and hinge detection. *Proteins*, Vol.48, No.2, (Aug 1), pp. 242-256, ISSN 0887-3585

Shimizu, K.; Hirose, S. & Noguchi, T. (2007) POODLE-S: web application for predicting protein disorder by using physicochemical features and reduced amino acid set of a position-specific scoring matrix. *Bioinformatics*, Vol.23, No.17, (Sep 1), pp. 2337-2338, ISSN 1460-2059

Shin, D.H.; Song, H.K.; Seong, I.S.; Lee, C.S.; Chung C.H & Suh, S.W. (1996) Crystal structure analyses of uncomplexed ecotin in two crystal forms: implications for its function and stability. *Protein Sci*, Vol.5, No.11, (Nov 1996), pp. 2236-2247, ISSN 0961-8368

Shoemaker, B.A. & Panchenko, A.R. (2007) Deciphering protein-protein interactions. Part II. Computational methods to predict protein and domain interaction partners. *PLoS Comput Biol*, Vol.3, No.4, (Apr 27), pp. e43, ISSN 1553-734X

Skrabanek, L.; Saini, H.K.; Bader, G.D. & Enright, A.J. (2008) Computational prediction of protein-protein interactions. *Mol Biotechnol*, Vol.38, No.1, (Jan 2008), pp. 1-17, ISSN 1073-6085

Snel, B.; Bork, P. & Huynen, M.A. (2002) The identification of functional modules from the genomic association of genes. *Proc Natl Acad Sci U S A*, Vol.99, No.9, (Apr 30), pp. 5890-5895, ISSN 0027-8424

Sprinzak, E. & Margalit, H. (2001) Correlated sequence-signatures as markers of protein-protein interaction. *J Mol Biol*, Vol.311, No.4, (Aug 24), pp.681-692, ISSN 0022-2836

Valencia, A. & Pozos, F. (2002) Computational methods for the prediction of protein interactions. *Curr Opin Struct Biol*, Vol.12, No.3, (Jun 2002), pp. 368-373, ISSN 0959-440X

Wako, H.; Kato, M. & Endo, S. (2004) ProMode: a database of normal mode analyses on protein molecules with a full-atom model. *Bioinformatics*, Vol.20, No.13, (Sep 1), pp. 2035-2043, ISSN 1460-2059

Williams, R.J. (1989) NMR studies of mobility within protein structure. *Eur J Biochem*, Vol.183, No.3, (Aug 15), pp. 479-497, ISSN 0014-2956

Yang, L.W.; Liu, X.; Jursa, C.J.; Holliman, M.; Rader, A.J.; Karimi, H.A. & Bahar, I. (2005)
    *i*GMN: a database of protein functional motions based on Gaussian Network
    Model. *Bioinformatics*, Vol.21, No.13, (Jul 1), pp. 2978-2987, ISSN 1460-2059

Ye, Y. & Godzik, A. (2004) Database searching by flexible protein structure alignment.
    Protein Sci, Vol.13, No.7, (Jul 2004), pp. 1841-1850, ISSN 0961-8368

8

Prediction of Protein Interaction Sites Using Mimotope Analysis

Jian Huang, Beibei Ru and Ping Dai
School of Life Science and Technology
University of Electronic Science and Technology of China
China

1. Introduction

Biological functions depend on all kinds of interaction networks; life is a miracle of all types of molecular interactions. Among them, proteins interacting with proteins, nucleic acids and small compounds play a central role (Barabasi & Oltvai, 2004; Przulj, 2011; Vidal, et al., 2011). To guide protein engineering studies for better enzymes, antibodies and drugs, structural and functional characterization of protein interaction sites at the residue or atom level is of great help. Experimental approaches such as X-ray diffraction of protein complex can define structural binding sites at the atomic level (Bickerton, et al., 2011; Higurashi, et al., 2009); mutagenesis and binding test are capable of identifying functional binding sites at the residue or group level (Moreira, et al., 2007; Peng, et al., 2011). However, these means are costly, time-consuming and sometimes technically difficult or even impossible. Moreover, they are not always applicable on a large scale. As a result, computer tools for the prediction of protein interaction sites have been increasingly popular for complementing experimental techniques (Fernández-Recio, 2011; Wass, et al., 2011).

The existing methods for the prediction of protein interaction sites can be grouped into three categories based on the main input data used. The first category consists of methods using protein sequence as the only input (Ofran & Rost, 2007; Res, et al., 2005). Methods in the second category such as molecular docking and simulation solely use structure data as input (Kozakov, et al., 2010; Mashiach, et al., 2010). Methods of the third category make use of a mimotope motif or a set of mimotope sequences together with protein sequence or structure as input (Huang, et al., 2011).

In this chapter, we review methods of the third category, focusing on their current statuses, discussing challenges and providing suggestions to advance this field.

2. Mapping protein-protein interaction sites using mimotope analysis

Mimotopes are peptides mimicking protein interaction sites; they are initially acquired from chemical synthesis (Geysen, et al., 1986). High-throughput obtainment of mimotopes has achieved since phage display and other surface display technologies became available (Smith, 1985; Smith & Petrenko, 1997). Taking phage display as an example, random DNA sequences can be inserted into genes coding for coat proteins of bacteriophage to make combinatorial libraries. As shown in Figure 1, the combinatorial library can be incubated and selected with an immobilized protein, termed as the target. The natural partner of the target is called as the template. Phages without affinity to the target are washed away with buffer. Then, bound phages are eluted with the target, the template or stronger buffer only. The bound phages are further amplified by infecting bacteria to form a secondary library, which is then used for the next round of incubating, washing, eluting and amplifying. After several rounds of such processes which are well known as biopanning, phage clones are picked randomly from the isolation of bound phages and sequenced. The affinities of these phage clones or corresponding peptides to the target are measured by surface plasmon resonance, enzyme-linked immunosorbent assay or other binding assays. The foreign inserts which enable corresponding phage clones to bind the target competitively with a template are considered as mimotopes.

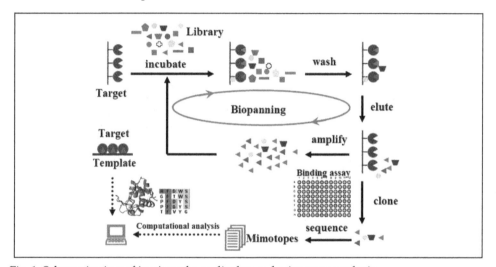

Fig. 1. Schematic view of in vitro phage display and mimotope analysis.

As described above, a set of mimotopes can be readily obtained via phage display. They are capable of binding to the target and blocking the interaction between the target and the template. Therefore, it implies that the information of protein interaction sites is encoded in mimotopes and can be predicted by decoding mimotopes. It is only natural to suppose that the mimotopes are similar to the binding site on the template at the sequential or structural level. Indeed, all approaches to prediction of protein interaction sites based on mimotopes depend on either the sequence or the structure of the template. Thus, the existing methods can be divided into the following two groups:

2.1 Methods based on template sequence

Various methods based on template sequence are summarized in Figure 2. In brief, a set of mimotopes are aligned with the corresponding template to find out the similar region in sequence, which is thought to be at least a part of the target-binding site on the template protein. Sometimes, sequences of paralogs or orthologs of the template are also aligned to help the identification of the protein interaction site. In some studies, consensus sequences or motifs are derived from the blocks of mimotope alignments. Then, consensus sequences are aligned to the template sequence; motifs are scanned along the template sequence. And the template segments similar to consensus sequences or matching the motifs are considered to be a part of the protein interaction sites. If the template itself is not determined, local alignment search with each mimotope or the consensus sequence against the protein database would help to predict reasonable candidates of template and its binding sites. The template and corresponding interaction sites can also be predicted through pattern search with mimotope motifs against the protein database.

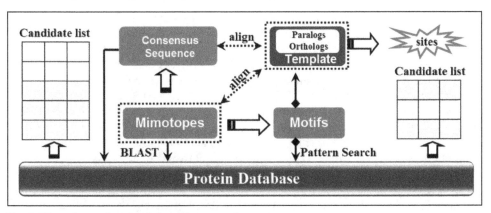

Fig. 2. Flow chart of methods based on template sequence.

As shown in Figure 2, methods based on template sequence involve several steps and tasks such as aligning sequence, inferring consensus sequence or motifs, searching local alignment or motif against the protein database. Among them, sequence alignment is undoubtedly the most important one. Methods based on template sequence can be fulfilled with visual inspection, general-purpose programs and tools specially designed for mimotope analysis.

2.1.1 Manual sequence analysis with visual inspection

Some mimotopes are very similar or even identical to some part of the template sequence every now and then, indicating the segment involving in binding the target protein. In this situation, the protein binding site can be easily depicted through aligning mimotope and template sequence manually by visual inspection. A 6mer random library was screened with the monoclonal antibody GDO5 raised against the Hantaan virus glycoprotein G2. After three rounds of panning, the mimotope obtained had the sequence LEYPWH, which was very similar to the template sequence 94YEYPWH99, implying the site where GDO5 bound (Fack, et al., 1997). The Ph.D.-7 random phage library was panned using the anti-SEB monoclonal antibody ab53981 (Urushibata, et al., 2010). Among the mimotopes obtained, SPDELHK was almost identical to 8PDELHK13 of the staphylococcal enterotoxin B. The ab53981 binding site was thus located. Four anti-HBsAg monoclonal antibodies, namely H5, H35, H53 and H166 were characterized using phage display (Chen, et al., 1996). Manually aligning 16 H166-binding mimotopes with the HBsAg sequences from subtype adw2, ayw2 and ayw3, Chen et al found that most mimotopes have the CRTC or CKTC subsequences by visual inspection, which were identical to the segments from 121 to 124 of the HBsAg. The epitope recognized by H166 was thus indicated. Nonetheless, sequence alignment programs are necessary when there are a lot of sequences to be aligned or the similarity between the mimotope and template sequence is not obvious.

2.1.2 General-purpose sequence analysis tools

General-purpose tools for sequence alignment, local alignment and pattern search have been widely used in the prediction of protein interaction sites based on mimotope and template sequences. As we described above, Chen et al identified the H166-binding site by visual inspection. However, the software GENEWORK was used in the left three cases (Chen, et al., 1996). Significant matches were found by manual analysis on the dot-matrix diagrams produced by GENEWORK. For example, ARARCEHRSGLSL as one part of an H35-selected mimotope was aligned to 166ASARFSWLSL175 of the HBsAg sequence from subtype ayw3, locating the H35-binding site (Chen, et al., 1996). MDM2-binding peptides were obtained using mRNA display (Shiheido, et al., 2011); the peptides were aligned using the ClustalW program (Larkin, et al., 2007). Compared with the sequence of P53, a similar segment 17-28 was found be the MDM2-P53 interaction site (Shiheido, et al., 2011). A monoclonal antibody against the West Nile virus capsid protein was generated and designated as 6D3 (E. C. Sun, et al., 2011). A 12mer peptide library was screened with 6D3 to produce a set of mimotopes. Alignment revealed a consensus segment KKPGGPG, which was same to the subsequence 3-9 of West Nile virus capsid protein. A monoclonal antibody 2A10G6 was raised against the heat-inactivated dengue virus and used to screen the Ph.D.-12 random phage library. Alignment of mimotopes revealed a consensus FFDRTWP, which corresponded well with 98DRGW101 located at the tip of the fusion loop of E protein of dengue virus (Deng, et al., 2011). In the studies of Sun and Deng, MegAlign software within the Lasergene suite was used to align the orthologs of the template. In the study of Urushibata, the ClustalW program was used to align the paralogs of the template (Urushibata, et al., 2010). These studies showed that orthologs or paralogs were helpful for locating the binding sites.

Unlike the interaction between an antigen and corresponding antibody, a protein may have quite a few partners sometimes or its natural partner may be unknown. In these situations, the sequence alignment between mimotope and template cannot be done directly. However, a local alignment search against the protein sequence database helps to identify candidate templates and binding sites. The AC3 protein of geminiviruses was characterized using phage display (Pasumarthy, et al., 2011). Each AC3-specific peptide sequence obtained was then searched for local alignment against the Arabidopsis non-redundant protein database at NCBI through BLASTP program adjusted for short sequence (Mount, 2007). Proteins from a few metabolic pathways were identified as putative AC3-interacting proteins. For example, YALKHLPESTIP was very similar with 704YALKHIRES712 of the Hua Enhancer 1 (HEN1). Thus HEN1 might interact with the AC3 protein around 704YALKHIRES712

(Pasumarthy, et al., 2011). LipL32, the major outer membrane protein of pathogenic Leptospira, was panned against the Ph.D.-7 phage library. For each mimotope obtained, a BLASTP search against the protein database was performed. Quite a few proteins expressed on the surface of target cells of pathogenic Leptospira were suggested to interact with LipL32. For example, the mimotope HLPPNHT is similar with the sequence PLPPEHT of Collagen XX, indicating LipL32 might bind to Collagen XX at that site (Chaemchuen, et al., 2011). The strategy of mimotope blast against protein databases has also been used to deduce small molecule binding sites and drug targets (Chen, et al., 2006; Takakusagi, et al., 2010; Takami, et al., 2011), infer proteins involving in cell interactions (Kanki, et al., 2011; Zhao, et al., 2010).

Besides tools for local alignment search, pattern search against the protein database has also been used to find possible templates and their target-binding sites. SdrC is important in the interactions between Staphylococcus aureus and its host. However, the host ligand interacting with SdrC was not previously identified. The Ph.D.-12 phage library was screened with SdrC and eight phage clones displayed significantly higher affinity to SdrC. These clones were sequenced, and an alignment revealed the consensus sequence HHIHHFH. It was then used for a pattern search against the human protein database allowing for zero, one and two residue mismatches. The results showed that human neurexin 1β, 2β, 3β and a T-type voltage-dependent calcium channel might be the host ligands interacting with SdrC. Among them, the subsequence 10-16 of human neurexin 1β was identical to the consensus sequence, which implied SdrC might bind to human neurexin 1β at the site 10HHIHHFH16 (Perosa, et al., 2010). Autoantibodies against centromere associated protein A (CENP-A) were purified from sera of eight systemic sclerosis patients with the immunodominant epitope of CENP-A (Ap17–30). These antibodies were used to screen a phage library. The binding phage clones were sequenced, and the inserted peptides were aligned with MULTALIN (Corpet, 1988) to derive antigenic motifs. Human proteins containing such motifs were searched in the SwissProt Protein Sequence Database using the ScanProsite tool. Taking the PTPxxGPxxR motif as an example, 20PTPTPGPSRR29 of human CENP-A was certainly found. However, 53PTPAPGPGRR62 of human Forkhead box protein E3 (FOXE3) also matched the motif, indicating those autoantibodies could interact with FOXE3 at the site around 53-62. Indeed, the peptide 53-62 of FOXE3 was confirmed to behave similarly in binding and inhibition assays with anti-Ap17–30 IgG (Barbu, et al., 2010).

2.1.3 Specially designed sequence analysis tools

Even very recently, general-purpose tools for sequence alignment, local alignment and pattern search remain popular in the study of mapping protein interaction sites based on mimotopes. One reason for this is that these tools are freely, stably and conveniently available. However, these general-purpose tools have their limits. For example, most of them are not good at aligning a very short sequence (mimotope) to quite a long sequence (template). Furthermore, they are less efficient to deduce conformational binding sites, which are made of segments far away in primary sequence but close on the surface of template structure. Specially designed tools are thus needed for sequence analyses of mimotopes and templates.

To delineate conformational binding sites on protein, the program FINDMAP was proposed (Mumey, et al., 2003). FINDMAP allowed any permutations (e.g. inversion) of the mimotope sequence to align its template sequence. Furthermore, gaps even large gaps were permitted in both mimotope and template sequences. Such alignment was proven to be NP-complete and a branch-and-bound algorithm was used to solve the problem in practice. As FINDMAP could deal with only one mimotope each time, an improved version called EPIMAP was introduced later. It was capable of aligning each mimotope to the template, producing a set of top-scoring alignments, selecting the most mutually compatible alignments and filtering out spurious alignments (Mumey, et al., 2006). MimAlign was a meta-method. It combined results from four multiple sequence alignments of the template and its mimotopes (Moreau, et al., 2006). In the RELIC suite, there were quite a few tools specially designed for analysis on mimotopes (Mandava, et al., 2004). For example, MOTIF1 and MOTIF2 were designed to identify weak sequence motifs within short peptide sequence; MATCH, FASTAcon and FASTAskan were designed for optimal sequence alignments between mimotopes and its template. Although the RELIC suite focused on the interaction between small molecule and protein, its sequence tools were often used in the analysis of protein-protein interaction sites. For instance, MMACHC-binding peptides were aligned to MMADHC with tools in RELIC and five MMACHC-binding sites on the protein MMADHC were predicted (Plesa, et al., 2011). Mouse monoclonal antibodies against the predominant VSGs LiTat 1.3 and LiTat 1.5 of T.b. gambiense were used to screen PhD.-12 and PhD.-C7C phage libraries. Epitopes were identified by sequence alignment performed manually and with RELIC suite (Van Nieuwenhove, et al., 2011). For example, ALLPFKDHLPYP selected with the monoclonal antibody H12H3 against VSG LiTat 1.5 was aligned to 269AQAVYKDHDPDQ280 of VSG LiTat 1.5. The following experiment did show that the binding of H12H3 to synthetic ALLPFKDHLPYP was inhibited by human African trypanosomiasis sera. Regretfully, all the special tools described here are now hard or impossible to access.

2.1.4 Methods based on template sequence: challenges and suggestions

Methods based on template sequence are of their advantages. For example, they can be used in any condition because no structural information is required during prediction. Even if the template sequence is not given, local alignment or pattern search against protein databases may fulfil the task of inferring possible templates and protein interaction sites. However, to evaluate the results of sequence alignment, local alignment search and pattern search is still a great challenge.

Two formulae have been proposed to compute the frequency of finding similar sequences in two random sequences with different lengths (Chen, et al., 1996). One formula is for a single sequence match; another is for nearby matches within a pair of two sequences. This was a good attempt to evaluate if a continuous or discontinuous match was significant or just by chance. Chen et al assumed that 20 different residues were with equal probability at each position of the two sequences. However, it is not true in real case. To be more reasonable, we suggest using the residue frequency of the corresponding phage library for mimotopes and the actual frequency for template with long sequence. For short or unknown template, use the amino acids frequency of SwissProt.

Although the BLAST program has its statistical means to evaluate a match, they are not fit for short peptides such as mimotopes. In the study of Pasumarthy et al, a lot of matches were found. Among them, the mimotope FPKAFHHHKIY was found to be similar with 317HKIY310 of the Retinoblastoma like protein (pRBR) with an E-value of 1250 and pRBR was known to be an AC3-interacting protein. The mimotope DAMIMKKHWHRF was found to be similar with 164MIMK167 of the Geminivirus Rep interacting kinase 1 (GRIK1) with an E-value of 517 and GRIK1 did interact with the AC1 protein. Thus, they used the E-value 1250 as the threshold to filter the blast results. The candidate list was further shortened with one of the following conditions: (1) at least two hits from the same or different peptides; (2) with E-value less than 517 (Pasumarthy, et al., 2011). In another study, only a tri-peptide or longer sequence match was considered (Kanki, et al., 2011). It seems that the evaluation of sequence matches found by sequence alignment, local alignment search and pattern search are rather arbitrary. As the standard is different case by case, the results from these tools are more like a kind of indication rather than a formal prediction. The results can be confirmed only when more background information is available. Results from sequence alignment, local alignment search and pattern search are same in nature: similarity matches between mimotope and a protein sequence. Thus, a general statistics model or method that evaluates the similarity match reasonably is urgently needed.

As described previously, methods based on template sequence have succeeded in many cases. However, it is more frequent that mimotopes show little similarities to the template, especially when the interaction sites are conformational. Thus methods based on template sequence often fail too. TSOL18 is a host-protective oncosphere antigen of Taenia solium, which is a cestode parasite causing cysticercosis in humans and pigs. The Ph.D.-12 phage library was screened with the anti-TSOL18 monoclonal antibody 17E1. The mimotopes were aligned to the TSOL18 protein sequence using ClustalW software. No significant match was found (Guo, et al., 2010). Intact oocytes surrounded by canine zona pellucida proteins were used to identify peptide sequences from phage display libraries that could recognize and bind to zona pellucida proteins (Samoylova, et al., 2010). The selection of a 12mer library resulted in identification of four sequences with the common NNXXPIL motif discovered by the MOTIF2 program in the RELIC suite. Among them, NNQSPILKLSIH was synthesized and immunized in dogs. The anti-NNQSPILKLSIH antibodies did bind to the acrosomal region of the canine sperm cell. However, BLAST search did not result in identification of homologies to known sperm proteins or other mammalian proteins. Thus, to predict protein interaction sites that are discontinuous using only sequences of mimotope and template is a great challenge. Though the FINDMAP program is a good attempt on this, it is still far from satisfactory. As the entry number of the PDB database increases exponentially, more and more protein structures become available to be used in the prediction of protein interaction sites based on mimotope analysis (Rose, et al., 2011).

2.2 Methods based on template structure

When sequence similarities are not found, it is very likely that mimotopes resemble a special region on the surface of template rather than a linear segment of template sequence. The prediction of protein interaction sites based on mimotope sequences and corresponding template structure is actually to identify and evaluate surface regions on the template that are similar to mimotopes.

2.2.1 Algorithms, programs and web servers

In 1995, Pizzi et al described the first method that predicted discontinuous antibody binding site based on mimotopes and the antigen structure (Pizzi, et al., 1995). Since then, quite a few algorithms, programs and web servers have been published by different teams around the world. All these methods can be divided into four groups. The first one is the motif-based group, which align a motif or consensus sequence to template structure. This group includes 3DEX (Schreiber, et al., 2005), MIMOX (Huang, et al., 2006) and the MimCons section of MIMOP program (Moreau, et al., 2006). The second group includes Mapitope (Bublil, et al., 2006; Bublil, et al., 2007; Enshell-Seijffers, et al., 2003; Tarnovitski, et al., 2006) and its derivatives (Denisov, et al., 2009; Denisova, et al., 2008; Denisova, et al., 2009; Denisova, et al., 2010). It can be called the pairs-based group because amino acid pairs on the template surface are considered to be simulated by amino acid pairs in the mimotope sequence. The third one is the patch-based group, which evaluates similarities between surface patches on template and mimotopes. SiteLight (Halperin, et al., 2003) and EpiSearch (Negi & Braun, 2009) belong to this group. The fourth is the graph-based group, which aligns a set of query peptides to a graph representing the template surface. Pepsurf (Mayrose, Shlomi, et al., 2007) and Pep-3D-Search (Huang, et al., 2008) belong to this group. To improve the performances of existing programs, hybrid methods such as MimoPro (Chen, et al., 2011) and meta-servers such as Pepitope (Mayrose, Penn, et al., 2007) were also proposed.

As tools mentioned above have been reviewed in detail recently (Huang, et al., 2011), here we only introduce LocaPep, a tool proposed very recently (Pacios, et al., 2011). For each mimotope, this program firstly scans the template surface to select seeds. Then it searches residues adjacent to each seed to form a cluster. For each residue in a cluster, its total score is the weighted sum of the area, exposure, contacts and distance score. At last, the final consensus cluster is calculated to form the binding site predicted. LocaPep is written with Fortran90 independent of any specific library and runs in command line mode. Its source code, manual and binaries are available at http://atenea.montes.upm.es.

2.2.2 Benchmarking tools of the trade

As described above, quite a few methods based on template structure are available for the phage display community to predict protein interaction sites. All these methods have succeeded in some case studies. These test cases were either compiled from published papers or from special databases such as the ASPD database (Valuev, et al., 2002) and the MimoDB database (Huang, et al., 2012; Ru, et al., 2010). However, no systematic evaluations were done when these methods were published. This is due to a relative lack of the type of data where the target-template complex is solved and the relevant mimotope data is available simultaneously.

As the protein structure and mimotope data increase rapidly (Huang, et al., 2012; Rose, et al., 2011), now it becomes possible to make benchmarks for the trade to evaluate its tools at a larger scale. Sun et al compiled a benchmark from the PDB database (Sun, et al., 2011) and the MimoDB database. It included 47 test cases in which 18 cases were with structures of the antigen-antibody complexes and 29 cases had structures of other protein-protein complexes. They further kept only one test case for each complex with the same template, which made a representative dataset with 30 test cases. Five popular tools, i.e. Mapitope, PepSurf, Pepitope, EpiSearch and Pep-3D-Search, were evaluated with the benchmark and the representative dataset. The results showed that performances of these tools were better than random predictions. However, their overall performances were still not satisfactory. Most tools were good at some cases but failed with other cases.

Our group has also compiled a benchmark called MimoBench (Huang, et al., 2012). It can be freely accessed from http://immunet.cn/mimodb/mimobench.php. Currently, MimoBench has 23, 23 and 27 sets of data for antibody–antigen complex, receptor–ligand complex and other protein-protein complex respectively. Using this benchmark, we have performed a preliminary evaluation on Mapitope, Episearch and MimoPro by their default parameters. Our results showed that performances of these tools were poor in many cases. However, they made quite accurate predictions in some cases. Taking the AUC value 0.8 as a cutoff, the three benchmarked tools succeeded in overlapping but different cases, which suggested that these tools complemented each other. Thus, it is recommended to use several tools together in the prediction of protein–protein interaction sites based on mimotopes.

2.2.3 Methods based on template structure: Challenges and suggestions

Methods based on template structure are capable of predicting the conformational sites of protein-protein interactions. However, the existing tools are not robust enough. Sun et al reported that many test cases in their benchmark dataset could not be applied to the five tools they evaluated due to software limitations (Sun, et al., 2011). We met the same problem when we compared Mapitope, Episearch and MimoPro using MimoBench. For example, four test cases were excluded from benchmarking because these tools did not work on the template with two or more chains. Another 10 cases were dropped because MimoPro returned no results for unknown reason (Huang, et al., 2012). Hence, tools in the future should be more robust. Furthermore, they should also be more convenient to access. It is hoped that web sites of these tools are stable and easy to access. No login is required. Thus, they can be utilized more conveniently whether they are standalone tools or web servers.

As described in the previous section, performances of the existing tools based on template structure are poor in many cases. To improve their performances is one of the greatest challenges in this field. We have suggested that the poor performance might partly due to information loss and noise inclusion during the experimental and computational process (Huang, et al., 2009). Considering the two points in mind, the accuracy of deciphering protein interaction sites using mimotopes might be improved. We will discuss on this issue in the following section.

2.3 Data cleaning tools

Due to the limitation of experiments, the biopanning results are noisy. They are usually a mixture of mimotopes (desired signal) and target-unrelated peptides (unwanted noise).

Target-unrelated peptides (TUPs) can be divided into two categories. One is called selection-related TUP. They appear in the biopanning results because they are selected by contaminants or other components of the screening system rather than the target (Menendez & Scott, 2005; Vodnik, et al., 2011). Propagation-related TUP makes another category (Brammer, et al., 2008; Derda, et al., 2011; Thomas, et al., 2010). They sneak into the output of biopanning because they have a higher infection rate or faster secretion rate. Phages with growth advantage can be not only noise but also decrease the library diversity and lead to a loss of useful mimotopes. Simulations and experiments showed that subtle differences in growth rate yielded drastic differences in clone abundances after rounds of amplifications (Derda, et al., 2011). Thus, propagation-related TUP may even dominate the biopanning results. As TUPs are peptides unrelated to the target, they undoubtedly interfere with the prediction of protein interaction sites based on mimotopes if a TUP is taken as a mimotope. Changing experimental conditions and improving experimental methods can decrease TUPs. For example, increasing the stringency of panning may reduce TUPs; subtractive procedures may decrease selection-related TUPs; amplification in isolated compartment can mitigate the growth advantage of propagation-related TUPs (Derda, et al., 2010). However, TUPs cannot be eradicated experimentally. To exclude TUPs from the biopanning with computational tools has become an alternative and more convenient choice.

2.3.1 Data cleaning tool based on information theory

Based on the information theory, the program INFO in the RELIC suite (Mandava, et al., 2004) calculates information content for each peptide of the panning result. Two input files are required. The first one is a text file with a minimum of 50 peptide sequences from clones randomly selected from a naive library. The second file is the query of users, one or more peptide sequences selected from that same library. INFO first uses AAFREQ to calculate the amino acid frequency distributions at each position of the inserted peptide sequences from the parent library. The probability of random occurrence of any peptide can be calculated by multiplying the probability of each amino acid occurring at each position. The natural logarithm of the probability of a peptide multiplied by -1 is defined as its information content. Obviously, if the query peptide has very high information content, it is less possible to appear in the panning result. If it does occur in the result, it is more likely to be the mimotope selected by specific binding to the target. On the contrary, when a peptide with very low information content is observed in the result, it is less confident of taking it as a mimotope because it may be a propagation-related TUP. The INFO program was also integrated in other tools in the RELIC suite such as MATCH, HETEROalign and FASTAskan. However, all these tools have regretfully been inaccessible for about one year, which makes the RELIC suite now a real relic.

2.3.2 Data cleaning tool based on TUP motif

We have developed a free web server called SAROTUP, which is short for scanner and reporter of target-unrelated peptides (Huang, et al., 2010). It can be used to scan and exclude possible target-unrelated peptides from biopanning result. SAROTUP is based on known TUP motifs and sequences. In the current version, a set of 26 TUP motifs and 27 known TUP sequences are collected from literature and compiled into the program. Among them, nine sequences are known or highly suspected to be propagation-related TUPs; the left 42 motifs or sequences are for selection-related TUPs, including 14 for albumin binders, six for unrelated antibody binders, five for immunoglobulin Fc region binders, five for streptavidin binders, five for plastic binders, four for bivalent metal ion binders and one for biotin binders, protein A binders and lipid A binders respectively. We had tested SAROTUP before the MimoDB database was constructed. The results showed that: (1) TUPs were often seen and taken as mimotopes; (2) epitope prediction based on mimotopes was greatly interfered if TUPs were used in the analysis; (3) SAROTUP improved performances of epitope mapping based on mimotopes through cleaning the input data; (4) SAROTUP also helped to explain experiment results. However, as a tool based on pattern search, SAROTUP cannot deal with TUPs without known motifs.

2.3.3 Data cleaning tool based on database search

The problem mentioned above was partly solved when the MimoDB database became available (Huang, et al., 2012; Ru, et al., 2010). With a lot of biopanning results and relevant background information collected, this database can be used as a virtual and comprehensive control for experimental biologists. In the MimoDB database version 2.0, a batched peptide search tool can be used for a set of peptides to search against the database. If a peptide has been reported by different groups with different targets, it may be a TUP rather than a mimotope. This is because the chance of obtaining an identical peptide from a library having millions or billions of different peptides with a completely different target is extremely small. If this happens, the peptide obtained may be due to some common factors in the biopanning systems rather than by the target. The MimoBlast tool of the MimoDB database can further find out those peptides not identical but highly similar to the query peptides. Such peptides may also be TUPs. New TUP motifs can be derived from analyzing the result of MimoBlast. With these tools, we studied the peptides in the MimoDB database and claimed confidently that GETRAPL, SILPYPY, LLADTTHHRPWT, TMGFTAPRFPHY, SAHGTSTGVPWP and HLPTSSLFDTTH are TUPs which were not reported before (Huang, et al., 2012).

2.3.4 Data cleaning tools: Challenges and suggestions

Although the data cleaning tools described in this section complement each other, none of them are real classifiers but rather reminders. Without a solid statistical estimation, they can only tell users that a peptide in the result may be a TUP rather than a mimotope. However, as the entries in the MimoDB database increases rapidly, it is now practical to construct various TUP predictors based on machine learning methods. Secondly, the data cleaning procedure was ignored by most existing tools for the prediction of protein interaction sites based on mimotopes. This situation should be changed in the future.

3. Conclusion

Identification of the protein interaction site is very important for basic and applied research. Computational analysis on mimotopes obtained from phage display or other surface display experiments is a relatively cheap, convenient and efficient strategy to locate a protein interaction site at the segment or residue level. Although used mostly in epitope prediction, this strategy can also be used to other types of protein interaction sites. Insights can be gained by methods based on template sequence, which find sequence similarities between mimotopes and template through sequence alignment, local alignment search and pattern search. Conformational sites can also be mapped by methods based on template structure. However, performances of all existing methods are not satisfactory enough. This is at least partly due to TUPs that crept into the biopanning result. Several tools are available to detect TUPs based on information theory, known TUP motifs or special database. With the rapid accumulation of experimental data and improvement of methods, an evidence-based virtual phage display platform is expected to be established and the performance of predicting protein interaction sites based on mimotopes will substantially be increased.

4. Acknowledgment

This work was supported by the National Natural Science Foundation of China (grant 61071177) and the Scientific Research Foundation of UESTC for Youth (grant JX0769).

5. References

Barabasi, A. L. & Oltvai, Z. N. (2004). Network biology: understanding the cell's functional organization. Nat Rev Genet, Vol. 5, No. 2, (Feb), pp. 101-113, ISSN 1471-0056

Barbu, E. M.; Ganesh, V. K.; Gurusiddappa, S.; Mackenzie, R. C.; Foster, T. J.; Sudhof, T. C. & Hook, M. (2010). beta-Neurexin is a ligand for the Staphylococcus aureus MSCRAMM SdrC. PLoS Pathog, Vol. 6, No. 1, (Jan), p. e1000726, ISSN 1553-7374

Bickerton, G. R.; Higueruelo, A. P. & Blundell, T. L. (2011). Comprehensive, atomic-level characterization of structurally characterized protein-protein interactions: the PICCOLO database. BMC Bioinformatics, Vol. 12, p. 313, ISSN 1471-2105

Brammer, L. A.; Bolduc, B.; Kass, J. L.; Felice, K. M.; Noren, C. J. & Hall, M. F. (2008). A target-unrelated peptide in an M13 phage display library traced to an advantageous mutation in the gene II ribosome-binding site. Anal Biochem, Vol. 373, No. 1, (Feb 1), pp. 88-98, ISSN 0003-2697

Bublil, E. M.; Yeger-Azuz, S. & Gershoni, J. M. (2006). Computational prediction of the cross-reactive neutralizing epitope corresponding to the monoclonal antibody b12 specific for HIV-1 gp120. FASEB J, Vol. 20, No. 11, (Sep), pp. 1762-1774, ISSN 1530-6860

Bublil, E. M.; Freund, N. T.; Mayrose, I.; Penn, O.; Roitburd-Berman, A.; Rubinstein, N. D.; Pupko, T. & Gershoni, J. M. (2007). Stepwise prediction of conformational discontinuous B-cell epitopes using the Mapitope algorithm. Proteins, Vol. 68, No. 1, (Jul 1), pp. 294-304, ISSN 1097-0134

Chaemchuen, S.; Rungpragayphan, S.; Poovorawan, Y. & Patarakul, K. (2011). Identification of candidate host proteins that interact with LipL32, the major outer membrane protein of pathogenic Leptospira, by random phage display peptide library. Vet Microbiol, Vol. 153, No. 1-2, (Nov 21), pp. 178-185, ISSN 1873-2542

Chen, T.; Cui, J.; Liang, Y.; Xin, X.; Owen Young, D.; Chen, C. & Shen, P. (2006). Identification of human liver mitochondrial aldehyde dehydrogenase as a potential target for microcystin-LR. Toxicology, Vol. 220, No. 1, (Mar 1), pp. 71-80, ISSN 0300-483X

Chen, W. H.; Sun, P. P.; Lu, Y.; Guo, W. W.; Huang, Y. X. & Ma, Z. Q. (2011). MimoPro: a more efficient Web-based tool for epitope prediction using phage display libraries. BMC Bioinformatics, Vol. 12, p. 199, ISSN 1471-2105

Chen, Y. C.; Delbrook, K.; Dealwis, C.; Mimms, L.; Mushahwar, I. K. & Mandecki, W. (1996). Discontinuous epitopes of hepatitis B surface antigen derived from a filamentous phage peptide library. Proc Natl Acad Sci U S A, Vol. 93, No. 5, (Mar 5), pp. 1997-2001, ISSN 0027-8424

Corpet, F. (1988). Multiple sequence alignment with hierarchical clustering. Nucleic Acids Res, Vol. 16, No. 22, (Nov 25), pp. 10881-10890, ISSN 0305-1048

Deng, Y. Q.; Dai, J. X.; Ji, G. H.; Jiang, T.; Wang, H. J.; Yang, H. O.; Tan, W. L.; Liu, R.; Yu, M.; Ge, B. X.; Zhu, Q. Y.; Qin, E. D.; Guo, Y. J. & Qin, C. F. (2011). A broadly flavivirus cross-neutralizing monoclonal antibody that recognizes a novel epitope within the fusion loop of E protein. PLoS One, Vol. 6, No. 1, p. e16059, ISSN 1932-6203

Denisov, D. A.; Denisova, G. F.; Lelic, A.; Loeb, M. B. & Bramson, J. L. (2009). Deciphering epitope specificities within polyserum using affinity selection of random peptides and a novel algorithm based on pattern recognition theory. Mol Immunol, Vol. 46, No. 3, (Jan), pp. 429-436, ISSN 0161-5890

Denisova, G.; Denisov, D.; Evelegh, C.; Weissgram, M.; Beck, J.; Foley, S. R. & Bramson, J. L. (2009). Characterizing complex polysera produced by antigen-specific immunization through the use of affinity-selected mimotopes. PLoS One, Vol. 4, No. 4, p. e5309, ISSN 1932-6203

Denisova, G. F.; Denisov, D. A.; Yeung, J.; Loeb, M. B.; Diamond, M. S. & Bramson, J. L. (2008). A novel computer algorithm improves antibody epitope prediction using affinity-selected mimotopes: a case study using monoclonal antibodies against the West Nile virus E protein. Mol Immunol, Vol. 46, No. 1, (Nov), pp. 125-134, ISSN 0161-5890

Denisova, G. F.; Denisov, D. A. & Bramson, J. L. (2010). Applying bioinformatics for antibody epitope prediction using affinity-selected mimotopes - relevance for vaccine design. Immunome Res, Vol. 6 Suppl 2, p. S6, ISSN 1745-7580

Derda, R.; Tang, S. K. & Whitesides, G. M. (2010). Uniform amplification of phage with different growth characteristics in individual compartments consisting of monodisperse droplets. Angew Chem Int Ed Engl, Vol. 49, No. 31, (Jul 19), pp. 5301-5304, ISSN 1521-3773

Derda, R.; Tang, S. K.; Li, S. C.; Ng, S.; Matochko, W. & Jafari, M. R. (2011). Diversity of phage-displayed libraries of peptides during panning and amplification. Molecules, Vol. 16, No. 2, pp. 1776-1803, ISSN 1420-3049

Enshell-Seijffers, D.; Denisov, D.; Groisman, B.; Smelyanski, L.; Meyuhas, R.; Gross, G.; Denisova, G. & Gershoni, J. M. (2003). The mapping and reconstitution of a conformational discontinuous B-cell epitope of HIV-1. J Mol Biol, Vol. 334, No. 1, (Nov 14), pp. 87-101, ISSN 0022-2836

Fack, F.; Hugle-Dorr, B.; Song, D.; Queitsch, I.; Petersen, G. & Bautz, E. K. (1997). Epitope mapping by phage display: random versus gene-fragment libraries. J Immunol Methods, Vol. 206, No. 1-2, (Aug 7), pp. 43-52, ISSN 0022-1759

Fernández-Recio, J. (2011). Prediction of protein binding sites and hot spots. WIREs Computational Molecular Science, Vol. 1, No. 5, pp. 680-698, ISSN 1759-0884

Geysen, H. M.; Rodda, S. J. & Mason, T. J. (1986). A priori delineation of a peptide which mimics a discontinuous antigenic determinant. Mol Immunol, Vol. 23, No. 7, (Jul), pp. 709-715, ISSN 0161-5890

Guo, A.; Cai, X.; Jia, W.; Liu, B.; Zhang, S.; Wang, P.; Yan, H. & Luo, X. (2010). Mapping of Taenia solium TSOL18 antigenic epitopes by phage display library. Parasitol Res, Vol. 106, No. 5, (Apr), pp. 1151-1157, ISSN 1432-1955

Halperin, I.; Wolfson, H. & Nussinov, R. (2003). SiteLight: binding-site prediction using phage display libraries. Protein Sci, Vol. 12, No. 7, (Jul), pp. 1344-1359, ISSN 0961-8368

Higurashi, M.; Ishida, T. & Kinoshita, K. (2009). PiSite: a database of protein interaction sites using multiple binding states in the PDB. Nucleic Acids Res, Vol. 37, No. Database issue, (Jan), pp. D360-364, ISSN 1362-4962

Huang, J.; Gutteridge, A.; Honda, W. & Kanehisa, M. (2006). MIMOX: a web tool for phage display based epitope mapping. BMC Bioinformatics, Vol. 7, p. 451, ISSN 1471-2105

Huang, J.; Xia, M.; Lin, H. & Guo, F.B. Information loss and noise inclusion risk in mimotope based epitope mapping, The 3rd International Conference on Bioinformatics and Biomedical Engineering(iCBBE2009), pp. 1-3, Beijing, 2009

Huang, J.; Ru, B.; Li, S.; Lin, H. & Guo, F. B. (2010). SAROTUP: scanner and reporter of target-unrelated peptides. J Biomed Biotechnol, Vol. 2010, p. 101932, ISSN 1110-7251

Huang, J.; Ru, B. & Dai, P. (2011). Bioinformatics resources and tools for phage display. Molecules, Vol. 16, No. 1, pp. 694-709, ISSN 1420-3049

Huang, J.; Ru, B.; Zhu, P.; Nie, F.; Yang, J.; Wang, X.; Dai, P.; Lin, H.; Guo, F. B. & Rao, N. (2012). MimoDB 2.0: a mimotope database and beyond. Nucleic Acids Res, Vol. 40, No. D1, pp. D271-D277, ISSN 1362-4962

Huang, Y. X.; Bao, Y. L.; Guo, S. Y.; Wang, Y.; Zhou, C. G. & Li, Y. X. (2008). Pep-3D-Search: a method for B-cell epitope prediction based on mimotope analysis. BMC Bioinformatics, Vol. 9, p. 538, ISSN 1471-2105

Kanki, S.; Jaalouk, D. E.; Lee, S.; Yu, A. Y.; Gannon, J. & Lee, R. T. (2011). Identification of targeting peptides for ischemic myocardium by in vivo phage display. J Mol Cell Cardiol, Vol. 50, No. 5, (May), pp. 841-848, ISSN 1095-8584

Kozakov, D.; Hall, D. R.; Beglov, D.; Brenke, R.; Comeau, S. R.; Shen, Y.; Li, K.; Zheng, J.; Vakili, P.; Paschalidis, I. & Vajda, S. (2010). Achieving reliability and high accuracy in automated protein docking: ClusPro, PIPER, SDU, and stability analysis in CAPRI rounds 13-19. Proteins, Vol. 78, No. 15, (Nov 15), pp. 3124-3130, ISSN 1097-0134

Larkin, M. A.; Blackshields, G.; Brown, N. P.; Chenna, R.; McGettigan, P. A.; McWilliam, H.; Valentin, F.; Wallace, I. M.; Wilm, A.; Lopez, R.; Thompson, J. D.; Gibson, T. J. &

Higgins, D. G. (2007). Clustal W and Clustal X version 2.0. Bioinformatics, Vol. 23, No. 21, (Nov 1), pp. 2947-2948, ISSN 1367-4811

Mandava, S.; Makowski, L.; Devarapalli, S.; Uzubell, J. & Rodi, D. J. (2004). RELIC--a bioinformatics server for combinatorial peptide analysis and identification of protein-ligand interaction sites. Proteomics, Vol. 4, No. 5, (May), pp. 1439-1460, ISSN 1615-9853

Mashiach, E.; Schneidman-Duhovny, D.; Peri, A.; Shavit, Y.; Nussinov, R. & Wolfson, H. J. (2010). An integrated suite of fast docking algorithms. Proteins, Vol. 78, No. 15, (Nov 15), pp. 3197-3204, ISSN 1097-0134

Mayrose, I.; Penn, O.; Erez, E.; Rubinstein, N. D.; Shlomi, T.; Freund, N. T.; Bublil, E. M.; Ruppin, E.; Sharan, R.; Gershoni, J. M.; Martz, E. & Pupko, T. (2007). Pepitope: epitope mapping from affinity-selected peptides. Bioinformatics, Vol. 23, No. 23, (Dec 1), pp. 3244-3246, ISSN 1367-4803

Mayrose, I.; Shlomi, T.; Rubinstein, N. D.; Gershoni, J. M.; Ruppin, E.; Sharan, R. & Pupko, T. (2007). Epitope mapping using combinatorial phage-display libraries: a graph-based algorithm. Nucleic Acids Res, Vol. 35, No. 1, pp. 69-78, ISSN 0305-1048

Menendez, A. & Scott, J. K. (2005). The nature of target-unrelated peptides recovered in the screening of phage-displayed random peptide libraries with antibodies. Anal Biochem, Vol. 336, No. 2, (Jan 15), pp. 145-157, ISSN 0003-2697

Moreau, V.; Granier, C.; Villard, S.; Laune, D. & Molina, F. (2006). Discontinuous epitope prediction based on mimotope analysis. Bioinformatics, Vol. 22, No. 9, (May 1), pp. 1088-1095, ISSN 1367-4803

Moreira, I. S.; Fernandes, P. A. & Ramos, M. J. (2007). Hot spots--a review of the protein-protein interface determinant amino-acid residues. Proteins, Vol. 68, No. 4, (Sep 1), pp. 803-812, ISSN 1097-0134

Mount, D. W. (2007). Using the Basic Local Alignment Search Tool (BLAST). CSH Protoc, Vol. 2007, p. pdb top17, ISSN 1559-6095

Mumey, B.; Ohler, N.; Angel, T.; Jesaitis, A. & Dratz, E. (2006). Filtering Epitope Alignments to Improve Protein Surface Prediction, In: Frontiers of High Performance Computing and Networking – ISPA 2006 Workshops, G. Min, B. Di Martino, L. Yang, M. Guo and G. Ruenger, (Eds.), 648-657, Springer,

Mumey, B. M.; Bailey, B. W.; Kirkpatrick, B.; Jesaitis, A. J.; Angel, T. & Dratz, E. A. (2003). A new method for mapping discontinuous antibody epitopes to reveal structural features of proteins. J Comput Biol, Vol. 10, No. 3-4, pp. 555-567, ISSN 1066-5277

Negi, S. S. & Braun, W. (2009). Automated detection of conformational epitopes using phage display Peptide sequences. Bioinform Biol Insights, Vol. 3, pp. 71-81, ISSN 1177-9322

Ofran, Y. & Rost, B. (2007). ISIS: interaction sites identified from sequence. Bioinformatics, Vol. 23, No. 2, (Jan 15), pp. e13-16, ISSN 1367-4811

Pacios, L. F.; Tordesillas, L.; Palacin, A.; Sanchez-Monge, R.; Salcedo, G. & Diaz-Perales, A. (2011). LocaPep: localization of epitopes on protein surfaces using peptides from phage display libraries. J Chem Inf Model, Vol. 51, No. 6, (Jun 27), pp. 1465-1473, ISSN 1549-960X

Pasumarthy, K. K.; Mukherjee, S. K. & Choudhury, N. R. (2011). The presence of tomato leaf curl Kerala virus AC3 protein enhances viral DNA replication and modulates virus induced gene-silencing mechanism in tomato plants. Virol J, Vol. 8, p. 178, ISSN 1743-422X

Peng, L.; Oganesyan, V.; Damschroder, M. M.; Wu, H. & Dall'acqua, W. F. (2011). Structural and Functional Characterization of an Agonistic Anti-Human EphA2 Monoclonal Antibody. J Mol Biol, Vol. 413, No. 2, (Oct 21), pp. 390-405, ISSN 1089-8638

Perosa, F.; Vicenti, C.; Racanelli, V.; Leone, P.; Valentini, G. & Dammacco, F. (2010). The immunodominant epitope of centromere-associated protein A displays homology with the transcription factor forkhead box E3 (FOXE3). Clin Immunol, Vol. 137, No. 1, (Oct), pp. 60-73, ISSN 1521-7035

Pizzi, E.; Cortese, R. & Tramontano, A. (1995). Mapping epitopes on protein surfaces. Biopolymers, Vol. 36, No. 5, (Nov), pp. 675-680, ISSN 0006-3525

Plesa, M.; Kim, J.; Paquette, S. G.; Gagnon, H.; Ng-Thow-Hing, C.; Gibbs, B. F.; Hancock, M. A.; Rosenblatt, D. S. & Coulton, J. W. (2011). Interaction between MMACHC and MMADHC, two human proteins participating in intracellular vitamin B metabolism. Mol Genet Metab, Vol. 102, No. 2, (Feb), pp. 139-148, ISSN 1096-7206

Przulj, N. (2011). Protein-protein interactions: making sense of networks via graph-theoretic modeling. Bioessays, Vol. 33, No. 2, (Feb), pp. 115-123, ISSN 1521-1878

Res, I.; Mihalek, I. & Lichtarge, O. (2005). An evolution based classifier for prediction of protein interfaces without using protein structures. Bioinformatics, Vol. 21, No. 10, (May 15), pp. 2496-2501, ISSN 1367-4803

Rose, P. W.; Beran, B.; Bi, C.; Bluhm, W. F.; Dimitropoulos, D.; Goodsell, D. S.; Prlic, A.; Quesada, M.; Quinn, G. B.; Westbrook, J. D.; Young, J.; Yukich, B.; Zardecki, C.; Berman, H. M. & Bourne, P. E. (2011). The RCSB Protein Data Bank: redesigned web site and web services. Nucleic Acids Res, Vol. 39, No. Database issue, (Jan), pp. D392-401, ISSN 1362-4962

Ru, B.; Huang, J.; Dai, P.; Li, S.; Xia, Z.; Ding, H.; Lin, H.; Guo, F. & Wang, X. (2010). MimoDB: a New Repository for Mimotope Data Derived from Phage Display Technology. Molecules, Vol. 15, No. 11, pp. 8279-8288, ISSN 1420-3049

Samoylova, T. I.; Cox, N. R.; Cochran, A. M.; Samoylov, A. M.; Griffin, B. & Baker, H. J. (2010). ZP-binding peptides identified via phage display stimulate production of sperm antibodies in dogs. Anim Reprod Sci, Vol. 120, No. 1-4, (Jul), pp. 151-157, ISSN 1873-2232

Schreiber, A.; Humbert, M.; Benz, A. & Dietrich, U. (2005). 3D-Epitope-Explorer (3DEX): localization of conformational epitopes within three-dimensional structures of proteins. J Comput Chem, Vol. 26, No. 9, (Jul 15), pp. 879-887, ISSN 0192-8651

Shiheido, H.; Takashima, H.; Doi, N. & Yanagawa, H. (2011). mRNA display selection of an optimized MDM2-binding peptide that potently inhibits MDM2-p53 interaction. PLoS One, Vol. 6, No. 3, p. e17898, ISSN 1932-6203

Smith, G. P. (1985). Filamentous fusion phage: novel expression vectors that display cloned antigens on the virion surface. Science, Vol. 228, No. 4705, (Jun 14), pp. 1315-1317, ISSN 0036-8075

Smith, G. P. & Petrenko, V. A. (1997). Phage Display. Chem Rev, Vol. 97, No. 2, (Apr 1), pp. 391-410, ISSN 1520-6890

Sun, E. C.; Zhao, J.; Yang, T.; Liu, N. H.; Geng, H. W.; Qin, Y. L.; Wang, L. F.; Bu, Z. G.; Yang, Y. H.; Lunt, R. A. & Wu, D. L. (2011). Identification of a conserved JEV serocomplex B-cell epitope by screening a phage-display peptide library with a mAb generated against West Nile virus capsid protein. Virol J, Vol. 8, p. 100, ISSN 1743-422X

Sun, P.; Chen, W.; Huang, Y.; Wang, H.; Ma, Z. & Lv, Y. (2011). Epitope prediction based on random peptide library screening: benchmark dataset and prediction tools evaluation. Molecules, Vol. 16, No. 6, pp. 4971-4993, ISSN 1420-3049

Takakusagi, Y.; Takakusagi, K.; Sugawara, F. & Sakaguchi, K. (2010). Use of phage display technology for the determination of the targets for small-molecule therapeutics. Expert Opinion on Drug Discovery, Vol. 5, No. 4, pp. 361-389, ISSN 1746-0441

Takami, M.; Takakusagi, Y.; Kuramochi, K.; Tsukuda, S.; Aoki, S.; Morohashi, K.; Ohta, K.; Kobayashi, S.; Sakaguchi, K. & Sugawara, F. (2011). A screening of a library of T7 phage-displayed peptide identifies E2F-4 as an etoposide-binding protein. Molecules, Vol. 16, No. 5, pp. 4278-4294, ISSN 1420-3049

Tarnovitski, N.; Matthews, L. J.; Sui, J.; Gershoni, J. M. & Marasco, W. A. (2006). Mapping a neutralizing epitope on the SARS coronavirus spike protein: computational prediction based on affinity-selected peptides. J Mol Biol, Vol. 359, No. 1, (May 26), pp. 190-201, ISSN 0022-2836

Thomas, W. D.; Golomb, M. & Smith, G. P. (2010). Corruption of phage display libraries by target-unrelated clones: diagnosis and countermeasures. Anal Biochem, Vol. 407, No. 2, (Dec 15), pp. 237-240, ISSN 1096-0309

Urushibata, Y.; Itoh, K.; Ohshima, M. & Seto, Y. (2010). Generation of Fab fragment-like molecular recognition proteins against staphylococcal enterotoxin B by phage display technology. Clin Vaccine Immunol, Vol. 17, No. 11, (Nov), pp. 1708-1717, ISSN 1556-679X

Valuev, V. P.; Afonnikov, D. A.; Ponomarenko, M. P.; Milanesi, L. & Kolchanov, N. A. (2002). ASPD (Artificially Selected Proteins/Peptides Database): a database of proteins and peptides evolved in vitro. Nucleic Acids Res, Vol. 30, No. 1, (Jan 1), pp. 200-202, ISSN 1362-4962

Van Nieuwenhove, L. C.; Roge, S.; Balharbi, F.; Dieltjens, T.; Laurent, T.; Guisez, Y.; Buscher, P. & Lejon, V. (2011). Identification of peptide mimotopes of Trypanosoma brucei gambiense variant surface glycoproteins. PLoS Negl Trop Dis, Vol. 5, No. 6, (Jun), p. e1189, ISSN 1935-2735

Vidal, M.; Cusick, M. E. & Barabasi, A. L. (2011). Interactome networks and human disease. Cell, Vol. 144, No. 6, (Mar 18), pp. 986-998, ISSN 1097-4172

Vodnik, M.; Zager, U.; Strukelj, B. & Lunder, M. (2011). Phage display: selecting straws instead of a needle from a haystack. Molecules, Vol. 16, No. 1, pp. 790-817, ISSN 1420-3049

Wass, M. N.; David, A. & Sternberg, M. J. (2011). Challenges for the prediction of macromolecular interactions. Curr Opin Struct Biol, Vol. 21, No. 3, (Jun), pp. 382-390, ISSN 1879-033X

Zhao, S.; Zhao, W. & Ma, L. (2010). Novel peptide ligands that bind specifically to mouse embryonic stem cells. Peptides, Vol. 31, No. 11, (Nov), pp. 2027-2034, ISSN 1873-5169

9

Slow Protein Conformational Change, Allostery and Network Dynamics

Fan Bai[1], Zhanghan Wu[2], Jianshi Jin[1],
Phillip Hochendoner[3] and Jianhua Xing[2]
[1]Biodynamic Optical Imaging Centre,
Peking University, People's Republic of China, Beijing
[2]Department of Biological Sciences, Virginia Tech,
[3]Department of Physics, Virginia Tech,
[1]China
[2,3]USA

1. Introduction

Macromolecules such as proteins contain a large number of atoms, which lead to complex dynamic behaviors not usually seen in simpler molecular systems with only a few to tens of atoms. Characterizing the biochemical and biophysical properties of macromolecules, including their interactions with other molecules, has been a central research theme for many decades. The field is especially accelerated by recent advances in experimental techniques, such as nuclear magnetic resonance (NMR) and single-molecule measurements, and computational powers that has been facilitated to simulate molecular dynamics at large scales.

Chemical kinetics has been well developed for simple molecular systems, and most of the small molecular reactions can be described accurately by kinetic equations. However, it's hard to describe a macromolecular system using simple mathematical equations, because reactions at the macromolecular level usually involve complicated processes and dynamic behaviors. Even so, biochemists have done many efforts to find a way to describe the biological systems. Many equations and models have been published by using approximate treatments or hypothesis.

If biochemists were asked what is the most important mathematical equation they know, most likely the answer you will hear is the Michaelis-Menten equation. Michaelis–Menten equation is one of the simplest and best-known equations describing enzyme kinetics (Menten and Michaelis, 1913). It is named after American biochemist Leonor Michaelis and Canadian physician Maud Menten. For a typical enzymatic reaction one often finds that the following scheme works reasonably well,

$$S + E \underset{\alpha_{-1}}{\overset{\alpha}{\rightleftarrows}} ES \overset{k}{\longrightarrow} E + P \tag{1}$$

with S, E, ES, P representing the substrate, the free enzyme, the enzyme-substrate complex, and the product. Then one has the rate of product formation: (after certain assumptions, such as the enzyme concentration being much less than the substrate concentration)

$$\frac{d[P]}{dt} = \frac{k[E]_t[S]}{(\alpha_{-1}+k)/\alpha+[S]} \tag{2}$$

In this model, the rate of product formation increases along with the substrate concentration [S] with the characteristic hyperbolic relationship, asymptotically approaching its maximum rate $V_{max} = k[E]_t$, ($[E]_t$ is the total enzyme concentration) attained when all enzymes are bound to substrates .We can use K_m to represent $(a_{-1}+k)/a$, named Michaelis constant. It is the substrate concentration at which the reaction rate is at half the maximum rate, and is a measure of the substrate's affinity for the enzyme. A small K_m indicates high affinity, meaning that the rate approaches V_{max} more quickly.

The Michaelis–Menten equation was first proposed for investigating the kinetics of an enzymatic (invertase) reaction mechanism in 1913 (Menten and Michaelis, 1913). Later, it has been widely used in a variety of biochemical transitions other than enzyme-substrate interaction, which includes antigen-antibody binding, DNA-DNA hybridization and protein-protein interaction. There is no exaggeration to say that the Michaelis–Menten model has greatly pushed forward our understanding of enzymatic reactions.

However, biochemists also found that many enzymes show kinetics are more complicated than the Michaelis-Menten kinetics. Frieden coined the name "hysteretic enzyme" referring to "those enzymes which respond slowly (in terms of some kinetic characteristic) to a rapid change in ligand, either substrate or modifier, concentration" (Frieden, 1970). Since then a sizable literature exists on the enzyme behavior. The list of hysteretic enzymes cover proteins working in many organisms from bacteria to mammalians (Frieden, 1979), with one of the latest examples related to the protein secreted by bacteria *Staphylococcus aureus* to induce host blood coagulation (Kroh et al., 2009). The kinetics, especially the enzymatic activity of a hysteretic enzyme, cannot adapt to new environmental conditions quickly. The delay time can be surprisingly long. For example, upon changing the solution's pH value, it takes more than two hours for alkaline phosphatase to relax to the enzymatic activity corresponding to the new pH value (Behzadi et al., 1999). The mnemonic behavior is another key example of slow conformational dynamic disorder advocated by Richard and his colleagues (Cornish-Bowden and Cardenas, 1987; Frieden, 1970; Frieden, 1979; Ricard and Cornish-Bowden, 1987). It refers to the phenomenon that "the free enzyme alone which undergoes the 'slow' transition…upon the desorption of the last product from the active site, the enzyme retains for a while the conformation stabilized by that product before relapsing to another conformation" (Ricard and Cornish-Bowden, 1987). Their observation revealed that Mnemonic enzymes show non-Michaelis-Menten (NMM) behaviors. The concepts of mnemonic and hysteretic enzymes emphasize the steady-state kinetics and the transient kinetics leading to the steady state, respectively. However, the conformational change in a protein is the rate limiting step in both enzymatic reactions which are slower than the actual chemical reaction step (chemical bond breaking and forming). To this end, a unified model exists (Ainslie et al., 1972).

A deeper understanding on the origin of the mnemonic and hysteretic behaviors comes from biophysical studies. A related phenomenon called dynamic disorder has been discussed extensively in the physical chemistry and biophysics communities. Dynamic disorder refers to the phenomena that the 'rate constant' of a process is actually a random function of time, and is affected by some slow protein conformational motions (Frauenfelder et al., 1999; Zwanzig, 1990). A molecule fluctuates constantly at finite temperature. The Reaction Coordinate (RC) is an important concept in chemical rate theories (Hanggi et al., 1990). The RC is a special coordinate in the configurational space (expanded by the spatial coordinates of all the atoms in the system), which leads the system from the reactant configuration to the product configuration. A fundamental assumption in most rate theories (such as the transition state theory) states that the dynamics along the RC is much slower than fluctuations along all other coordinates. Consequently, for any given RC position, one may assume other degrees of freedom approaches approximately equilibrium. This is the so-called adiabatic approximation. Deviation from this assumption is treated as secondary correction (Grote and Hynes, 1980). Chemical rate theories based on this assumption are remarkably successful in explaining the dynamics involving small molecules. The dynamics of a system can be well characterized by a rate constant. However, the situation is much more complicated in macromolecules like proteins, RNAs, and DNAs. Macromolecules have a large number of atoms and possible conformations. The conformational fluctuation time scales of macromolecules span from tens of femtoseconds to hundreds of seconds (McCammon and Harvey, 1987). Consequently, conformational fluctuations can be comparable or even slower than the process involving chemical bond breaking and formation. The adiabatic approximation seriously breaks down at this regime. If one focuses on the dynamics of processes involving chemical reactions, the canonical concept of "rate constant" no longer holds. Since the pioneering work of Frauenfelder and coworkers on ligand binding to myoglobin (Austin et al., 1975), extensive experimental and theoretical studies have been performed on this subject (see for example ref. (Zwanzig, 1990) for further references). Additionally, the conformational fluctuation of a macromolecule is an individual behaviour, many dynamic processes were hidden under the ensemble measurements. Fortunately, recent advances in room-temperature single-molecule fluorescence techniques gave us an opportunity to investigate the conformational dynamics on the single-molecule level. Hence, the dynamic disorders in an individual macromolecule has been demonstrated directly through single molecule enzymology measurements recently (English et al., 2006; Min et al., 2005b; Xie and Lu, 1999). For example, Xie and coworkers showed that both enzymes' conformation and catalytic activity fluctuate over time, especially the turnover time distribution of one β-galactosidase molecule spans several orders of magnitude (10^{-3} s to 10 s). Their results revealed that although a fluctuating enzyme still exhibits MM steady-state kinetics in a large region of time scales, the apparent Michaelis and catalytic rate constants do have different microscopic interpretations. It is also shown that at certain conditions dynamic disorder results in Non-Michaelis-Menten kinetics (Min et al., 2006). Single molecule measurements on several enzymes suggested that the existence of dynamic disorder in biomolecules is a rule rather than exception (Min et al., 2005a). So if problems arise, when there are only a few copies of a particular enzyme in a living cell, do these fluctuations result in a noticeable physiological effect?

Therefore, an important question we need to ask is: What is the biological consequence of dynamic disorder? Frieden insightfully noticed that "*it is of interest that the majority of enzymes exhibiting this type of (hysteretic) behavior can be classed as regulatory enzymes*" (Frieden, 1979). A series of important questions emerge naturally: Is the existence of complex enzymatic kinetic behaviors an evolutional byproduct or selected trait? Is there any biological function for it? How can such diverse and complex enzymatic kinetic behaviors affect our understanding of regulatory protein interaction networks?

In recent years, studying interactions of molecules in a cell from a systems perspective has been gaining popularity. Researchers in this newly formed field "systems biology" emphasize that to characterize a complex system, it is insufficient to take the reductionist's view. Combining several reactions together, one can form reaction networks with emerging dynamic behaviors such as switches, oscillators, etc, and ultimately the life form (Alon, 2007; Kholodenko, 2006; Tyson et al., 2001). In the new era of systems biology, a modeler may deal with hundreds to thousands ordinary differential rate equations describing various biological processes. The hope is that by knowing the network topology and associated rate constants (which requires daunting experimental efforts), one can reveal the secret of life and even synthesize life.

On modeling such regulatory protein interaction networks, it is common practice to assume that each enzymatic reaction can be described by a simple rate process, especially by the Michaelis-Menten kinetics. In our opinion, most contemporary researches on biological network dynamics emphasize the effect of network topology without giving sufficient consideration of the biochemical/biophysical properties of each composing macromolecule. One of the reasons that account for the current state of affair is due to a lack of experimental data and theoretical understanding in the "intermediate regime" between single-molecule studies of individual enzymes (relatively simple) and cellular dynamics (too complex). Recent advances in single-molecule techniques give us hope to study larger systems. One of its unique advantages is the ability to study macromolecular dynamics under room temperature and nonequilibrium state, which well mimics physiological conditions of a living cell. Using these single-molecule experimental results to build the cellular dynamics model will be a promising and significative research field.

In this chapter, we will present a unified mathematical formalism describing both conformational change and chemical reactions. Then we will discuss some implications of slow conformational changes in protein allostery and network dynamics.

2. Coarse grained mathematical description of conformational changes

Substrate binding often induces considerable changes of the protein conformation, especially in the binding pocket. This is the so-called induced-fit model. To explicitly take into account the induced conformational change, one can generalize the scheme given in Equation 1 to what shown in Fig. 1A. The substrate and protein form a loosely bound complex first. Their mutual interactions drive further conformational change of the binding pocket to form a tight bound complex, where atoms are properly aligned for chemical bond breaking and forming to take place. Next the binding pocket opens to release the product and is ready for another cycle. Mathematically one can write a set of ordinary differential based rate equations to describe the dynamics, or perform stochastic simulations of the process.

For a more complete description of the continuous nature of conformational changes, one can reduce the conformational complexity of the system to a few well defined degrees of freedom with slow dynamics (Xing, 2007). For example, let's denote x to represent the conformational coordinate of the enzyme from open to close of the binding pocket, and $U(x)$ the potential of mean force along x. In general $U(x)$ is affected by substrate binding. Therefore, in a minimal model the chemical state of the binding pocket (the catalytic site) can be: Emp (empty), Rec (reactant bound), or Prod (product bound). As shown in Figure 1B, each state is described by a potential curve $U_i(x)$ along the conformational coordinate, and localized transitions can happen between two potentials. For an enzymatic cycle, a reactant molecule first binds onto the catalytic site (Emp→Rec), then forms a more compact complex, next the chemical reaction happens (Rec→Prod), and finally the catalytic site is open and the product is released (Prod→Emp). Notice that binding molecules may shift both the curve shape and minimum position, and some conformational motion is necessary during the cycle. The harmonic shape of the curves shown in Figure 1B is only illustrative. A more complete description is to use the two (or higher) dimensional potential surfaces plotted in Figure 1C. The plot should be only viewed as illustrative. Within an enzymatic cycle, the system zigzags through the potential surface, with motions along both the conformational and reaction coordinates coupled. Figure 1D gives projection of the potential surface along the reaction coordinate at two conformational coordinate values. The curves have the characteristic double well shape. For barrier crossing processes, a system spends most of the time at potential wells, and the actual barrier-crossing time is transient and fast. Therefore, one can reduce the two-dimensional surface (Figure 1C) to one-dimensional projections along the conformational coordinate (Figure 1B), and approximate transitions along the reaction coordinate by rate processes among the one-dimensional potential curves.

With the above introduction of potential curves, we can now formulate the governing dynamic equations by a set of over-damped Langevin equations coupled to Markov chemical transitions (Xing, 2007; Zwanzig, 2001),

$$\zeta_i \frac{dx(t)}{dt} = -\frac{dU_i(x)}{dx} + f_i(t) , \qquad (3)$$

where x and U_i as defined above, ζ_i is the drag coefficient along the molecular conformational coordinate, and f is the random fluctuation force with the property $<f(t)f(t')>$ = $2\ k_BT\zeta\delta(t-t')$, with k_B the Boltamann's constant, T the temperature. Chemical transitions accompany motions along the conformational coordinate with x-dependent transition rates. In general the dynamics may be non-Markovian and contain a memory effect (Zwanzig, 2001). Min et al. observed a power law memory kernel for single protein conformational fluctuations (Min et al., 2005b). Xing and Kim showed that the observation can be well reproduced using a coarse-grained protein fluctuation model, with both of two adjustable parameters agree with other independent studies (Xing and Kim, 2006). However here we will assume Markovian dynamics for simplicity. The Langevin dynamics described by Equation 3 can be equally described by a set of coupled Fokker-Planck equations,

$$\frac{\partial}{\partial t}\rho_i(x) = -\frac{D_i}{k_BT} \cdot \frac{\partial}{\partial x}\left(-\frac{\partial U_i(x)}{\partial x}\rho_i(x)\right) + D_i\frac{\partial^2 \rho_i(x)}{\partial x^2} + \sum_{j\neq i}\left(K_{ij}(x)\rho_j(x) - K_{ji}(x)\rho_i(x)\right) \qquad (4)$$

Where $D_i = k_BT/\zeta_i$ is the diffusion constant, K_{ij} is the transition matrix element, and $\rho_i(x)$ is the probability density to find the system at position x and state i.

The formalism given by Equations 3 and 4 is widely used to model systems such as electron transfer reactions, protein motors (Bustamante et al., 2001; Julicher et al., 1997; Wang and Oster, 1998; Xing et al., 2006; Xing et al., 2005), as well as enzymatic reactions here (Gopich and Szabo, 2006; Min et al., 2008; Qian et al., 2009; Xing, 2007).

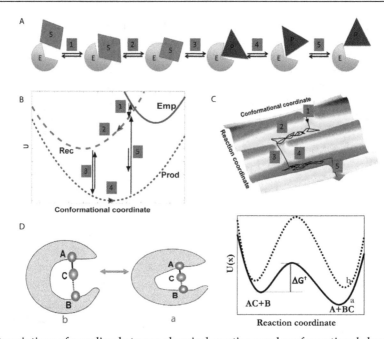

Fig. 1. Descriptions of coupling between chemical reactions and conformational changes.
(A) A discrete enzymatic cycle model with conformational changes. (B) A minimal continuous
model representing three potentials of mean force along a conformational coordinate.
(C) A continuous model with explicit reaction and conformational coordinates.
(D) Two protein conformations and the corresponding potentials of mean force along the
reaction coordinate.

The continuous form of Equation 4 can also be discretized to a form more familiar to
biochemists[1],

$$
\begin{array}{ccccc}
E_1 + S \underset{\alpha_{-1}}{\overset{\alpha_1}{\rightleftharpoons}} E_1 S \underset{k_{-1}}{\overset{k_1}{\rightleftharpoons}} E_1 P \underset{\beta_{-1}}{\overset{\beta_1}{\rightleftharpoons}} E_1 + P \\
\updownarrow \qquad \updownarrow \qquad \updownarrow \qquad \updownarrow \\
E_2 + S \underset{\alpha_{-2}}{\overset{\alpha_2}{\rightleftharpoons}} E_2 S \underset{k_{-2}}{\overset{k_2}{\rightleftharpoons}} E_2 P \underset{\beta_{-2}}{\overset{\beta_2}{\rightleftharpoons}} E_2 + P \\
\updownarrow \qquad \updownarrow \qquad \updownarrow \qquad \updownarrow \\
\cdots \\
\updownarrow \qquad \updownarrow \qquad \updownarrow \qquad \updownarrow \\
E_N + S \underset{\alpha_{-N}}{\overset{\alpha_N}{\rightleftharpoons}} E_N S \underset{k_{-N}}{\overset{k_N}{\rightleftharpoons}} E_N P \underset{\beta_{-N}}{\overset{\beta_N}{\rightleftharpoons}} E_N + P
\end{array}
\tag{5}
$$

---

[1] A mathematical procedure for the discretization is given in Xing, J., Wang, H.-Y., and Oster, G. (2005).
From continuum Fokker-Planck models to discrete kinetic models. Biophys J 89, 1551-1563.

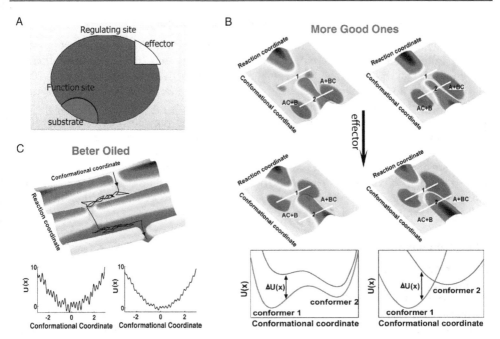

Fig. 2. Different models for allostery. (A) Schematic illustration of allosteric regulation. (B) Schematic potentials of mean force illustrating the MWC (left) and the KNF (right) models. (C) A nonequilibrium dynamic model.

Equations 3-5 describe richer physics than the simple induced fit model does. The conformational changes include contributions from binding induction as well as enzyme spontaneous fluctuations. There may be a number of parallel pathways for an enzymatic reaction corresponding to different protein conformations. An optimal conformation for one step of the reaction may not be the optimal conformation for another step. If an enzyme can transit among these conformations faster than a chemical transition event (including substrate/product binding and release), then the system can mainly follow the tortuous optimal pathway involving different conformations shown in Figure 1B and C. If the conformational change is comparable or slower than chemical events, multiple pathways may contribute significantly to the dynamics, and one observes time varying enzyme activity at the single molecule level, which leads to the phenomenon "dynamic disorder". One origin of the slow dynamics of intramolecular dynamics comes from diffusion along rugged potential surfaces with numerous potential barriers (Frauenfelder et al., 1991). Zwanzig shows that the effectic diffusion constant is greatly reduced along a rugged potential (Zwanzig, 1988). For example, for a rugged potential with a gaussian distributed barrier height, and root-mean-square ε, the so-called roughness parameter, the effective diffusion constant is scaled as,

$$D_{effective} = D \exp\left[-\left(\varepsilon / k_B T\right)^2\right] \tag{6}$$

which can be greatly reduced from the bare value of D.

3. Thermodynamic versus dynamic models for allostery

A cell needs to adjust its metabolic, transcriptional, and translational activities to respond to changes in the external and internal environment. Allostery and covalent modification are two fundamental mechanisms for regulating protein activities (Alberts et al., 2002). Allostery refers to the phenomenon that binding of an effector molecule to a protein's allosteric site affects the protein activity at its active site, which is usually physically distinct from where the effector binds. The discovery of allosteric regulations was in the 1950s, followed by a general description of allostery in the early 1960s, has been regarded as revolutionary at that time (Alberts et al., 2002). Not surprisingly, to understand the mechanism of allosteric regulation is an important topic in structural biology. Below we will focus on allosteric enzymes. For simplicity, we will restrict our discussions to positive allosteric effect, i.e., effector binding increases enzymatic activity. The discussions can be easily generalized to negative allosteric effects.

3.1 Conventional models of allostery

There are two popular models proposed to explain the allosteric effects. The concerted MWC model by Monod, Wyman, and Changeux, assumes that an allosteric protein can exist in two (or more) conformations with different reactivity, and effector binding modifies the thermal equilibrium distribution of the conformers (Monod et al., 1965). Recent population shift models re-emphasize the idea of preexisting populations (Goodey and Benkovic, 2008; Kern and Zuiderweg, 2003; Pan et al., 2000; Volkman et al., 2001). The sequential model described by Koshland, Nemethy, and Filmer is based on the induced-fit mechanism, and assumes that effector binding results in (slight) structural change at another site and affects the substrate affinity (Koshland et al., 1966). While different in details, both of the above models assume that the allosteric mechanism is through modification of the equilibrium conformation distribution of the allosteric protein by effector binding. For later discussions, we denote the mechanisms as "thermodynamic regulation".

The mechanisms of thermodynamic regulation impose strong requirements on the mechanical properties of an allosteric protein. The distance between the two binding sites of an allosteric protein can be far. For example, the bacterial chemotaxis receptor has the two reaction regions separated as far as 15 nm (Kim et al., 2002). In this case, signal propagation requires a network of mechanical strain relaying residues with mechanical properties distinguishing them well from the surroundings to minimize thermal dissipation – Notice that distortion of a soft donut at one side has negligible effect on another side of the donut. Mechanical stresses due to effector molecule binding irradiate from the binding site, propagate through the relaying network, and con-focus on the reaction region at the other side of the protein (Amaro et al., 2009; Amaro et al., 2007; Balabin et al., 2009; Cecchini et al., 2008; Cui and Karplus, 2008; Horovitz and Willison, 2005; Ranson et al., 2006). However, it is challenging to transmit the mechanical energy faithfully against thermal dissipation over a long distance. A possible solution is the attraction shift model proposed by Yu and Koshland(Yu and Koshland, 2001).

From a chemical physics perspective, current existing models on allosteric effects differ in some details of the potential shapes. The MWC and the recent population-shift model emphasizes that there are pre-existing populations for all the possible forms, as exemplified by the double well shaped potentials and the two corresponding conformers in the left panel of

Figure 2C. Effector binding only shifts their relative populations. The KNF model emphasizes that without the effector the protein exists mainly in one form (conformer 2 in the right panel of Figure 2C). Effector binding shifts the protein to another form (conformer 1) with different reactivity. The functions $U(x)$ are potentials of mean force, which suggests that the effect of effector binding can be enthalpic or entropic (Cooper and Dryden, 1984). Therefore in some sense there is no fundamental difference between the KNF and MWC models. They differ only in the extent of each conformer being populated, which is related to the free energy difference between conformers ΔU (in Figure 2C) through the Boltzman factor.

3.2 Possibly neglected dynamic aspect of allostery

The above allosteric models focus on the conformational changes decoupled from those changes associated with an enzymatic cycle. Consequently, the distribution along the conformational coordinate can be described as thermodynamic equilibrium. However, as discussed in section 2, an enzymatic cycle usually inevitably involves enzyme conformational changes, so the distribution of the latter is in general driven out of equilibrium due to coupling to the nonequilibrium chemical reactions. In many cases, as Frieden wrote, "conformational changes after substrate addition but preceding the chemical transformation, or after the chemical transformation but preceding product release may be rate-limiting" (Frieden, 1979). Recent NMR studies further demonstrate conformational changes as rate-limiting steps (Boehr et al., 2006; Cole and Loria, 2002). Based on these experimental observations, Xing proposed that the conformational change dynamics within an enzymatic cycle can be subject to allosteric modulation (Xing, 2007).

Enzyme conformational changes can be thermally activated barrier crossing events, and effectors function by modifying the height of the dominant barrier. Alternatively, effectors may accelerate conformational changes through decreasing the potential roughness (see Figure 2D). Intuitively, for the latter mechanism effectors transform rusty engines (enzymes) into better-oiled ones.

Figure 3 schematically summarizes possible effector binding induced changes of the potentials of mean force along a conformational coordinate, which then affects the

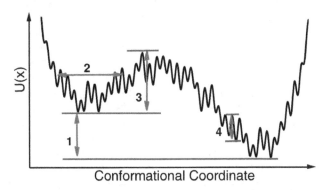

Fig. 3. Summary of effects of effector binding on the potential of mean force: (1) relative free energy difference of the two conformers; (2) Width of the potential well; (3) Barrier height; (4) Potential roughness.

enzymatic reaction dynamics. The changes can be the relative height of the potential wells representing different conformers (labelled 1 in Figure 3, enthalpic), the widths of potential wells (labelled 2, entropic), the barrier height (labelled 3) and the potential roughness (labelled 4) (dynamic). For a given enzyme subject to a given effector regulation, one or more effects may play the dominant role.

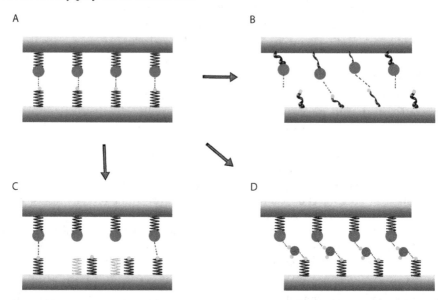

Fig. 4. Possible scenarios of modifying potential roughness. Relative motion between two protein surfaces (A) can be modulated through changing the linkage stiffness (B) or the arrangement of surface residues (C), or solvent accessibility (D).

Further experimental and theoretical studies are necessary to reveal the detailed molecular mechanisms for the proposed potential roughness regulation. Figure 4 gives some possible scenarios. Suppose during the process of conformational change, two protein surfaces need to move along each other, with numerous residues dangling on the surfaces forming and breaking noncovalent interaction pairs, e.g., hydrogen bonds. If these residues are rigidly connected to the protein body, one can treat the process as two rigid bodies moving relative to each other. At a given instance moving of the two surfaces requires breaking of all the previously formed interaction pairs (see Figure 4A). The repetitive breaking and forming interaction pairs result in rugged potentials along the moving coordinate. Effector binding may increase the elasticity of the residue linkages or the protein body. Then the two surfaces can move with some of the existing interaction pairs being stretched but not necessarily broken (see Figure 4B). Formation of new interaction pairs may energetically facilitate eventual broken of these bonds. This increased elasticity effectively smoothen the potential of mean force. Similarly, effector binding induced displacement of some residues may also reduce the average number of interaction pairs formed at a given relative position of the two surfaces. Effector binding may also increase solvent (water) molecule accessibility to the protein interface. Water molecules are effective on bridging interactions between displaced residues, and thus stabilizing the intermediate configurations (see Figure 4D).

3.3 Allosteric regulation of bacterial flagellar motor switching

Here we specifically discuss allosteric regulation in the bacterial flagellar motor system. Although the flagellar motor switching process does not involve enzymatic cycles directly, the process shares some features common to what we discussed in section 3.2.

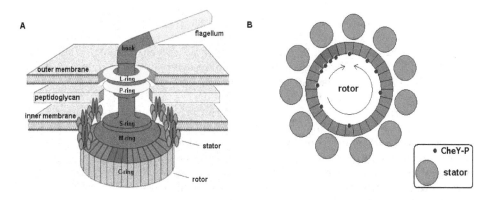

Fig. 5. Cartoon illustrations of the BFM torque generation/switching structure and the concept of conformational spread on the rotor ring (A) Schematic plot of the main structural components of the BFM. In this figure some rotor units (red) are in CW state against majority of the rotor units (blue) driving the motor rotating along CCW direction. (B) Top-view of the rotor ring complex with putative binding positions of the CheY-P molecules.

The bacterial flagellar motor (BFM) is a molecular device most bacteria use to rotate their flagella when swimming in aqueous environment. Using the transmembrane electrochemical proton (or sodium) motive force as the power source, the bacterial flagellar motor can rotate at an impressive high speed of a few hundred Hz and consequently, free-swimming bacteria can propel their cell body at a speed of 15-100 $\mu m/s$, or up to 100 cell body lengths per second (Berg, 2003, 2004; Sowa and Berry, 2008). Figure 5A shows a schematic cartoon plot of the major components of the *E. coli* BFM derived from previous research of electron microscopy, sequencing and mutational studies. These structural components can be categorized into two groups according to their function: the rotor and the stators. In the center of the motor, a long extracellular flagellum (about 5 or 10 times the length of the cell body) is connected to the basal body of the motor through a flexible hook domain. The basal body consists of a few protein rings, functioning as the rotor of the machine, and spans across the outer membrane, peptidoglycan and inner membrane into the cytoplasm of the cell (Berg, 2004). Around the periphery of the rotor, a circular array of 8-11 stator complexes are located. Each stator complex functions independently as a torque generation unit. When ions (proton or sodium) flow from periplasm to cytoplasm through an ion channel on the stator complex, conformational changes are triggered by ion binding on/off events, and therefore deliver torque to the rotor at the interface between the cytoplasmic domain of the stator complex and C-terminal domain of one of the 26 copies of FliG monomers on the rotor (Sowa et al., 2005). A series of mathematical models haven been proposed to explain the working mechanism of the BFM (Bai et al., 2009; Meacci and Tu, 2009; Mora et al., 2009; Xing et al., 2006).

The bacterial flagellar motor is not only important for the propulsion of the cell, but also crucial for bacterial chemotaxis. In the *E. coli* chemotaxis system, chemical gradients (attractant or repellent) are sensed through multiple transmembrane methyl-accepting chemotaxis proteins (MCPs) (Berg, 2004). When extracellular chemotactic attractants (or repellents) bind to MCPs, conformational changes through the membrane inhibit (or trigger) the autophosphorylation in the histidine kinase, CheA. CheA in turn transfers phosphoryl groups to conserved aspartate residues in the response regulators CheY. The phosphorylated form of CheY, CheY-P, diffuses away across the cytoplasm of the cell and binds to the bottom of the FliM/FliN complex of the flagellar motor. When attractant gradient is sensed, CheY-P concentration is low in the cytoplasm and therefore less CheY-P molecules bind to the flagellar motor, which favours counter-clockwise (CCW) rotation of the motor. When most of the motors on the membrane spin CCW, flagellar filaments form a bundle and propel the cell steadily forward. When repellent gradient is sensed, CheY-P concentration is raised and more CheY-P binds to the flagellar motor, which leads to clockwise (CW) rotation of the motor. When a few motors (can be as few as one) spin CW, flagellar filaments fly apart and the cell tumbles. The bacterial flagellar motor (BFM) switches stochastically between CCW and CW states and therefore the cell repeats a 'run'-'tumble'-'run' pattern. This enables a chemotactic navigation in a low Reynolds number environment (reviewed in Berg, book E. coli in motion). The ratio of the rotation direction CCW/CW is tuned by the concentration of the signalling protein, CheY-P.

The problem of BFM switching response to cytoplasmic CheY-P concentration is essentially a protein allosteric regulation. When the effector (CheY-P) binds to the bottom of each rotor unit (a protein complex formed by roughly 1:1:1 of FliG, FliM, FliN protein), it makes CW rotation more favourable (Figure 5B). However, a careful examination of the BFM switching shows that the allosteric regulation here has distinct features: 1) in previous *in vivo* experiment (Cluzel et al., 2000), Cluzel et al. monitored in real time the relationship between BFM switching bias and CheY-P concentration in the cell and found that the response curve is ultrasensitive with a Hill coefficient of ~ 10. Later FRET experiment further showed that binding of CheY-P to FliM is much less cooperative than motor switching response (Sourjik and Berg, 2002). The molecular mechanism of this high cooperativity in BFM switching response remains unknown. 2) the BFM rotor has a ring structure, which is a large multisubunit protein complexes formed by 26 identical rotor units. For such a large multisubunit protein complex, an absolute coupling between subunits as the MWC model requires seems very unlikely. 3) the BFM rotates in full speed stably in CCW or CW directions, and transitions between these two states are brief and fast. This indicates that the 26 rotor units on the basal body of the BFM are in a coherent conformation for most of the time and switching of the whole ring can finish within a very short time period. The above facts also put the KNF model in doubt. As in the KNF model, coupling between effector binding and conformation is absolute: When an effector binds a rotor unit, that rotor unit switches direction.

Therefore a new type of model is in needed to explain the molecular mechanism of the BFM switching. Duke et al. constructed a mathematical model of the general allosteric scheme based on the idea proposed by Eigen (Eigen, 1968) in which both types of coupling are probabilistic (Duke et al., 2001; Duke and Bray, 1999). This model encompasses the classical mechanisms at its limits and introduces the mechanism of conformational spread, with domains of a particular conformational state growing or shrinking faster than ligand binding. Particular regions in the parameter space of the conformational spread model reproduce the classical WMC and KNF model (Duke et al., 2001).

Here we introduce the conformational spread model modified for studying the BFM switching mechanism. In this model, first we assumed that each rotor unit can take two conformations: CCW and CW. The rotor unit in CCW state generates torque along CCW direction when interacting with a stator unit; the rotor unit in CW state generates torque along CW direction when interacting with a stator unit. Each rotor unit undergoes rapid flipping between these two conformations and may also bind a single CheY-P molecule. On the free energy diagram, we further assumed that for each rotor unit the CCW state is energetically favoured by Ea while the binding of CheY-P stabilizes the CW state. As shown in Figure 6A, the free energy of the CW state (red) changes from $+E_A$ to $-E_A$ relative to the CCW state (blue), when a rotor unit binds CheY-P.

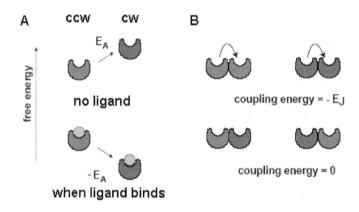

Fig. 6. Energy states of a rotor unit in the BFM switch complex. (A) The free energy of the CW state (red) changes from $+E_A$ to $-E_A$ relative to the CCW state (blue), when a rotor unit binds CheY-P. (B) The rotor unit is stabilized by E_J if the adjacent neighbor is in the same conformation.

In order to reproduce the ultrasensitivity of the BFM switching, a coupling energy E_J between adjacent neighbors in the ring is introduced. The free energy of a rotor unit is further stabilized by a coupling energy E_J when each neighboring rotor unit is in the same conformational state (Figure 6B), an idea inspired by the classical Ising phase transition theory from condensed matter physics.

In this conformational spread model, the rotor ring shows distinct features upon increasing of E_J. Below a critical coupling energy, the ring exhibits a random pattern of states as the rotor units flip independently of each other. Above the critical coupling energy, switch-like behaviour emerges: the ring spends the majority of time in a coherent configuration, either all in CCW or CW states, with abrupt stochastic switching between these two states. Unlike the MWC model, the conformational spread model allows the existence of an intermediate (or mixed) configuration of the rotor units on the ring; and unlike the KNF model, the conformational spread model also allows rotor units stay in its original conformation without being switched by effector binding events. By implementing parallel Monte Carlo processes, one can simulate BFM switching response to CheY-P concentration. In each iteration, each rotor unit on the ring is visited and polled to determine whether to stay in the old state or jump to a new state according to the free energy difference between the two states as a function of 1) free energy of the rotor unit itself 2) binding condition of the regulator molecule CheY-P 3) energy coupling of adjacent neighboring subunits.

The conformational spread model has successfully reproduced previous experimental observations that 1) the BFM switching bias responses ultrasensitively to changes in CheY-P concentration 2) the motor rotates stably in CCW and CW states with occasional fast transitions from one coherent state to the other. The model also made several new predictions: 3) creation of domains of the opposite conformation is frequent due to fast flipping of single rotor unit, but most of them shrink and disappear, failing to occupy the whole ring. However, some big fluctuations can still produce obvious slowdowns and pausing of the motor. Therefore, speed traces of the BFM should have frequent transient speed slowdowns and pauses. 4) the switch interval (the time that the motor spends in the CCW or CW state) follows a single exponential distribution. 5) the switch time, the time that the motor takes to complete a switch, is non-instantaneous. It can be modeled as a biased random walk along the ring. The characteristic switching time depends on the size of the ring and flipping rate of each rotor unit in a complicated manner. Due to the stochastic nature of this conformational spread, we expect to see a wide distribution of switching times.

With the cutting-edge single molecule detection technique, the above predictions of the conformational spread model has recently been confirmed (Bai et al., 2010). Instead of instant transition , switches between CCW and CW rotor states were found to follow a broad distribution, with switching time ranging from less than 2 milliseconds to several hundred milliseconds, and transient intermediate states containing a mixture of CW/CCW rotor units have been observed. The conformational spread model has provided a molecular mechanism for the BFM switching, and more importantly, it sheds light on allosteric regulation in large protein complexes. In addition to the canonical MWC and KNF models, the conformational spread model provides a new comprehensive approach to allostery, and is consistent with the discussion in section 3.2 that both kinetic and thermodynamic aspects should be considered.

4. Coupling between slow conformational change and network dynamics

A biological network usually functions in a noisy ever-changing environment. Therefore, the network should be: 1) robust — functioning normally despite environmental noises; 2) adaptive — the tendency to function optimally by adjusting to the environmental changes; 3) sensitive — sharp response to the regulating signals. It is not-fully understood how a biological network can achieve these requirements simultaneously. Contemporary researches emphasize that the dynamic properties of a network is closely related to its topology.

Many *in vivo* biological processes involve only a small number of substrate molecules. When this number is in the range of hundreds or even smaller, stochastic effect becomes predominant. Chemical reactions take place in a stochastic rather than deterministic way. Therefore one should track the discrete numbers of individual species explicitly in the rate equation formalism. So far, many studies have shown that one might make erroneous conclusions without considering the stochastic effect (Samoilov et al., 2005; Wylie et al., 2007). Noise propagation through a network is currently an important research topic (Levine et al., 2007; Paulsson et al., 2000; Pedraza and van Oudenaarden, 2005; Rao et al., 2002; Rosenfeld et al., 2005; Samoilov et al., 2005; Shibata and Fujimoto, 2005; Suel et al., 2007; Swain et al., 2002). One usually assumes that the stochastic effect mainly arises from small number of identical molecules, and rate constants are still assumed well defined.

With the existence of dynamic disorder, the activity of a single enzyme (and so of a small number of enzymes) is a varying quantity. This adds another noise source with unique (multi-time scale, non-white noise) properties (Min and Xie, 2006; Xing and Kim, 2006). For bulk concentrations, fluctuations due to dynamic disorder are suppressed by averaging over a large number of molecules. However, existence of NMM kinetics can still manifest itself in a network. If there are only a small number of protein molecules, as in many *in vivo* processes, dynamic disorder will greatly affect the network dynamics. The conventionally considered stochastic effect is mainly due to number variations of identical molecules. Here a new source of stochastic effect arises from small numbers of molecules with the same chemical structure but different conformations. Dynamic disorder induced stochastic effect has some unique properties, which require special theoretical treatment, and may result in novel dynamic behaviors. First, direct fluctuation of the rate constants over several orders of magnitude may have dramatic effects on the network dynamics. Second, the associated time scales have broad range. The Gaussian white noise approximation is widely used in stochastic modeling of network dynamics with the assumption that some processes are much faster than others (Gillespie, 2000). Existence of broad time scale distribution makes the situation more complicated. Furthermore, a biological system may actively utilize this new source of noise. Noises from different sources may not necessarily add up. Instead they may cancel each other and result in smaller overall fluctuations (Paulsson et al., 2000; Samoilov et al., 2005). We expect that the existence of dynamic disorder not only further complicates the situation, but may also provide additional degrees of freedom for regulation since the rates can be continuously tuned. Especially we expect that existence of dynamic disorder may require dramatic modification on our understanding of signal transduction networks. Many of these processes involve a small number of molecules, and are featured by short reaction time scales (within minutes), high sensitivity and specificity (responding to specific molecules only).

Wu et al. examined the coupling between enzyme conformational fluctuations and a phosphorylation-dephosphorylation cycle (PdPC) (Wu et al., 2009). The PdPC is a common protein interaction network structure found in biological systems. In a PdPC, the substrate can be in phosphorylated and dephosphorylated forms with distinct chemical properties. The conversions are catalyzed by a kinase (E1 in Figure 7A) and a phosphatase (E2 in Figure 7A) at the expense of ATP hydrolysis. Under the condition that the enzymes are saturated by the substrates, the system shows ultrasensitivity (Goldbeter and Koshland, 1981). As shown in Figure 7B, the fraction of the phosphorylated substrate form, f(W-P), is close to zero if the ratio between E1 and E2 enzymatic activities $\theta < 1$, but close to 1 if $\theta > 1$. Now consider a system with a finite size, e.g. 50 E1 molecules, 50 E2 molecules, and a total of 1500 substrate molecules as used to generate results in Figure 7C & D. Enzyme activities fluctuate due to conformational fluctuations. For simplicity let us assume that E1 can stochastically convert between an active and a less active forms. While the average value of θ = 1.1, it fluctuates within the range [0.7, 1.5], depending on the number of E1 molecules in the active form. For convenience of discussion, let us also define the response time of the PdPC, τ, as the time it takes for the fraction of W-P reaching half given at time 0 the system jumps from $\theta < 1$ to $\theta > 1$ due to enzyme conformational fluctuations. The response time clearly related to the enzymatic turnover rate. As the trajectories in Figure 7C show, for slow θ fluctuation $\Delta\theta$ is amplified to ΔW-P due to the ultrasensitivity of the PdPC and the much larger number of substrate molecules compared to enzymes. However, with θ fluctuation much faster than τ, the PdPC only responses to the average value of θ. Therefore depending on the relative time scale between θ fluctuation and the response time of the PdPC to θ change, fluctuations of θ can be either amplified or suppressed.

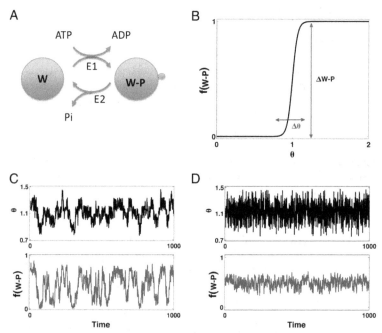

Fig. 7. Coupling between enzyme conformational fluctuations and a phosphorylation-dephosphorylation cycle. (A) A phosphorylation-dephosphorylation cycle (PdPC). (B) Ultrasensitivity of a PdPC. (C) Trajectories of enzyme activity due to slow conformational fluctuations and the corresponding substrate fluctuation. (D) Similar to C but with fast conformational fluctuations.

5. Conclusion

Slow conformational motions in macromolecules play crucial roles in their unique function in enzymatic reactions as well as biological networks. We suggest that these motions are of great functional importance, which can only be fully appreciated in the context of regulatory networks. Collaborative researches from molecular and cellular level studies are urgently needed for this largely unexplored area.

6. Acknowledgment

ZW and JX were supported by an NSF grant (EF-1038636) and a grant from the William and Mary Jeffress Memorial Trust.

7. References

Ainslie, G.R., Jr., Shill, J.P., and Neet, K.E. (1972). Transients and Cooperativity. A slow transition model for relating transients and cooperative kinetics of enzymes. J Biol Chem 247, 7088-7096.

Alberts, B., Johnson, A., Lewis, J., Raff, M., Roberts, K., and Walter, P. (2002). Molecular Biology of the Cell, 4d edn (New York, Garland).

Alon, U. (2007). An introduction to systems biology: Design principles of biological circuits, 1 edn (Chapman and Hall/CRC).

Amaro, R.E., Cheng, X., Ivanov, I., Xu, D., and McCammon, J.A. (2009). Characterizing Loop Dynamics and Ligand Recognition in Human- and Avian-Type Influenza Neuraminidases via Generalized Born Molecular Dynamics and End-Point Free Energy Calculations. J Am Chem Soc 131, 4702-4709.

Amaro, R.E., Sethi, A., Myers, R.S., Davisson, V.J., and Luthey-Schulten, Z.A. (2007). A Network of Conserved Interactions Regulates the Allosteric Signal in a Glutamine Amidotransferase,Ä†. Biochemistry 46, 2156-2173.

Austin, R.H., Beeson, K.W., Eisenstein, L., Frauenfelder, H., and Gunsalus, I.C. (1975). Dynamics of Ligand-Binding to Myoglobin. Biochemistry 14, 5355-5373.

Bai, F., Branch, R.W., Nicolau, Dan V.,,, Jr., Pilizota, T., Steel, B.C., Maini, P.K., Berry, R.M. (2010). Conformational Spread as a Mechanism for Cooperativity in the Bacterial Flagellar Switch. Science 327, 685-689.

Bai, F., Lo, C.-J., Berry, R.M., and Xing, J. (2009). Model Studies of the Dynamics of Bacterial Flagellar Motors. Biophys J 96, 3154-3167.

Balabin, I.A., Yang, W., and Beratan, D.N. (2009). Coarse-grained modeling of allosteric regulation in protein receptors. Proceedings of the National Academy of Sciences 106, 14253-14258.

Behzadi, A., Hatleskog, R., and Ruoff, P. (1999). Hysteretic enzyme adaptation to environmental pH: change in storage pH of alkaline phosphatase leads to a pH-optimum in the opposite direction to the applied change. Biophys Chem 77, 99-109.

Berg, H.C. (2003). The Rotary Motor of Bacterial Flagella. Annu Rev Biochem 72, 19-54.

Berg, H.C. (2004). E. Coli in Motion (New York, Springer-Verlag Press).

Boehr, D.D., McElheny, D., Dyson, H.J., and Wright, P.E. (2006). The Dynamic Energy Landscape of Dihydrofolate Reductase Catalysis. Science 313, 1638-1642.

Bustamante, C., Keller, D., and Oster, G. (2001). The physics of molecular motors. Acc Chem Res 34, 412-420.

Cecchini, M., Houdusse, A., and Karplus, M. (2008). Allosteric Communication in Myosin V: From Small Conformational Changes to Large Directed Movements. PLoS Comput Biol 4, e1000129.

Cluzel, P., Surette, M., and Leibler, S. (2000). An Ultrasensitive Bacterial Motor Revealed by Monitoring Signaling Proteins in Single Cells. Science 287, 1652-1655.

Cole, R., and Loria, J.P. (2002). Evidence for Flexibility in the Function of Ribonuclease A,Ä†. Biochemistry 41, 6072-6081.

Cooper, A., and Dryden, D.T.F. (1984). Allostery without Conformational Change - a Plausible Model. Eur Biophys J Biophys Lett 11, 103-109.

Cornish-Bowden, A., and Cardenas, M.L. (1987). Co-operativity in monomeric enzymes. J Theor Biol 124, 1-23.

Cui, Q., and Karplus, M. (2008). Allostery and cooperativity revisited. Protein Science 17, 1295-1307.

Duke, T., Novere, N.L., and Bray, D. (2001). Conformational spread in a ring of proteins: A stochastic approach to allostery. J Mol Biol 308.

Duke, T.A.J., and Bray, D. (1999). Heightened sensitivity of a lattice of membrane receptors. Proc Nat Acad Sci USA 96, 10104-10108.

Eigen, M. (1968). Kinetics of reaction control and information transfer in enzymes and nucleic acids. In Nobel Symp 5 on Fast Reactions and Primary Processes, Chem Kinetics, S. Claesson, ed. (Stockholm, Almquist & Wiksell).

English, B.P., Min, W., van Oijen, A.M., Lee, K.T., Luo, G.B., Sun, H.Y., Cherayil, B.J., Kou, S.C., and Xie, X.S. (2006). Ever-fluctuating single enzyme molecules: Michaelis-Menten equation revisited. Nat Chem Biol 2, 87-94.

Frauenfelder, H., Sligar, S.G., and Wolynes, P.G. (1991). The Energy Landscapes and Motions of Proteins. Science 254, 1598-1603.

Frauenfelder, H., Wolynes, P.G., and Austin, R.H. (1999). Biological physics. Rev Mod Phys 71, S419-S430.

Frieden, C. (1970). Kinetic Aspects of Regulation of Metabolic Processes. The hysteretic enzyme concept. J Biol Chem 245, 5788-5799.

Frieden, C. (1979). Slow Transitions and Hysteretic Behavior in Enzymes. Ann Rev Biochem 48, 471-489.

Gillespie, D.T. (2000). The chemical Langevin Equation. J Chem Phys 113, 297-306.

Goldbeter, A., and Koshland, D.E. (1981). An Amplified Sensitivity Arising from Covalent Modification in Biological-Systems. Proc Natl Acad Sci USA 78, 6840-6844.

Goodey, N.M., and Benkovic, S.J. (2008). Allosteric regulation and catalysis emerge via a common route. Nat Chem Biol 4, 474-482.

Gopich, I.V., and Szabo, A. (2006). Theory of the statistics of kinetic transitions with application to single-molecule enzyme catalysis. J Chem Phys 124, 154712.

Grote, R.F., and Hynes, J.T. (1980). The stable states picture of chemical-reactions 2: Rate constants for condensed and gas-phase reaction models. J Chem Phys 73, 2715-2732.

Hanggi, P., Talkner, P., and Borkovec, M. (1990). Reaction-rate theory: 50 years after Kramers. Rev Mod Phys 62, 254-341.

Horovitz, A., and Willison, K.R. (2005). Allosteric regulation of chaperonins. Current Opinion in Structural Biology 15, 646-651.

Julicher, F., Ajdari, A., and Prost, J. (1997). Modeling molecular motors. Rev Mod Phys 69, 1269-1281.

Kern, D., and Zuiderweg, E.R.P. (2003). The role of dynamics in allosteric regulation. Curr Opin Struc Biol 13, 748-757.

Kholodenko, B.N. (2006). Cell-signalling dynamics in time and space. Nat Rev Mol Cell Biol 7, 165-176.

Kim, S.H., Wang, W.R., and Kim, K.K. (2002). Dynamic and clustering model of bacterial chemotaxis receptors: Structural basis for signaling and high sensitivity. Proc Natl Acad Sci USA 99, 11611-11615.

Koshland, D.E., Nemethy, G., and Filmer, D. (1966). Comparison of Experimental Binding Data and Theoretical Models in Proteins Containing Subunits. Biochemistry 5, 365-385.

Kroh, H.K., Panizzi, P., and Bock, P.E. (2009). Von Willebrand factor-binding protein is a hysteretic conformational activator of prothrombin. Proc Natl Acad Sci U S A 106, 7786-7791.

Levine, J., Kueh, H.Y., and Mirny, L. (2007). Intrinsic fluctuations, robustness, and tunability in signaling cycles. Biophys J 92, 4473-4481.

McCammon, J.A., and Harvey, S.C. (1987). Dynamics of Proteins and Nucleic Acids (New York, Cambridge Univ Press).

Meacci, G., and Tu, Y. (2009). Dynamics of the bacterial flagellar motor with multiple stators. Proc Natl Acad Sci USA 106, 3746-3751.

Menten, L., and Michaelis, M.I. (1913). Die Kinetik der Invertinwirkung. Biochem Z 49, 333-369.

Min, W., English, B.P., Luo, G.B., Cherayil, B.J., Kou, S.C., and Xie, X.S. (2005a). Fluctuating enzymes: Lessons from single-molecule studies. Acc Chem Res 38, 923-931.

Min, W., Gopich, I.V., English, B.P., Kou, S.C., Xie, X.S., and Szabo, A. (2006). When Does the Michaelis-Menten Equation Hold for Fluctuating Enzymes? J Phys Chem B 110, 20093-20097.

Min, W., Luo, G.B., Cherayil, B.J., Kou, S.C., and Xie, X.S. (2005b). Observation of a power-law memory kernel for fluctuations within a single protein molecule. Phys Rev Lett 94, 198302.

Min, W., and Xie, X.S. (2006). Kramers model with a power-law friction kernel: Dispersed kinetics and dynamic disorder of biochemical reactions. Phys Rev E 73, -.

Min, W., Xie, X.S., and Bagchi, B. (2008). Two-Dimensional Reaction Free Energy Surfaces of Catalytic Reaction: Effects of Protein Conformational Dynamics on Enzyme Catalysis. J Phys Chem B 112, 454-466.

Monod, J., Wyman, J., and Changeux, J.P. (1965). On Nature of Allosteric Transitions - a Plausible Model. J Mol Biol 12, 88-118.

Mora, T., Yu, H., and Wingreen, N.S. (2009). Modeling Torque Versus Speed, Shot Noise, and Rotational Diffusion of the Bacterial Flagellar Motor. Phys Rev Lett 103, 248102.

Pan, H., Lee, J.C., and Hilser, V.J. (2000). Binding sites in Escherichia coli dihydrofolate reductase communicate by modulating the conformational ensemble. Proceedings of the National Academy of Sciences 97, 12020-12025.

Paulsson, J., Berg, O.G., and Ehrenberg, M. (2000). Stochastic focusing: Fluctuation-enhanced sensitivity of intracellular regulation. Proc Natl Acad Sci USA 97, 7148-7153.

Pedraza, J.M., and van Oudenaarden, A. (2005). Noise Propagation in Gene Networks. Science 307, 1965-1969.

Qian, H., Shi, P.-Z., and Xing, J. (2009). Stochastic bifurcation, slow fluctuations, and bistability as an origin of biochemical complexity. Phys Chem Chem Phys 11, 4861-4870.

Ranson, N.A., Clare, D.K., Farr, G.W., Houldershaw, D., Horwich, A.L., and Saibil, H.R. (2006). Allosteric signaling of ATP hydrolysis in GroEL-GroES complexes. Nat Struct Mol Biol 13, 147-152.

Rao, C.V., Wolf, D.M., and Arkin, A.P. (2002). Control, exploitation and tolerance of intracellular noise. Nature 420, 231-237.

Ricard, J., and Cornish-Bowden, A. (1987). Co-operative and allosteric enzymes: 20 years on. Eur J Biochem 166, 255-272.

Rosenfeld, N., Young, J.W., Alon, U., Swain, P.S., and Elowitz, M.B. (2005). Gene Regulation at the Single-Cell Level. Science 307, 1962-1965.

Samoilov, M., Plyasunov, S., and Arkin, A.P. (2005). Stochastic amplification and signaling in enzymatic futile cycles through noise-induced bistability with oscillations. Proc Natl Acad Sci USA 102, 2310-2315.

Shibata, T., and Fujimoto, K. (2005). Noisy signal amplification in ultrasensitive signal transduction. Proc Natl Acad Sci USA 102, 331-336.

Sourjik, V., and Berg, H.C. (2002). Binding of the Escherichia coli response regulator CheY to its target measured in vivo by fluorescence resonance energy transfer. Proc Natl Acad Sci USA 99, 12669-12674.

Sowa, Y., and Berry, R.M. (2008). Bacterial flagellar motor. Quart Rev Biophys 41, 103-132.

Sowa, Y., Rowe, A.D., Leake, M.C., Yakushi, T., Homma, M., Ishijima, A., and Berry, R.M. (2005). Direct observation of steps in rotation of the bacterial flagellar motor. Nature 437, 916-919.

Suel, G.M., Kulkarni, R.P., Dworkin, J., Garcia-Ojalvo, J., and Elowitz, M.B. (2007). Tunability and noise dependence in differentiation dynamics. Science 315, 1716-1719.

Swain, P.S., Elowitz, M.B., and Siggia, E.D. (2002). Intrinsic and extrinsic contributions to stochasticity in gene expression. Proc Natl Acad Sci USA 99, 12795-12800.

Tyson, J.J., Chen, K., and Novak, B. (2001). Network dynamics and cell physiology. Nat Rev Mol Cell Biol 2, 908-916.

Volkman, B.F., Lipson, D., Wemmer, D.E., and Kern, D. (2001). Two-state allosteric behavior in a single-domain signaling protein. Science 291, 2429-2433.

Wang, H., and Oster, G. (1998). Energy transduction in the F1 motor of ATP synthase. Nature 396, 279-282.

Wu, Z., Elgart, V., Qian, H., and Xing, J. (2009). Amplification and Detection of Single-Molecule Conformational Fluctuation through a Protein Interaction Network with Bimodal Distributions. J Phys Chem B 113, 12375-12381.

Wylie, D.C., Das, J., and Chakraborty, A.K. (2007). Sensitivity of T cells to antigen and antagonism emerges from differential regulation of the same molecular signaling module. Proc Natl Acad Sci USA, 0611482104.

Xie, X.S., and Lu, H.P. (1999). Single-molecule enzymology. J Biol Chem 274, 15967-15970.

Xing, J. (2007). Nonequilibrium dynamic mechanism for allosteric effect. Phys Rev Lett 99, 168103.

Xing, J., Bai, F., Berry, R., and Oster, G. (2006). Torque-speed relationship for the bacterial flagellar motor. Proc Natl Acad Sci USA 103, 1260-1265.

Xing, J., and Kim, K.S. (2006). Protein fluctuations and breakdown of time-scale separation in rate theories. Phys Rev E 74, 061911.

Xing, J., Wang, H.-Y., and Oster, G. (2005). From continuum Fokker-Planck models to discrete kinetic models. Biophys J 89, 1551-1563.

Yu, E.W., and Koshland, D.E. (2001). Propagating conformational changes over long (and short) distances in proteins. Proc Natl Acad Sci USA 98, 9517-9520.

Zwanzig, R. (1988). Diffusion in a Rough Potential. Proc Natl Acad Sci USA 85, 2029-2030.

Zwanzig, R. (1990). Rate-Processes with Dynamic Disorder. Acc Chem Res 23, 148-152.

Zwanzig, R. (2001). Nonequilibrium statistical mechanics (Oxford, Oxford University Press).

10

Structural Bioinformatics of Proteins: Predicting the Tertiary and Quaternary Structure of Proteins from Sequence

J. Planas-Iglesias, J. Bonet,
M.A. Marín-López, E. Feliu, A. Gursoy and B. Oliva
Structural Bioinformatics Lab. Universitat Pompeu Fabra, Catalunya
Spain

1. Introduction

Although the regulatory role of non-coding nucleic acids is currently being unraveled, the role of proteins is still a major issue as they mediate most biological functions. Thus, understanding how proteins fulfill their intricate functions is one of the most relevant current challenges in biology. It is well known that a protein's function is determined by its three-dimensional (3D) structure known as tertiary structure, which in turn is mainly dictated by its sequence (Thornton and cols. reviewed this issue in detail in (Watson et al., 2005)). Despite the exponential increase of available sequences and 3D structures, the number of sequences highly exceeds that of 3D structures. This difference in numbers is proportional to the disparity of the costs for experimentally obtaining either the sequence or the structure of a protein. Therefore, covering the gap between sequence and structure becomes a compelling requirement to achieve a molecular understanding of the protein function. Theoretical methods can help to bridge this gap by inferring the 3D structure from the sequence. These methods are classified into three different groups: comparative modeling, fold recognition and new fold or *ab initio* methods.

Besides the tertiary structure of a protein, other contextual factors may modulate its function. Among these, the ability of the proteins to interact with others and the particular partners with which they form complexes are one of the most important. This is because proteins rarely act alone; they rather constitute a mingled network of physical interactions, some times to form large macro-complexes and sometimes to produce transient interactions. In this context, understanding the function of a protein implies to recognize its partners and to grasp how they associate, even at the atomic level. The structure of these complexes is known as quaternary structure. To this end, computational techniques have been developed to dock one protein onto another (Janin, 2010; Vajda and Kozakov, 2009), and can help to infer 3D structure of a protein from the knowledge of the protein interactions (Fornes et al., 2009) and vice-versa (Stein et al., 2005). Furthermore, the combined use of data from multiple resources allows us to obtain an accurate model of large molecular complexes such as nucleopore (Alber et al., 2007).

There are two strategies for modeling the interaction between two proteins from sequence data. The first one is to model the unbound interactors and to dock them into the final complex (i.e. solving first the tertiary structure of the proteins and afterwards the quaternary). The second is to model the interacting pair or complex from the scratch, using as template the structural knowledge of an available homologous interacting pair (interolog, (Matthews et al., 2001)). When the template is not available the strategy can only play with the docking of the unbound partners. Figure 1 summarizes these possibilities. Here, we will cover these strategies and methods to infer and assess the 3D structure of binary protein interactions, and we will review the existing techniques to model large cellular macro-complexes.

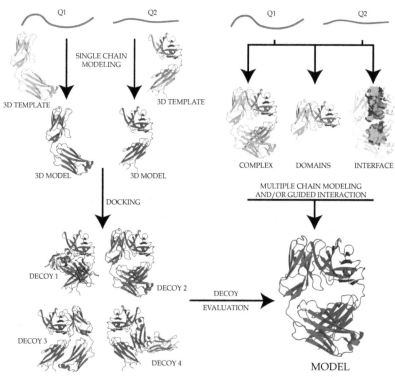

Fig. 1. Different strategies for modeling a protein interaction: The 3D structure of a binary protein interaction can be inferred by modeling individual interacting partners apart and subsequently docking them (left side) or modeling the interaction with one template, taking advantage of the available information of homologous complexes. Templates can be obtained from structural resources of information containing the full complex, a partial complex (in general formed by interacting domains) or with only the interacting interface (right side).

2. Modeling the tertiary structure of proteins

In order to obtain a complete model of a protein interaction, the interacting partners can be modeled separately and then docked into a functional complex. The first step of this approach is to obtain the 3D structure of each of the interacting partners. Comparative modeling, fold recognition, and *ab initio* (for new folds) are computational methods that may overcome the lack of experimental structural data. Models obtained using these approaches may be further assessed, in order to ensure that the inferred 3D structure contains no errors (see section 4). In case of persistence, such errors would damper the deduction of further biological conclusions such as the mode the modeled protein can interact with others. Figure 2 summarizes the different steps and strategies that can be exploited to achieve these objectives.

2.1 Homology modeling

Homology or comparative modeling techniques are those devoted to the prediction and construction of the 3D conformation of proteins. These methods are based on the assumption that structural features in proteins are more conserved than its sequences. Thus, two proteins with enough sequence similarity will fold in a similar way and share the same conformation in space. The process through which a tertiary structure is assigned to a given sequence is carried out in three steps, namely: template identification, template alignment, and model building. Finally, the produced model should be assessed (see section 4).

Template identification is the key step in the molecular modeling process. Templates are defined as the set of known structures used to build the tertiary structure of the query (target or problem protein). Known 3D data of proteins is stored in the Protein Data Bank (PDB) (Berman et al., 2000). Thus, the identification of the template refers to the process of identifying the structure of the PDB whose sequence is the closest homolog of the target. Such sequence homology search can be performed using sequence alignment tools like BLAST and PSI-BLAST (Altschul et al., 1997), or Hidden Markov Model (HMM) profile methods like HMMER (Eddy, 1998). While BLAST will reveal if there is any relatively close homolog to our query, PSI-BLAST and HMMER will also reveal the possibility of remote homologues. The homology threshold that can be used to define whether or not a template assignation is correct may be fuzzy. Those templates assigned with low percentage of identity, low homology, or in short parts of the sequence fall into what is known as the twilight zone (Rost, 1999). Some rules have been described to shed some light into that twilight region in order to better describe the viability of a template for a given query (Fornes et al., 2009).

Provided that a good template has been selected, the sequence alignment between the query and the template can be directly extracted or easily inferred (in case of the HMM) from the template search. Depending on specific requirements, the alignments can be redone with other sequence alignment methods such as CLUSTALW (Chenna et al., 2003) or T-COFFE (Notredame et al., 2000). Additionally, some methods optimize the sequence alignment through a genetic algorithm protocol that iterates the alignment, model building and model evaluation in order to obtain the best possible alignment (Fernandez-Fuentes et al., 2007) .

Model building is the process by which the three-dimensional data of the template(s) is applied on the query sequence. MODELLER is one of the most used and comprehensive pieces of modeling software (Sali et al., 1995). Provided the sequence alignment the modeling process is practically automatic. As many other modeling tools, it is based on satisfying a set of spatial constraints. Specifically it satisfies three spatial constraints being (1) homology-derived constraints, (2) stereochemical constraints such as bond angles, and (3) statistical preferences for dihedral angles and non-bonded interatomic distances.

Optionally, manually curated restraints from secondary structure packing to site-directed mutagenesis can be added to the modeling process.

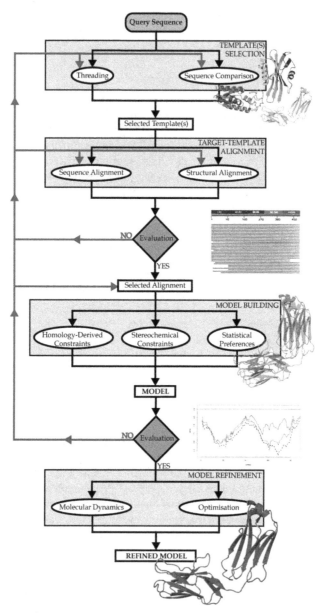

Fig. 2. Flowchart for single protein modeling: Scheme of the methods used for modeling, comprising template(s) selection, template-target alignment, model building, model evaluation, and model refinement steps.

2.2 Threading

In the same way as homology modeling, threading is a method to determine the tertiary structure of a protein based on the fairly small number of different folds contained in nature. The main difference resides in the fact that threading does not use specific protein 3D structures as templates but uses statistical knowledge extracted from all structures in PDB. Thus, threading is especially useful when no suitable template for the protein can be found. Basically, the prediction is made by aligning each amino acid of a given query sequence to a position in a set of structural templates. Once the optimal template is selected through this method the structural model is built according to the alignment between the query and the template. The threading process can be divided in four different steps: the template database construction, the scoring function, the threading alignment and the threading prediction. Among the most used programs on threading and fold prediction are GeneThreader (Jones, 1999) and Phyre (Kelley and Sternberg, 2009).

Gathering representative structures for the different folds avoiding redundancy creates the template database. This means extracting all the structures from PDB and picking one representative of each known fold, eliminating redundancy and sequence homology. The homology filtering is key to ensure that the predictions over the query sequence are not going to be biased because of the database composition.

Designing an optimal function to score the suitability of the templates for modeling the query protein is determinant. The scoring function should be based on the known relationships between structure and sequence. A good scoring function should contain as much information as possible such as pairwise potential, secondary structure compatibilities, environment fitness potential, and gap penalties. The accuracy of the alignment and the prediction will be directly related to the quality of the scoring function. During the threading alignment the query sequence is going to be tested against each given possible template. This part of the process is, by far, the most computationally costly as it takes into account pairwise contact potentials (see section 4.2) and cannot be substituted by the classic dynamic programming algorithm for sequence alignment. Finally the threading prediction uses the scoring function and all the provided alignments to select the better template and build the protein model by placing the backbone atoms of the query according to the position of their aligned counterparts in the template.

2.3 *Ab Initio* methods

Ab initio or *de novo* protein structure prediction tries to predict the tertiary structure of proteins directly from its sequence properties. The idea is that the structure of proteins can be determined without any explicit templates by means of applying the general principles that govern protein folding and the statistical tendencies of conformational features gathered from structural knowledge. Those predictions involve sampling the conformational space, which means that a large set of decoys (structural candidates) is likely to be generated. Scoring functions, either physics-based or knowledge based, are then used to select those decoys that can be identified as more native-like conformations. Optionally, high-resolution refinement is used to optimize those native-like structures. Few programs have been successful in this task, and among them the most flourishing are ROSETTA (Leaver-Fay et al., 2011) and TASSER (Chen and Skolnick, 2008).

3. Modeling the quaternary structure of proteins

As the structural data on protein complexes keep increasing steadily, using known protein complex structures has become an important approach for modeling protein interactions. This increase of structural knowledge of protein complexes (even if it is only partial) opens a new window of possibilities to infer the quaternary structure of proteins. However, for a large quantity of protein complexes this knowledge is still limited, and alternative techniques are required to infer their 3D structure. Docking methods surmount this lack of data providing predictions of the quaternary structure of the complex based on the physical, chemical, and biological known properties of protein complexes. New approaches have introduced the possibility to integrate different sources of experimental information, such as high-resolution electron-microscopy, SAXS, NMR, yeast-two-hybrid, and affinity purifications to extract restraints that can be applied to model the quaternary structure of macro-complexes (Alber et al., 2007).

3.1 Comparative modeling of protein binary complexes

Provided that a homolog structure of an interaction is known, homology modeling can be used to model a protein-protein interaction of interest. Two different approaches can be taken: (1) direct interaction modeling or (2) protein modeling and reorientation (see Figure 1).

When directly modeling a protein interaction, it has to be taken into account that both query proteins need to have a couple of acceptable templates that share the same crystal structure. If that is the case, MODELLER is able to directly model the protein-protein interaction. However, in not only each separate structure needs to be evaluated but also the interface created between them (see section 4.2).

An alternative is to model each protein separately and afterwards use a known interaction as a guide to reposition each structure in the way the interaction is supposed to be taking place. To do so it is required to perform a structural alignment between the model and the template for the interaction. That can be done with strictly devoted tools such as STAMP (Russell and Barton, 1992) or through a variety of protein structure graphical interfaces such as PYMOL (http://www.pymol.org). This approach should be selected if the resolution of the structure of the templates in the interaction is largely worse than the unbound templates. However, it has to be taken into account the need to introduce some structural flexibility produced to construct the interaction. Considering the principal motions and intrinsic fluctuations to accommodate the unbound structures (Dobbins et al., 2008) may help to this purpose. The final structure needs to be refined, and additional restraints are applied to keep the partners on the orientation defined by the template of the binary complex.

3.2 Modeling of protein binary complexes from partial structural information

The sequence and structural homology methods described in the previous section require global similarity (sequence or structure). However, recent research shows that the binding sites of proteins are somewhat more distinguishable from the rest of the protein surface. The binding site of two interacting proteins is called a protein-protein interface. If the structure of a protein complex is available, determining the interface is fairly simple. The interface can be found either by finding contacting residues (distance based) or by calculating the accessible surface area of the residues. Since proteins interact through interfaces, physico-chemical properties of interfaces are important to study protein interactions. Statistical studies of known protein complexes have revealed general characteristics of interfaces. Interfaces in general have electrostatic and shape complementary. Compared to the rest of the protein surface, interfaces are found to be slightly more conserved (Caffrey et al., 2004). Depending on the interaction type, properties of interfaces display variation. Homo-oligomeric complexes have more hydrophobic and larger interfaces than the hetero complexes. Homo-oligomers are usually permanent and their interfaces resemble interior of globular proteins. Transient interactions, on the other hand, are mediated by smaller interfaces (less than 1500 Å^2) and have more polar and charged amino acids than the interfaces of permanent interactions (Nooren and Thornton, 2003). The small surface-area of transient interfaces are partly due to requirement of individual partners of the interaction to fold independently and to be soluble. The secondary structure content of interfaces shows differences between permanent and transient interfaces as well. For example, turns are observed more frequently in non-obligatory interfaces since flexibility is required to repeatedly associate/disassociate (De et al., 2005). Even within an interface, the properties and organization of residues are not uniform. The interface area may be dissected into regions where a set of buried residues forming a core region is surrounded by a rim of residues that are partially solvent accessible. The composition of residues are distinct between these two regions (Guharoy and Chakrabarti, 2005). Alanine scanning mutagenesis of interface residues has also revealed that some residues contribute more to the binding energy (Clackson and Wells, 1995). These areas, called hot spots, are particularly enriched in Trp, Tyr, and Arg residues and are structurally conserved, which can be used to differentiate binding sites from the rest of the surface (Ma et al., 2003).

All these characteristics can be used to identify binding sites of proteins either from sequence (Ofran and Rost, 2007) or from unbound structures (Neuvirth et al., 2004), and potentially for modeling protein interactions. Therefore, a systematic collection and categorization of protein interfaces play important role. Several databases of interfaces have been compiled along this direction, including PiBASE (Davis and Sali, 2005), SCOWLP (Teyra et al., 2006), SCOPPI (Winter et al., 2006) and PRINT (Tuncbag et al., 2008). These databases, in general, present interfaces extracted form known protein complexes together with features of the interfaces such as change in accessible surface area, conservation, or residue composition (reviewed in (Tuncbag et al., 2009)). PRINT database presents all interfaces from PDB (as of 2006) clustered by structural similarity where each cluster represents a different interface architecture. Some interface architectures are observed to be more favorable and reused frequently. These interface architectures are found to be similar to domain folds, consistent with earlier studies indicating that the folding and binding are similar processes (Tsai et al., 1997).

The analysis of protein interactions and interfaces has suggested that the number of possible interfaces is much smaller than the possible number of protein interactions (Aloy and Russell, 2004; Tuncbag et al., 2008). In addition, interfaces are observed to be reused in different protein interactions that are not globally similar (the same interface used by proteins with different fold architectures) (Keskin et al., 2004). This information can be used to overcome the global similarity of requirement of the homology based modeling methods.

That is, modeling protein interactions using only the similarity of protein interfaces. PRISM (Aytuna et al., 2005) is one of the first approaches that has used interface similarity along this direction. It was originally developed to predict protein interactions between proteins (target set) from a set of known protein interfaces (template set). If the two complementary sides of a template interface are found to be structurally similar to two target proteins (one side on one protein, the other on another protein), then the proteins are predicted to be interacting and modeled using the binding site dictated by the template interface. A schematic description of the method is illustrated in Figure 3. Putative interactions are then re-ordered by flexible refinement. PRISM protocol is a collection of scripts that performs a) preparing a set of template interfaces from known complexes, b) preparing surfaces of target proteins that interactions among them to be predicted, c) structural alignment of templates to targets, d) scoring with flexible refinement. The method can be used to model a protein interaction by selecting the two potentially interacting proteins as targets, and using all non-redundant interfaces as the template set. Although the method is limited by the availability and coverage of known protein-protein interfaces from PDB, the continuous growth of the PDB database will increase the applicability of the method. In fact, a recent study on the structural coverage of known protein interfaces already points out that the coverage is close to complete (Gao and Skolnick, 2010).

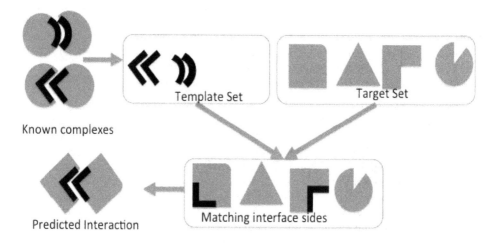

Fig. 3. Schematic representation of PRISM: Two target proteins are predicted to interact if the two complementary sides of a template interface are found to be structurally similar to them (a different side on each protein).

3.3 Protein-protein docking

In contrast to previously described methods (which are based on the structural knowledge of the interaction), protein docking is one of the computational techniques for elucidating the structures of binary bio-molecules (e.g. two proteins) when experimental data regarding the structure of the complex is lacking but the structures of the interacting proteins are known. Docking techniques sample the orientation of two unbound protein structures to produce several predictions about their interaction, followed by a scoring step to rank the predictions. These methods were introduced in 1978 (Wodak and Janin, 1978). Since then, docking algorithms have substantially improved, with a breakthrough in algorithm speed given by the introduction of the Fast Fourier Transform (FFT) (Katchalski-Katzir et al., 1992) (e.g. FTDock (Gabb et al., 1997), ZDock (Mintseris et al., 2007), PIPER (Kozakov et al., 2006)), and by some other very successful geometry-based methods (e.g. FRODOCK (Garzon et al., 2009), Hex (Ritchie and Kemp, 2000), MolFit (Katchalski-Katzir et al., 1992)). A docking procedure usually involves several steps (Vajda and Kozakov, 2009). First, a rigid-docking search is performed by treating the two proteins as rigid bodies. One of the proteins, called the receptor, is kept fixed while the other protein, the ligand, is rotated and translated around the first. Next, further refinement of some structures takes place, allowing changes in conformation of the two unbound structures upon binding (Dobbins et al., 2008; Shen et al., 2008); this step may or may not be supported by experimental evidence.

3.3.1 The ranking problem

Docking methods yield a large number of output conformations (ranging from 10000 to more than 50000), which include a large number of false positives. Thus, a crucial point after a rigid-docking search is the discrimination of near-native structures for further consideration and refinement. The number of selected conformations typically spans from 10 to 2000. There are two non-excluding strategies to perform such selection. The first strategy is to re-rank the docked conformations with a scoring function, which is supposed to rank near-native structures at the top (i.e. describe the molecular environment of the molecular interaction). Scoring functions are usually built upon different properties of protein-protein interactions observed in known binary complexes. These properties include physical and chemical characteristics of the binding site, at the level of residue or atomic contacts (Z-rank (Pierce and Weng, 2007), Fold X (Guerois et al., 2002)). Among these scoring functions, statistical potential is a term that refers to a knowledge-based scoring function that depends on specific properties of known protein-protein interactions stored in some database. Initially, statistical potentials were derived in order to distinguish a correct protein fold (i.e. near-native) of a model from a plethora of generated solutions (see section 4.2). In contrast to atomistic-detailed scoring functions, statistical potentials represent a much faster approach to solve this problem. It has been recently shown that the performance of split statistical potentials to rank docking poses (see following sections) may surpass that of scoring functions encoding atomistic energy terms or other statistical potentials (Feliu et al., 2011).

The atomistic scoring potentials of Z-rank and FoldX split the score into a linear combination of energetic terms and further obtained the best parameterization. In FoldX (1) the energy terms were split in the van der Waals (Gvdw), electrostatic (Gel), solvation (Gsolv) and hydrogen bonding (GHbond) contributions, and the entropy was also included. Some of these terms were split with different weights (i.e. the solvation of hydrophobic residues had a different weight than the solvation of polar residues, and the entropy of the main-chain had different weight than the entropy of side-chains). The parameters optimizing the final score were obtained using single-point mutations of nine different proteins and the corresponding free energies obtained with their 3D conformations.

$$\Delta G = \alpha_{vdw}\Delta G_{vdw} + \Delta G_{solv} + \Delta G_{Hbond} + \Delta G_{el} + T\Delta S$$

$$\Delta G_{solv} = \alpha_{sh}\Delta G_{hydrophobic}^{solv} + \alpha_{sp}\Delta G_{polar}^{solv}$$

$$\Delta G_{Hbond} = \Delta G_{water-bridge} + \left(G_{Hbond}^{prot} - G_{Hbond}^{water}\right)$$ (1)

$$\Delta S = \alpha_{mc}\Delta S_{main-chain} + \alpha_{sc}\Delta S_{side-chain}$$

In Z-rank the energies were also split in van der Waals, electrostatic and solvation terms, but the weights of van der Waals and electrostatic interactions were different for attractive (a) and repulsive (r) interactions, and also different for short-range (<5Å) and long-range (>5 Å) interactions (sr and lr, for short and long ranges, respectively):

$$score = E_{vdw} + E_{el} + E_{solv}$$

$$E_{vdw} = \alpha_{vdw}^{lrr}E_{vdw}^{lrr} + \alpha_{vdw}^{lra}E_{vdw}^{lra} + \alpha_{vdw}^{srr}E_{vdw}^{srr} + \alpha_{vdw}^{sra}E_{vdw}^{sra}$$ (2)

$$E_{el} = \alpha_{el}^{lrr}E_{el}^{lrr} + \alpha_{el}^{lra}E_{el}^{lra} + \alpha_{el}^{srr}E_{el}^{srr} + \alpha_{el}^{sra}E_{el}^{sra}$$

The second strategy follows the rationale that near-native structures will show a broader and deeper well in the energy landscape compared to non–near-native structures. This assumption is the basis of clustering a collection of output conformations (around 1000–2000) as a function of the number of similar structures. Clustering is performed using as the similarity measure either the Cα binding site root mean square deviation (named I-RMSD) (Comeau et al., 2004) or the ligand Cα RMSD (Ritchie and Kemp, 2000). Selection based on the clustering methodology has proved to be better for determining near-native conformations than selection based solely on scoring functions (Ritchie and Kemp, 2000; Vajda and Kozakov, 2009). Consequently, the clustering method has become popular, mainly in combination with a re-ranking given by a scoring function that guides the selection of structures to cluster (Comeau et al., 2004; Shen et al., 2008).

3.3.2 Knowledge based potentials

In knowledge-based potentials, also named statistical potentials, the interaction between two residues is scored by the potential of mean force (PMF) obtained from the probability of finding a pair of residues at a given distance (Sippl, 1990). Let k_B denote the Boltzmann constant and let T be the standard temperature (300K). If A and B are the two interacting chains and a,b are two residues in chains A and B (respectively) at distance d, the potential of mean force is given by:

$$PMF(a,b,d) = PMF_{std}(d) - k_B T \log\left(\frac{P(a,b\mid d)}{P(a)P(b)}\right)$$ (3)

where $PFM_{std}(d)=k_B T\log(P(d))$; $P(a)$, $P(b)$ are the individual probabilities of residues a, b; $P(a,b\mid d)$ is the conditional probability of residues a,b at distance smaller or equal to d and $P(d)$ the probability of any pairs of residues at distance smaller or equal to d. All probabilities correspond to the observed frequencies of the events in the reference database (i.e. 3DID (Stein et al., 2005))

The score of the interaction is then defined as the sum over all interacting pairs of the pair residue scores. Formally, if $a_1,...a_s$ is the residue sequence of chain A, $b_1,...,b_r$ is the residue sequence of chain B, Γ is the set of pair position indices (i,j) of interacting residues a_i,b_j at distance d_{ij}, then the statistical potential E_{pair} is:

$$E_{pair} = \sum_{(i,j)\in\Gamma} PMF(a_i,b_j,d_{ij}) \qquad (4)$$

As energy can usually be split in independent terms from which different forces are derived, the statistical potential can also be split in terms that would describe the different parts of the interaction as particular forces. Particularly, considering a residue *condition* θ as the triplet formed by (secondary structure, polarity, degree of exposure), then the PMF in (3) can be decomposed using:

$$PMF_{pair}(a,b) = -k_B T \log\left(\frac{P(a,b\mid d_{ab})}{P(a)P(b)P(d_{ab})}\right)$$

$$PMF_{local}(a,b) = k_B T \log\left(\frac{P(a\mid\theta_a)P(\theta_a)}{P(a)}\right) + k_B T \log\left(\frac{P(b\mid\theta_b)P(\theta_b)}{P(b)}\right)$$

$$PMF_{3D}(a,b) = k_B T \log(P(d_{ab})) \qquad (5)$$

$$PMF_{3DC}(a,b) = k_B T \log\left(\frac{P(\theta_a,\theta_b\mid d_{ab})}{P(\theta_a,\theta_b)}\right)$$

$$PMF_{S3DC}(a,b) = -k_B T \log\left(\frac{P(a,b\mid d_{ab},\theta_a,\theta_b)P(\theta_a,\theta_b)}{P(a,b\mid\theta_a,\theta_b)P(\theta_a,\theta_b\mid d_{ab})}\right)$$

Finally, the split statistical potentials E_{pair}, E_{local}, E_{3D}, E_{3DC}, and E_{S3DC} can be obtained by applying the formula (4) to the decomposed PMFs (5), with corresponding subindexes between $E_\_$ and $PMF_\_$. It was shown (Aloy and Oliva, 2009) that Epair admits a decomposition of the form:

$$E_{pair} = E_{S3DC} - E_{3DC} + E_{3D} - E_{Local} + E_{cmp} \qquad (6)$$

where E_{cmp} is a residual energy term depending only on the conditions of the interacting residues and accounts for the reference state (first term in PMF equations). This equation was initially derived for the scoring of protein folds, but it remains valid when applied to the residues in the interface between two interacting proteins (Feliu et al., 2011).

Note that the statistical potential E_{S3DC} is a refinement of the residue-pair statistical potential E_{pair}, in the sense that it takes into account not only the residues that interact but also the condition in which each of them sits. On the contrary, the statistical potential E_{3DC} depends on the occurrence of interacting conditions, disregarding the specific interacting residues. The score E_{local} reflects the probability of placing a residue on a specific condition. Moreover, it splits into two terms, each of them depending only on the probability of placing a certain residue in some condition for each chain separately. The energy term E_{3D} concerns only the distance at which pairs of residues interact, and increases together with the number of interacting residue-pairs, thus being proportional to the number of residues implied in the interface.

3.3.3 Using split statistical potentials to rank docking poses

To test the scoring functions, the benchmark decoy dataset of Weng and cols. (Hwang et al., 2008) is widely used as gold standard. This dataset is based on a set of non-redundant real interactions for which both the complex 3D structure and the individual chain structures are available. It consists of a collection of binary complexes (124) with known structure (named targets) and a set of decoys for each of them (named target set). The 54,000 decoys generated using the rigid-body docking algorithm ZDock3.0 (Mintseris et al., 2007) from the individual chain structures were considered. The set of binary-complex conformations of a rigid-body prediction are classified according to the expected difficulties to obtain a near-native solution of the target. They deal with three types named: easy, medium and difficult cases. In total, the dataset consists of 124 cases, 88 of which are straight forward for rigid-body docking, 19 are medium and 17 are difficult cases for which further conformational changes are required upon binding. Only 97 of them (88 rigid-body and 9 medium) fit into the common near-native decoy criterion of structures differing from the native one at most 2.5Å (computed in terms of I-RMSD from the native structure). For difficult cases it is not possible to have near-native poses because of the deformation suffered by one or two of the protein partners. Thus, a different definition of a successful prediction is required in these cases. A selected pose was considered good if its I-RMSD differs less than 0.5Å from the lowest I-RMSD among all the decoys in the target set. This measure enables to determine if the scoring function top-ranks the best available decoys of the set. Figure 4 shows the success curves for the split potentials, revealing the relative importance of E_{local} and E_{3D} in the composition of the residue-pair statistical potential E_{pair}.

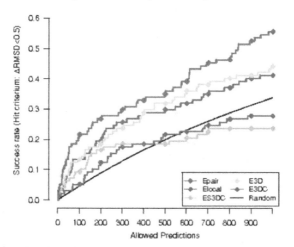

Fig. 4. Success curves for the split potentials: Success curves on the whole benchmark dataset are plotted for the five statistical potentials E_{pair} (red), E_{S3DC} (orange), E_{local} (blue), E_{3D} (light green) and E_{3DC} (purple), plus the success curve expected by random (black).

Based on the observation that E_{pair} and E_{S3DC} provided a fairly amount of non-overlapping hits, a new ranking strategy was defined: "MixRank". This strategy consists of first considering the lists of decoys ranked by different scoring functions separately, and then alternatively selecting one decoy from each list. Then, in order to avoid repetitions, we apply a removal of redundant predictions (Feliu and Oliva, 2010). That is, we do not include decoys that are less than 5Å of I-RMSD from an already selected decoy. This way of removal of redundancies was analyzed (Feliu and Oliva, 2010) and was proved to provide better selection of near-native decoys. This ranking strategy proved to be able to compete with other ranking strategies based on atomistic-detailed scoring functions if large conformational changes of the interacting partners are required for the interaction. These are the cases typically included in the medium and difficult categories of the benchmark data set. This is shown in Figure 5, where E_{pair} and MixRank surpass ranking system based either on a reference statistical potential (RPScore (Moont et al., 1999)) or on an atomistic-detailed scoring function (ZRank) when predicting near-native poses within the medium and difficult categories of the benchmark data set.

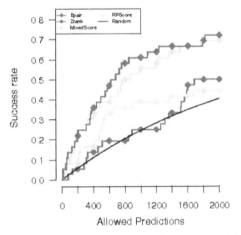

Fig. 5. Different ranking approaches compared for difficult cases of the benchmark data set: Success curves are plotted after removal of redundant solutions for the MixRank strategy (light green), Epair (red), Zrank (dark green) and RPScore (blue) scoring functions, and also compared with the success curve expected by random (black), only with the medium and difficult cases of the benchmark dataset.

4. Errors in models

The quality of the obtained model establishes the limits of the biological information that can be safely extracted from it. Although all structural models may enclose errors, these become less of a problem if correctly detected and assessed: once an error is identified, it is possible to discriminate whether it affects key structural or functional regions. Therefore an essential step in any structural modeling process is the detection of the wrongly modeled regions.

4.1 Sources of errors

In comparative modeling (homology modeling and threading), wrongly modeled regions are expected to be more frequent as the sequence identity between the query protein and the template decreases. Errors can be expected to occur at any step of the process, thus, they can be catalogued according to the step in which they can be found and, therefore, the step in which they can be corrected or compensated. Docking techniques may incorporate similar errors during the step of molecular refinement.

Wrong template selection is the most costly error that can be found in a modeling process. Being a key step in the process, the selection of a wrong template cannot be overcome at any other part of the process and will inevitable yield to a wrong model. Correcting such error implies going back to the beginning of the modeling process and start all over again. The selection of a wrong template usually derives from the lack of a sequence homologous enough to sequence the query protein. A lot of effort is being put into trying to describe the optimal thresholds of identity and similarity to decide whether or not a sequence can be chosen as template.

Misalignment errors tend to appear under a 40% sequence identity. Their abundance rapidly increases below 30% of identity, as the occurrence of local regions with very low sequence identity makes wrong alignments more feasible. These errors are specially focused on gap misplacements in the alignment, and are one of the major sources of problems in homology modeling. As with the detection of the correct template, the sequence-template alignment is key, and correcting it requires redoing the alignment.

Structural distortions can be found both in well-aligned and in unaligned regions. Those in aligned regions appear when the sequence identity is too low in a local region and the sequence does, in fact, acquire a different secondary structure than that of the template. This problem can be overcome by using several templates in low identity regions in order to explore the possibilities. The regions that, even with multiple templates, are not aligned to any template have to be predicted by energy-based methods of database searching. The alignments at the sequence boundaries and 3D boundaries of such regions will determine the accuracy of the prediction.

Finally, side chain packing needs to be optimized especially as sequence identity decreases. Such optimization can be a major issue, specifically when it involves residues implicated in the protein's function and mostly in the interface of interacting proteins.

4.2 Detecting the errors

Automated methods for detecting errors in 3D models rely on the knowledge of previously solved structures in the PDB. This knowledge has lead to identify stereochemical and energy-related restrictions in the final 3D conformation of a protein. Considering stereochemical restrictions, perhaps the most obvious is that two amino-acids cannot clash (i.e. they cannot occupy the same spatial region). In addition, not all possible relative orientations of two correlative amino-acids in the protein sequence are allowed. These orientations are defined by the Φ and Ψ angles of the amino-acidic bond and the applicable restrictions are summarized in the Ramachandran diagram (Ramachandran et al., 1963), which represents the allowed conformations as a function of the Φ and Ψ angles. PROCHECK program (Laskowski et al., 1996) assess the overall quality of a protein model based on these parameters.

Besides stereochemistry, there are other protein spatial features in the proteins that could be used as indicators of errors in the models: packing, creation of a hydrophobic core, residue and atomic solvent accessibilities, spatial distribution of charged groups, distribution of atom-atom distances, and main-chain hydrogen bonding structures (Sali, 1995). These are key features to understand the mechanisms by which a protein finds its native state. This mechanism is known as the folding pathway and the possibilities space for the folding of a protein is vast (Levinthal, 1968). Solving this problem requires an accurate potential describing the interactions among different amino-acid residues (Dinner et al., 2000). However, the use of such atomistic-detailed potentials (Brooks et al., 2009) is quasi-prohibitive and it does not ensure the native and biologically active conformation.

An alternative approach to the full atomistic description is to construct a coarse grained potential. The aim of such potential would be to approximate the function: a) whose global minimum corresponds to the native structure (Sippl, 1990), and b) capable to drive the structure from incorrect folding states toward native-like conformations (i.e. the having a correlation with native structure similarity (Keasar and Levitt, 2003)) describing a funnel-like energy surface. This scoring function, termed knowledge-based or statistical potential, works as a coarse-grained descriptor of the environment of the protein, and can be used to assess the quality of a protein 3D model. Based on this approach PROSAII (Sippl, 1993) is probably the most widely used program to assess the quality of a protein 3D model. Similarly, specific potentials have been derived for the interaction between macromolecules in order to assess protein-protein interactions (e.g., M-TASSER (Chen and Skolnick, 2008), MULTIPROSPECTOR (Lu et al., 2002) or InterPreTS (Aloy and Russell, 2003). Nevertheless, a funneling theory such as the Levinthal paradox in protein folding is still under development and some explanations are recently found (Wass et al., 2011).

5. Integrative modeling

The previous detailed methods could be useful in small complexes, where the docking of few subunits can solve the quaternary structure. However, the assembly of large macromolecular complexes such as the nucleopore complex, which contains more than 450 proteins, is unaffordable. In these cases, the presence of such amount of subunits forces the necessity to find methods that could manage the assembly problem in terms of costs and time.

During the last years, the integration of the maximum amount of structural information available about the structurally unknown macromolecular complex has become the state of the art solution to this problem. The main idea of this methodology is to use particular characteristics of the complex that can be synergistically combined in order to restrict the possible solutions to only those consistent with these features.

Electron microscopy has been established as a crucial technique for studying the structure of macromolecular assemblies (Alber et al., 2007). The resolution is insufficient to construct an atomic model but reveals insights into the shape and size of the whole complex. Thus, fitting atomic-resolution structures into the electron density maps is a suitable method for determining not only large macromolecular assemblies but also small ones.

Several methods have been developed for simultaneously fitting the individual protein subunits into the density map of their assembly. MultiFit (Lasker et al., 2010b) solve the position and orientation of each component within a protein structure using a function score that maximizes the quality of fit in the electron density map, the protrusion from the density map envelope, and the complementary shape between subunits. An optimizer algorithm DOMINO (Discrete Optimization of Multiple Interacting Objects) (Lasker et al., 2009) searches like a puzzle the positions of the subunits within a discrete sampling space. Each subunit is placed in a particular position inside the density map, conditioning the position of the rest of the subunits. This algorithm is used to efficiently find the global minimum in an affordable way.

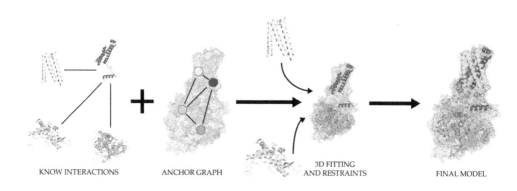

Fig. 6. Schematic representation of integrative modeling.

Often, the electron density map or the high-resolution structure of the subunits is not available. In these cases, it is not possible to apply the fitting procedure mentioned above. However, the integrative approach is not restricted to this data. There are different techniques that provide different types of information that can be used to understand particular features of the assembled complex. Table 1 highlights a list of proteomics, biophysical and computational methods used to obtain this valuable data.

In this way, Sali and collaborators developed an integrative modeling platform (IMP) (Lasker et al., 2010a) that collect this information and consider them simultaneously to generate models consistent with the data. This platform was used to describe the nuclear pore complex (Alber et al., 2007) and the structure of chromatin at megabase scale (Bau et al., 2010). Moreover, this platform can be used to solve any kind of 3D structure when enough data is provided.

IMP performs its function in an iterative series of four different steps. Below, a brief description of each step gave us an insight into how this heterogeneous data can be combined to deliver such large complex models.

Type of Structural Information	Techniques
Composition	Mass spectrometry and quantitative immunoblotting
Interactions	Genetic interactions and bioinformatics predictions
Connectivity	Affinity purification and surface plasmon resonance (SPR)
Interaction partners	Yeast to hybrid, protein microarrays, protein-fragment complementation assay (PCA) and calorimetry
Interaction distances	Fluorescence resonance energy transfer (FRET), bioluminescence resonance energy transfer (BRET) and cross-linking
Complex shape	X-ray scattering (SAXS) Cryo-electron microscopy, Cryo-electron tomography and Negative stain electron microscopy
Protein positions	High resolution electron microscopy, gold-labelling, green fluorescence protein (GFP) labelling and Docking
Residue positions	Crosslinking, hydrogen/deuterium exchange, Limited Proteolysis and Footprinting
Atomic positions	X-rays crystallography and nuclear magnetic resonance (NMR)

Table 1. Proteomics, biophysical and computational methods used to obtain information for modelling macromolecular complexes.

5.1 Data gathering

The collection of structural information is the first requirement needed to start the assembly process. The techniques listed in table 1 are appropriate generators of this data. In addition, a large amount of biological information is available through databases. Table 2 lists some databases with structural relevant information.

5.2 System representation and data translation into spatial restraints

One of the most characteristic features of the integrative modeling process is the ability to use structures that are not solved in high-resolution. In those cases, it is necessary to find an appropriate representation of the system. For example, on one hand, an atomic-resolution structure can be represented with particles corresponding to atoms and, on the other hand, in a low-resolution structure a single particle can represent a sphere corresponding to a group of atoms, residues or domains. Consequently, the resolution of the final complex is dictated by the resolution of the available data.

The raw data gathered in the first step must to be translated into spatial restraints, which specify values for the encoded data in order to decide if the model satisfies or not the experimental information about it. A restraint is a scoring function that reaches its minimum if the feature is consistent with the experimental data. A 0 indicates a model that is perfectly consistent with the restraint, whereas the result of the function is higher when the restraint is violated. In Table 3, most common types of restrains are reviewed.

Database	Description
PDB	PDB (Protein Data Bank) is the worldwide repository of information of 3D biological molecules structures
ModBase	ModBase is a relational database of protein structure models calculated by comparative homology modelling of known structures
SCOWLP	SCOWLP (Structural Characterization Of Water, Ligands and Proteins) is a relational database for detailed structural analysis of PDB protein interfaces at atomic level. The SCOWLP includes proteins, ligands and water as descriptors of interfaces
3DID	3did (3D interacting domains) is a collection of domain-domain interactions extracted from atomic-resolution structures. Each domain is associated to a Pfam domain and the database is GO term functional annotated
EMdataBank	EM Data Bank is a database of cryo-electron microscopy maps, models and associated metadata
BioGRID	BioGRID (Biological General for Interaction Datasets) is a database that archives genetic and proteomic interactions curated from high-throughput datasets and individual studies
PRISM	PRISM (Protein Interactions by Structural Matching) is a web-served compilation of protein-protein interaction interfaces
SCOPPI	SCOPPI (Structural Classification of Protein-Protein Interactions) is a database of all domain-domain interactions and their interfaces derived from PDB structure files and SCOP domain definitions

Table 2. Databases of structural information suitable for the integrative modelling process.

Type of restraint	Description of the restraint
Distance restraints	Restraints the distance between two particles
Connectivity restraints	Restraints all proteins in a set to interact or not.
Quality of fit restraint	Restraints the overlapping of the particles in an electron density map
Excluded volume	Restraint steric clashes
Geometric complementary	Maintains the geometric complementary between two particles interfaces
Statistical potential restraint	Restraint depending on the frequencies of contacts in previous solved complexes
Angle restraint	Restraint the angle between three particles
Protein localization restraint	Restraints a particle in a specific position
Complex diameter restraint	Restraints the maximum distance between the two most distance particles
Symmetry restraint	Maintains the same configuration of equivalent particles across multiple symmetry units
Radial distribution function restraint	Restraints the correlation between experimentally measured and computed radial distribution functions

Table 3. Most common types of spatial restrains obtained from structural data.

5.3 Calculation of an ensemble consistent with the restraints

At this point, the different restraints are combined into a final scoring function, which is commonly the sum of the singular scoring functions corresponding to each restraint. Then, the configuration of the constituent protein beads is determined by optimizing this scoring function.

The optimization process consists of searching through the configuration space the positions and orientations of the structural subunits that minimizes this function. It starts from random positions and iteratively moves them to minimize the violation of the restraints. In essence, a kind of 'force' pulls the proteins together to the native complex configuration. For this task, it is possible to use methods that explore the scoring function landscape in an efficient manner, such as conjugate gradient, molecular dynamics with simulated annealing or personalized optimizers, such as DOMINO (Lasker et al., 2009).

5.4 Analysis of the ensemble

Assuming a unique native state of the complex, the optimization process it is supposed to give a single model that satisfies all restraints. However, if the data used to encode the restraints is insufficient, more than one solution might be obtained. This problem could be solved introducing new restrains and running the process again. Conversely, in case of incorrect restraints, it is possible that no solution is obtained because there is not a model that satisfies all the restraints. In conclusion, the integrative method is a very powerful tool but it is clearly conditioned by the quality of the gathered information. Finally, the structure of the complex needs to be evaluated using similar approaches as in modelling, but adding the quality of the accomplishment of the restraints applied to construct the macro-complex.

6. Conclusions

Protein sequences are totally valueless if meaningful information about their biological function is not reported. In the past 30 years clear relationships between proteins sequence, structure, and function have been proven. Thus, the knowledge of a protein's 3D structure is normally required to completely understand its function. Since many proteins act in association with others, the knowledge of the structure of the complex formed by this association (named quaternary structure) is crucial to understand how proteins perform their functions. In this chapter we have attempted to establish the capabilities and limitations of currently available computational methods for predicting the tertiary and quaternary structure of proteins. Different strategies can be followed depending on the data available, and this review hopefully could serve as a practical guide for modelling the tertiary structures of proteins and its association into complexes.

When known structures of homologous proteins are available, these can be used as a template to model the structure of a target protein or a protein complex in a process termed comparative modelling. Being the current knowledge on the structure of protein complexes much more limited than that of single proteins, several databases of protein-protein interfaces such as PRISM (Aytuna et al., 2005) have been developed to overcome this problem. In comparative modelling, the percentage of sequence identity between the problem proteins and the templates is crucial. Below a certain threshold of sequence identity

(~30%) comparative modelling becomes a difficult task even for experts. In any case, models must be critically evaluated to be sure that they are correct, devoting most efforts to the region involved in the function (usually implying its interaction with other proteins or compounds).

On the lack of experimental data of the structure of a complex of proteins, protein docking is one of the computational techniques for elucidating the structures of binary interactions. We have shown that the use of split knowledge-based statistical potentials to score and rank the different docking solutions can be as accurate as atomistic-detailed potentials (Feliu et al., 2011) in any type of docking. Furthermore, these statistical poteintals surpass atomistic detailed scores when the complex requires large conformational changes of the interacting partners upon the interaction and we apply a rigid docking protocol.

Finally, we reviewed how different sources of experimental data are synergistically used to model large macromolecular complexes by the Integrative Modelling Platform (Lasker et al., 2010a). This approach has successfully been used to elucidate the structure of the nucleopore complex or the structure of chromatin at megabase scale.

7. References

Alber, F., Dokudovskaya, S., Veenhoff, L.M., Zhang, W., Kipper, J., Devos, D., Suprapto, A., Karni-Schmidt, O., Williams, R., Chait, B.T., et al. (2007). Determining the architectures of macromolecular assemblies. Nature 450, 683-694.

Aloy, P., and Oliva, B. (2009). Splitting statistical potentials into meaningful scoring functions: testing the prediction of near-native structures from decoy conformations. BMC Struct Biol 9, 71.

Aloy, P., and Russell, R.B. (2003). InterPreTS: protein interaction prediction through tertiary structure. Bioinformatics 19, 161-162.

Aloy, P., and Russell, R.B. (2004). Ten thousand interactions for the molecular biologist. Nat Biotechnol 22, 1317-1321.

Altschul, S.F., Madden, T.L., Schaffer, A.A., Zhang, J., Zhang, Z., Miller, W., and Lipman, D.J. (1997). Gapped BLAST and PSI-BLAST: a new generation of protein database search programs. Nucleic Acids Res 25, 3389-3402.

Aytuna, A.S., Gursoy, A., and Keskin, O. (2005). Prediction of protein-protein interactions by combining structure and sequence conservation in protein interfaces. Bioinformatics 21, 2850-2855.

Bau, D., Sanyal, A., Lajoie, B.R., Capriotti, E., Byron, M., Lawrence, J.B., Dekker, J., and Marti-Renom, M.A. (2010). The three-dimensional folding of the alpha-globin gene domain reveals formation of chromatin globules. Nat Struct Mol Biol 18, 107-114.

Berman, H.M., Westbrook, J., Feng, Z., Gilliland, G., Bhat, T.N., Weissig, H., Shindyalov, I.N., and Bourne, P.E. (2000). The Protein Data Bank. Nucleic Acids Res 28, 235-242.

Brooks, B.R., Brooks, C.L., 3rd, Mackerell, A.D., Jr., Nilsson, L., Petrella, R.J., Roux, B., Won, Y., Archontis, G., Bartels, C., Boresch, S., et al. (2009). CHARMM: the biomolecular simulation program. J Comput Chem 30, 1545-1614.

Caffrey, D.R., Somaroo, S., Hughes, J.D., Mintseris, J., and Huang, E.S. (2004). Are protein-protein interfaces more conserved in sequence than the rest of the protein surface? Protein Sci 13, 190-202.

Chen, H., and Skolnick, J. (2008). M-TASSER: an algorithm for protein quaternary structure prediction. *Biophys J 94*, 918-928.

Chenna, R., Sugawara, H., Koike, T., Lopez, R., Gibson, T.J., Higgins, D.G., and Thompson, J.D. (2003). Multiple sequence alignment with the Clustal series of programs. *Nucleic Acids Res 31*, 3497-3500.

Clackson, T., and Wells, J.A. (1995). A hot spot of binding energy in a hormone-receptor interface. *Science 267*, 383-386.

Comeau, S.R., Gatchell, D.W., Vajda, S., and Camacho, C.J. (2004). ClusPro: an automated docking and discrimination method for the prediction of protein complexes. *Bioinformatics 20*, 45-50.

Davis, F.P., and Sali, A. (2005). PIBASE: a comprehensive database of structurally defined protein interfaces. *Bioinformatics 21*, 1901-1907.

De, S., Krishnadev, O., Srinivasan, N., and Rekha, N. (2005). Interaction preferences across protein-protein interfaces of obligatory and non-obligatory components are different. *BMC Struct Biol 5*, 15.

Dinner, A.R., Sali, A., Smith, L.J., Dobson, C.M., and Karplus, M. (2000). Understanding protein folding via free-energy surfaces from theory and experiment. *Trends Biochem Sci 25*, 331-339.

Dobbins, S.E., Lesk, V.I., and Sternberg, M.J. (2008). Insights into protein flexibility: The relationship between normal modes and conformational change upon protein-protein docking. *Proc Natl Acad Sci U S A 105*, 10390-10395.

Eddy, S.R. (1998). Profile hidden Markov models. *Bioinformatics 14*, 755-763.

Feliu, E., Aloy, P., and Oliva, B. (2011). On the analysis of protein-protein interactions via knowledge-based potentials for the prediction of protein-protein docking. *Protein Sci*.

Feliu, E., and Oliva, B. (2010). How different from random are docking predictions when ranked by scoring functions? *Proteins 78*, 3376-3385.

Fernandez-Fuentes, N., Rai, B.K., Madrid-Aliste, C.J., Fajardo, J.E., and Fiser, A. (2007). Comparative protein structure modeling by combining multiple templates and optimizing sequence-to-structure alignments. *Bioinformatics 23*, 2558-2565.

Fornes, O., Aragues, R., Espadaler, J., Marti-Renom, M.A., Sali, A., and Oliva, B. (2009). ModLink+: improving fold recognition by using protein-protein interactions. *Bioinformatics 25*, 1506-1512.

Gabb, H.A., Jackson, R.M., and Sternberg, M.J. (1997). Modelling protein docking using shape complementarity, electrostatics and biochemical information. *J Mol Biol 272*, 106-120.

Gao, M., and Skolnick, J. (2010). Structural space of protein-protein interfaces is degenerate, close to complete, and highly connected. *Proc Natl Acad Sci U S A 107*, 22517-22522.

Garzon, J.I., Lopez-Blanco, J.R., Pons, C., Kovacs, J., Abagyan, R., Fernandez-Recio, J., and Chacon, P. (2009). FRODOCK: a new approach for fast rotational protein-protein docking. *Bioinformatics 25*, 2544-2551.

Guerois, R., Nielsen, J.E., and Serrano, L. (2002). Predicting changes in the stability of proteins and protein complexes: a study of more than 1000 mutations. *J Mol Biol 320*, 369-387.

Guharoy, M., and Chakrabarti, P. (2005). Conservation and relative importance of residues across protein-protein interfaces. *Proc Natl Acad Sci U S A 102*, 15447-15452.

Hwang, H., Pierce, B., Mintseris, J., Janin, J., and Weng, Z. (2008). Protein-protein docking benchmark version 3.0. *Proteins 73*, 705-709.

Janin, J. (2010). Protein-protein docking tested in blind predictions: the CAPRI experiment. *Mol Biosyst 6*, 2351-2362.

Jones, D.T. (1999). GenTHREADER: an efficient and reliable protein fold recognition method for genomic sequences. *J Mol Biol 287*, 797-815.

Katchalski-Katzir, E., Shariv, I., Eisenstein, M., Friesem, A.A., Aflalo, C., and Vakser, I.A. (1992). Molecular surface recognition: determination of geometric fit between proteins and their ligands by correlation techniques. *Proc Natl Acad Sci U S A 89*, 2195-2199.

Keasar, C., and Levitt, M. (2003). A novel approach to decoy set generation: designing a physical energy function having local minima with native structure characteristics. *J Mol Biol 329*, 159-174.

Kelley, L.A., and Sternberg, M.J. (2009). Protein structure prediction on the Web: a case study using the Phyre server. *Nat Protoc 4*, 363-371.

Keskin, O., Tsai, C.J., Wolfson, H., and Nussinov, R. (2004). A new, structurally nonredundant, diverse data set of protein-protein interfaces and its implications. *Protein Sci 13*, 1043-1055.

Kozakov, D., Brenke, R., Comeau, S.R., and Vajda, S. (2006). PIPER: an FFT-based protein docking program with pairwise potentials. *Proteins 65*, 392-406.

Lasker, K., Phillips, J.L., Russel, D., Velazquez-Muriel, J., Schneidman-Duhovny, D., Tjioe, E., Webb, B., Schlessinger, A., and Sali, A. (2010a). Integrative structure modeling of macromolecular assemblies from proteomics data. *Mol Cell Proteomics 9*, 1689-1702.

Lasker, K., Sali, A., and Wolfson, H.J. (2010b). Determining macromolecular assembly structures by molecular docking and fitting into an electron density map. *Proteins 78*, 3205-3211.

Lasker, K., Topf, M., Sali, A., and Wolfson, H.J. (2009). Inferential optimization for simultaneous fitting of multiple components into a CryoEM map of their assembly. *J Mol Biol 388*, 180-194.

Laskowski, R.A., Rullmannn, J.A., MacArthur, M.W., Kaptein, R., and Thornton, J.M. (1996). AQUA and PROCHECK-NMR: programs for checking the quality of protein structures solved by NMR. *J Biomol NMR 8*, 477-486.

Leaver-Fay, A., Tyka, M., Lewis, S.M., Lange, O.F., Thompson, J., Jacak, R., Kaufman, K., Renfrew, P.D., Smith, C.A., Sheffler, W., et al. (2011). ROSETTA3: an object-oriented software suite for the simulation and design of macromolecules. *Methods Enzymol 487*, 545-574.

Levinthal, C. (1968). Are there pathways for protein folding? *J Chem Phys*, 44-45.

Lu, L., Lu, H., and Skolnick, J. (2002). MULTIPROSPECTOR: an algorithm for the prediction of protein-protein interactions by multimeric threading. *Proteins 49*, 350-364.

Ma, B., Elkayam, T., Wolfson, H., and Nussinov, R. (2003). Protein-protein interactions: structurally conserved residues distinguish between binding sites and exposed protein surfaces. *Proc Natl Acad Sci U S A 100*, 5772-5777.

Matthews, L.R., Vaglio, P., Reboul, J., Ge, H., Davis, B.P., Garrels, J., Vincent, S., and Vidal, M. (2001). Identification of potential interaction networks using sequence-based searches for conserved protein-protein interactions or "interologs". *Genome Res 11*, 2120-2126.

Mintseris, J., Pierce, B., Wiehe, K., Anderson, R., Chen, R., and Weng, Z. (2007). Integrating statistical pair potentials into protein complex prediction. *Proteins 69*, 511-520.

Moont, G., Gabb, H.A., and Sternberg, M.J. (1999). Use of pair potentials across protein interfaces in screening predicted docked complexes. *Proteins 35*, 364-373.

Neuvirth, H., Raz, R., and Schreiber, G. (2004). ProMate: a structure based prediction program to identify the location of protein-protein binding sites. *J Mol Biol 338*, 181-199.

Nooren, I.M., and Thornton, J.M. (2003). Structural characterisation and functional significance of transient protein-protein interactions. *J Mol Biol 325*, 991-1018.

Notredame, C., Higgins, D.G., and Heringa, J. (2000). T-Coffee: A novel method for fast and accurate multiple sequence alignment. *J Mol Biol 302*, 205-217.

Ofran, Y., and Rost, B. (2007). ISIS: interaction sites identified from sequence. *Bioinformatics 23*, e13-16.

Pierce, B., and Weng, Z. (2007). ZRANK: reranking protein docking predictions with an optimized energy function. *Proteins 67*, 1078-1086.

Ramachandran, G.N., Ramakrishnan, C., and Sasisekharan, V. (1963). Stereochemistry of polypeptide chain configurations. *J Mol Biol 7*, 95-99.

Ritchie, D.W., and Kemp, G.J. (2000). Protein docking using spherical polar Fourier correlations. *Proteins 39*, 178-194.

Rost, B. (1999). Twilight zone of protein sequence alignments. *Protein Eng 12*, 85-94.

Russell, R.B., and Barton, G.J. (1992). Multiple protein sequence alignment from tertiary structure comparison: assignment of global and residue confidence levels. *Proteins 14*, 309-323.

Sali, A. (1995). Modeling mutations and homologous proteins. *Curr Opin Biotechnol 6*, 437-451.

Sali, A., Potterton, L., Yuan, F., van Vlijmen, H., and Karplus, M. (1995). Evaluation of comparative protein modeling by MODELLER. *Proteins 23*, 318-326.

Shen, Y., Paschalidis, I., Vakili, P., and Vajda, S. (2008). Protein docking by the underestimation of free energy funnels in the space of encounter complexes. *PLoS Comput Biol 4*, e1000191.

Sippl, M.J. (1990). Calculation of conformational ensembles from potentials of mean force. An approach to the knowledge-based prediction of local structures in globular proteins. *J Mol Biol 213*, 859-883.

Sippl, M.J. (1993). Recognition of errors in three-dimensional structures of proteins. *Proteins 17*, 355-362.

Stein, A., Russell, R.B., and Aloy, P. (2005). 3did: interacting protein domains of known three-dimensional structure. *Nucleic Acids Res 33*, D413-417.

Teyra, J., Doms, A., Schroeder, M., and Pisabarro, M.T. (2006). SCOWLP: a web-based database for detailed characterization and visualization of protein interfaces. *BMC Bioinformatics 7*, 104.

Tsai, C.J., Xu, D., and Nussinov, R. (1997). Structural motifs at protein-protein interfaces: protein cores versus two-state and three-state model complexes. *Protein Sci 6*, 1793-1805.

Tuncbag, N., Gursoy, A., Guney, E., Nussinov, R., and Keskin, O. (2008). Architectures and functional coverage of protein-protein interfaces. *J Mol Biol 381*, 785-802.

Tuncbag, N., Kar, G., Keskin, O., Gursoy, A., and Nussinov, R. (2009). A survey of available tools and web servers for analysis of protein-protein interactions and interfaces. *Brief Bioinform 10*, 217-232.

Vajda, S., and Kozakov, D. (2009). Convergence and combination of methods in protein-protein docking. *Curr Opin Struct Biol 19*, 164-170.

Wass, M.N., Fuentes, G., Pons, C., Pazos, F., and Valencia, A. (2011). Towards the prediction of protein interaction partners using physical docking. *Molecular systems biology 7*, 469.

Watson, J.D., Laskowski, R.A., and Thornton, J.M. (2005). Predicting protein function from sequence and structural data. *Curr Opin Struct Biol 15*, 275-284.

Winter, C., Henschel, A., Kim, W.K., and Schroeder, M. (2006). SCOPPI: a structural classification of protein-protein interfaces. *Nucleic Acids Res 34*, D310-314.

Wodak, S.J., and Janin, J. (1978). Computer analysis of protein-protein interaction. *J Mol Biol 124*, 323-342.

11

Computational Approaches to Predict Protein Interaction

Darby Tien-Hao Chang

National Cheng Kung University

Taiwan

1. Introduction

Recently there have been large advances in high-throughput experimental approaches to identifying protein interactions. However, these experimental verified interactions still account for a small proportion of the complete interaction network. For example, based on current understanding (Stumpf, Thorne et al. 2008), less than 10% of interactions of the human protein interaction network (PIN) are identified and collected in the Human Protein Reference Database (HPRD) (Peri, Navarro et al. 2003; Stumpf, Thorne et al. 2008). The low coverage can be complemented by the computational approaches methods to predict protein interaction. This chapter describes approaches based on different biological observations and/or different computational techniques. Another focus of this chapter is to highlight the importance of creating a benchmark - especially negative samples since there are very limited techniques developed to confirm that two proteins do not interact (Doerr 2010; Smialowski, Pagel et al. 2010) - in evaluating computational approaches.

Computational methods can be roughly divided into two categories. Methods in the first category utilize the observation that functionally related proteins have patterns of co-occurrence, such as co-evolution or co-expression; while methods in the second category compile proteins into features potentially related to protein interaction, such as protein surface area, and resort to machine learning (ML) techniques for prediction. Different co-occurrence-based methods are distinct in where, namely which biological properties, the co-occurrence is observed and in the implementation details to record the co-occurrence. For example, Salgado et al. suggested that some related genes are close in genome to make the transcription more efficient (Salgado, Moreno-Hagelsieb et al. 2000). Methods based on this observation utilize co-localization in genome and could, for example, use the distance between two genes to record the co-occurrence. Section 2 will introduce seven categories of co-occurrence as follows.

1. **Genomic location**—some genes producing proteins that will interact are close in genome to facilitate transcription;
2. **Cellular compartment**—interacting proteins should appear in the same area in a cell to interact, another co-localization pattern which consider the cellular location;

3. **Phylogenetic tree** — if one protein was mutated in evolution, its cooperating protein should have a corresponding mutation to keep their interaction/function and thus the species survival, i.e. cooperating proteins should have similar phylogenetic trees;

4. **Existence in close species** — if two proteins co-work for a function to a species, then the species will have both of them, otherwise the species will have none of them, i.e. some related proteins are present in/absent from species together;

5. **Interacting domains** — interacting proteins usually have complementary parts of a interacting domain pair;

6. **Literature** — related proteins, since there must be some papers describing their relations, are prone to be mentioned together in literature, as opposed to other proteins existing only in articles describing their individual characteristics;

7. **Gene fusion** — some interacting proteins whose homologues form a fused protein chain, a special biological phenomenon named Rosetta Stone protein.

Different ML-based (or feature-based) methods, however, may share partial features to previous studies but develop new features at the same time. This led to more complicated relations than that among co-occurrence-based methods. For example, Shen et al. (Shen, Zhang et al. 2007) proposed to use a composition of short sequences as protein features and a following work by Chang et al. (Chang, Syu et al. 2010) combined these features with protein surface information. In addition to the overlap of features among different ML-based methods, they may use identical or different ML techniques. Using the two ML-based methods as an example, Shen et al. chose the widely used support vector machine (SVM) (Vapnik and Vapnik 1998), while Chang et al. used a relaxed variable kernel density estimator (RVKDE) (Oyang, Hwang et al. 2005) developed by their group. Thus to keep the description structure compact, we will focus on the features in section 3. We will provide only a minimum introduction to several well-known ML techniques in section 4 since they are beyond the scope of this chapter. Knowing the concepts of these ML techniques may help to understand the design of different ML-based PPI predictors and to select appropriate features. This chapter roughly divides features into four categories.

1. **Sequence information** — many studies extracted features only from protein sequences. Methods using only such features are very challenging but provide much applicability. Some derived features such as protein polarity (by summing the polarity index of its amino acids) are also included in this category.

2. **Evolution information** — features involving alignment with other sequences fall into this category. Methods using such features usually require a collection of protein sequences of many species.

3. **Structure information** — methods of this category can perform geometry and even energy analyses. Many useful features such as protein surface, secondary structure and binding affinity can be derived. These methods are usually time-consuming, where researchers will expect to obtain extremely accurate predictions.

4. **Auxiliary information** — some studies used auxiliary information such as function annotation. These studies usually used such features to analyze rather than to predict protein interactions, since some features were manually curated. It was hard to perform a fair comparison with other methods not using such features.

After the features used in recent ML-based methods there is an introduction to three well-known ML techniques.

1. **Decision tree** — a time-honored tool, which is less accurate than modern ML tools but preferred by many biologists because its learning model is more interpretable to human;
2. **SVM** — a state-of-the-art tool that overwhelmingly prevails in the field of computational biology because of its accuracy;
3. **RVKDE** — another modern ML tool that solves the most critical problem of SVM, unacceptable execution time on large data, by slightly sacrificing accuracy.

This chapter ends up with the important issue of computational approaches - evaluation. Computational approaches of identifying protein interactions have a fateful difference to experimental approaches. That is, their results are considered as "prediction" rather than the answer. So it is an inevitable step for the studies of computational methods that they must test their algorithms and report the prediction accuracy compared to a benchmark with the answers already known.

As a summary, this chapter will first introduce the concept of co-occurrence pattern and the implementation details of some co-occurrence-based methods. For the ML-based methods, this chapter focuses on the features and a little on the ML techniques. Finally, three contradictions are used to describe to readers the importance of evaluating these computational methods and explain how to interpret the accuracy they see in literature.

2. Co-occurrence-based approaches

This section introduces seven concepts of co-occurrence patterns that have been adopted to predict protein interactions. An identical concept, based on the available materials, may have different implementation details. In this section, the concept of each co-occurrence pattern is first introduced followed by the implementation details of several methods as examples of that co-occurrence pattern.

2.1 Genomic location

The advance of sequencing leads to the opportunity not only of identifying the genomic locations of genes, but also of analyzing genomic context to predict interactions between genes (Huynen and Snel 2000). The genomic location, also known as genomic context, co-occurrence pattern relies on the fact that operons and some adjacent genes are likely to encode functionally related proteins (Rogozin, Makarova et al. 2002). Huynen and Snel proposed a method to assess the probability that two genes occur as neighbours in a genome only by chance (Huynen and Snel 2000). They randomized the genes in each genome over the loci in that genome. The expected number of the co-occurrences of two genes, namely those that occur as neighbours, in the randomized genomes was less than one. A functional interaction between genes was inferred if the observed number of co-occurrences is significantly higher than the expectation. Rogozin et al. proposed a procedure to compare the orders of orthologous genes (Rogozin, Makarova et al. 2002). They clustered genes into orthologous groups which were then projected onto genomes to identify the neighborhood genes. The results show that the gene neighbours have good functional coherence.

2.2 Cellular compartment

Proteins that occur in different cellular compartments are, in principle, considered not to interact since they do not have the chance to meet. However, some *in vitro* experiments such as tandem affinity purification-mass spectroscopy (TAP-MS) method (Krogan, Cagney et al. 2006), might report such interactions of two proteins in different cellular compartments. It is difficult to determine if these *in vitro* interactions are correct. Thus, this co-occurrence pattern is usually used to increase the prediction reliability of another method or to generate a reliable benchmark rather than an individual interaction predictor. For example, the Eukaryotic Linear Motif (ELM) server used cellular compartment information as a filter to double-check gene function (Davey, Van Roey et al. 2011). The gene function, represented by its Gene Ontology (GO) terms (Ashburner, Ball et al. 2000), was required to be consistent with its cellular compartment. Guo et al. used cellular compartment information to build the negative data of protein interaction (Guo, Yu et al. 2008). They assumed that proteins that occur in different cellular compartments do not interact. They grouped proteins into eight subsets based on the eight main types of cellular compartment—cytoplasm, nucleus, mitochondrion, endoplasmic reticulum, golgi apparatus, peroxisome, vacuole and cytoplasm and nucleus. The negative samples of non-interacting pairs were generated by pairing proteins from different subsets.

2.3 Phylogenetic tree

The phylogenetic tree was proposed to reflect the evolution information. Thus, the similarity between phylogenetic tress provides a good measure of gene co-evolution. Interacting proteins usually co-evolve since mutations in one protein led to the loss of function or a compensation mutation of the other protein to preserve the interaction (Walhout, Sordella et al. 2000). Jothi et al. proposed the MORPH, an algorithm to search the best superimposition between evolutionary trees based on the tree *automorphism* group in 2005 (Jothi, Kann et al. 2005). The search was done by Monte Carlo algorithm that probes the search space of all possible superimpositions, which is computationally intensive. In graph theory, two trees are isomorphic if there is a one-to-one mapping between their vertices (genes) and edges (interactions). Jothi et al. extended this definition to automorphic whereby a tree is isomorphic to itself. The search space was largely reduced to the automorphism group of a phylogenetic tree. The same group proposed another method to assess the degree of co-evolution of domain pairs in interacting proteins in 2006 (Jothi, Cherukuri et al. 2006). Multiple sequence alignments of two proteins/domains to a reference set of genomes were used to construct phylogenetic trees and similarity matrices. The degree of co-evolution of two domains was then estimated by the correlation coefficient of the two corresponding similarity matrices.

2.4 Existence in close species

The co-occurrence pattern of the existence in close species, known as phylogenetic profile, is based on the fact that functionally related proteins usually co-evolve and have homologues in the close genomes (Snitkin, Gustafson et al. 2006). A phylogenetic profile of a gene is a vector, representing the presence or absence of homologues to that gene across a collection of reference organisms. There are two major components in a phylogenetic profile-based method: i) how to construct a phylogenetic profile of a given gene and ii) how to determine the similarity of two phylogenetic profiles. First, the presence or absence of homologues can be determined by sequence alignment scores, such as a BLAST (Altschul, Madden et al. 1997) E-value, with a threshold of presence (Sun, Xu et al. 2005). Such binary vectors were improved as real valued vectors of normalized alignment scores without arbitrarily determining a score threshold (Enault, Suhre et al. 2003). Second, any similarity or distance function between two vectors can be used to define the similarity of two phylogenetic profile vectors. Enault et al. have examined two Euclidean-like distance funtions and another two correlation coefficient variants (Enault, Suhre et al. 2003). They concluded that inner product, shown as follows, is a good indicator in predicting *Escherichia coli* protein interactions.

$$Sim(i, j) = \frac{\sum_{k=1}^{n} R_{ik} \times R_{jk}}{\left[\left(\sum_{k=1}^{n} R_{ij}^2 \right) \times \left(\sum_{k=1}^{n} R_{jk}^2 \right) \right]^{1/2}}$$

2.5 Interacting domains

Proteins usually depend on a short sequence of residues to perform interactions with other molecules. The functional short sequences between two interacting proteins form the contact interfaces, also known as interaction sites (Sheu, Lancia et al. 2005). These interaction sites are usually represented by domains/motifs. Li et al. proposed a method to detect interaction sites, which required only protein sequences (Li, Li et al. 2006). They developed an efficient itemset mining algorithm that can identify the most conserved motifs within two interacting protein groups. Here interacting protein groups indicate two groups, A and B, of proteins where all proteins in group A interact with all proteins in group B, denoted an *all-versus-all* interaction network. The conserved motifs within group A were considered, in principle, related to the conserved motifs within group B. The identified interacting motif pairs can be then used to predict novel interacting proteins. Tan et al. proposed a method, D-STAR to find correlated motifs that were overrepresented in interacting protein pairs (Tan, Hugo et al. 2006). The basic idea of D-STAR is to check all possible (l, d)-motif pairs, where (l, d) indicate an alignment of length l with at most d mismatches. Tan et al. speeded up the brute force procedure by transforming the problem into a clique-finding problem (Pevzner and Sze 2000).

2.6 Literature

Owing to the advance of Internet technologies, the scale of public accessible biomedical literature has increased astonishingly in the last decade. Text mining tools are critical to maximize the usage of such a large-scale knowledge base. Extracting protein interactions from literature is generally categorized as *relationship* mining, which aims to detect co-occurrences of a pair of entities of specific types (such as gene, protein, drug or disease) to a pre-specified relationship (such as interact, regulate, activate or inhibit) in the same article (Cohen and Hersh 2005). Albert et al. proposed a method to retrieve abstracts reporting nuclear receptors (NRs) (Albert, Gaudan et al. 2003). The retrieved data were reviewed manually. Albert et al. generated a dictionary focusing on NRs, cofactors and other NR-binding proteins of human, mouse and rat. The extraction process as follows was performed on MEDLINE abstracts: i) identify abstracts with at least one NR in the generated dictionary, ii) tag entities (proteins) and relationships (interactions) according to the generated dictionary and iii) extract sentences contains two tagged proteins and a tagged interaction. In the current genomic era, the text-minded information is widely applied in database annotation. Many popular protein interaction databases such the Database of Interacting Proteins (DIP) database (Salwinski, Miller et al. 2004) and the Search Tool for the Retrieval of Interacting Genes (STRING) database (Szklarczyk, Franceschini et al. 2011) included automatically extracted literature information as an additional line of evidence.

2.7 Gene fusion

Gene fusion is a special genomic organization whereby some interacting proteins have orthologues in the close genomes fused as a single protein (Enright, Iliopoulos et al. 1999). The fused protein is usually called a Rosetta Stone protein, thus this method is sometime called the Rosetta Stone method. This genomic organization of gene fusion is formed for efficiently transcribing related genes together, thus it is preserved evolutionarily. Marcotte et al. applied the gene fusion method on *Escherichia coli* (Marcotte, Pellegrini et al. 1999). They identified 6,809 protein pairs of which both protein sequences were significantly similar to the same protein sequence of at least a genome. More than half of these 6,809 protein pairs have been shown to be related. This method, unlike previous co-occurrence patterns, is a very specific genomic organization rather than a concept of co-occurrence. Thus, there is very limited space for the algorithm development and implementation details. For any new genome, researchers can always search for Rosetta Stone proteins first. But other methods are required since many interacting proteins are not Rosetta Stone proteins. For example, in the DIP database that deposits experimentally confirmed protein interactions, only 6.4% interacting protein pairs formed Rosetta Stone proteins (Shoemaker and Panchenko 2007).

3. Machine learning-based approaches

This chapter roughly divides features into four categories: sequential, evolutionary, structural and other. Note that the power of ML tools allows researchers to submit any features, with or without obvious biological glues to protein interaction, into a magical black box and wait for the prediction without knowing how the prediction was made. For example, amino acid composition (20 features) and number of search results in PubMed can be used as features. Namely, each co-occurrence pattern can be used as a feature — the only thing to do is designing a rule to record the pattern with one or more real numbers. So in this chapter we only demonstrate several features that have been shown to help the prediction accuracy in published articles, but cannot list all features in a category.

3.1 Sequence information

One of the most widely used data to encode proteins is their primary sequence. Methods that only rely on protein sequences have a great advantage of the wide applicability. Because such methods do not rely on other information, they are sometime called *de novo* (*ab initio*) predictors of protein interaction. Yu et al. proposed a method that encoded protein sequences as feature vectors by considering the amino acid triads observed in it (Yu, Chou et al. 2010). An amino acid triad regards three continuous residues as a unit. However, considering all 20^3 amino acid triads requires an 8000-dimensional feature vector to represent a protein, which is too large for contemporary machine learning tools. Thus, the 20 amino acid types were clustered into seven groups based on their dipole strength and side chain volumes to reduce the dimensions of the feature vector (Shen, Zhang et al. 2007). The frequencies of the $7^3 = 343$ triads can be used to encode a protein sequence. However, such a frequency is highly correlated to the distribution of amino acids. To overcome this problem, Yu et al. proposed a significance calculation by answering the question: how rare is the number of observed occurrences considering the amino acid composition of the protein? The significances of all triads were used to encode protein sequences.

Methods based on sequence motifs/domains also fall into this category since sequence motifs are mined from protein sequences. One may notice that the co-occurrence-based methods mentioned in subsection 2.5 used similar features. In this regard, the co-occurrence-based methods use domain as features with a straightforward rule: if two proteins have interacting domains, then they are predicted as interacting. On the other hand, ML-based methods use domain as features but resort to ML tools for the final decision/prediction. Depending on the ML tools used, the decision rules could be very complicated models of, for example, non-linear equations or a combination of multiple individual components (it could be a 'sum' of multiple functions). Dijk et al. proposed a ML-based method to select relevant motifs from a set of pre-mined motifs (Van Dijk, Ter Braak et al. 2008). They first invoked the D-STAR (Tan, Hugo et al. 2006) algorithm to identify correlated motifs that overrepresented in interacting protein pairs. The vector of the presence or absence of the identified motif pairs were used to encode proteins.

3.2 Evolution information

Methods that require not only the sequences of the query protein pairs but also a collection of supporting sequences fall into this category. The supporting collection is usually from other species for calculating the conservation score. Position-specific scoring matrix (PSSM) is a widely used scheme to encode a protein sequence while considering its orthologues. For a protein sequence, PSSM describes the likelihood of a particular residue substitution at a specific position based on evolutionary information (Altschul, Madden et al. 1997). It is outputted by BLAST when aligning the query protein seuqence to a seqeucne database, e.g. the non-redundant (NR) database from National Center for Biotechnology Information (NCBI). The likelihood values are scaled to [0,1] using the following logistic function:

$$x' = \frac{1}{1 - \exp(-x)},$$

where x is the raw value in PSSM profile and x' is the value corresponding to x after scaling. Each position of a protein sequence is represented by a 21-dimensional vector where 20 elements take the likelihood values of 20 amino acid types from the scaled PSSM profile and the last element is a terminal flag. Finally, the feature vector of a residue comprises a window of positions. Chang et al. proposed a method based on the assumption that protein interactions are more related to amino acids at the surface than those at the core (Chang, Syu et al.). They first used PSSM to encode protein sequences for surface prediction and then used the surface sequence for interaction prediction.

Espadaler et al. proposed a method that made use of conservation of protein pairs (Espadaler, Romero-Isart et al. 2005). They first collected 855 protein complexes with known three-dimensional structure with <80% sequence identity. The 855 complexes were further classified into 16 groups. In a protein complex, the distance between a residue pair from two proteins was defined as the distance of the nearest heavy atoms of the two residues. Via setting a cut-off of the contact distance, one can identify the interface of two proteins in these complexes. These identified interfaces were actually unordered sequence fragments, among which Espadaler et al. defined more than five contiguous residues a *patch*. The conservation of the patches obtained by multiple sequence alignment was considered to select the final patches. These conserved structural patch pairs can be used to predict novel protein interactions. Notice that this method proposed by Espadaler et al. also used the structure information which will be introduced in the next subsection. This also reveals that with the ML tools, combining multiple resources becomes relatively easy since it is no longer dependent on a single co-occurrence pattern.

3.3 Structure information

The most critical problem of sequence-based methods is the reliability. Conversely, researchers usually resort to structure-based methods for verification since the results delivered by structure-based methods can be visualized. Aloy and Russell proposed a method to detect interactions based on protein tertiary structures. They used empirical potentials to compute the fitness between two protein structures. Thus, success of such a method is highly dependent on the performance of the underlying potential function. The adopted potential function did not rely on model proteins, which enlarges its applicability. Aloy and Russell defined interacting residues as those having at least one i) hydrogen bonds ($N-O$ distances ≤ 3.5 Å), salt bridges ($N-O$ distances ≤ 5.5 Å), or van de Waals interactions ($C-C$ distances ≤ 5 Å). Buried side-chains were excluded by filtering out residues with relative accessibility $\geq 10\%$. The identified interacting residues were used to train the empirical potentials based on a molar-fraction random state model as follows:

$$S_{ab} = \log_{10}\left(\frac{O_{ab}}{E_{ab}}\right) E_{ab} = N \frac{n_a}{\sum_{a=1}^{20} n_a} \frac{n_b}{\sum_{b=1}^{20} n_b},$$

where a and b are amino acid types, O_{ab} and E_{ab} are the number of observed/expected contacts, N is the number of analyzed residue pairs and n_a and n_b are number of residues of the corresponding types. The method of Aloy and Russell provided ranks of analyzed protein pairs so that researchers can pick the most promising prediction for further biological experiments.

3.4 Auxiliary information

Important data that is not mentioned above is microarray data, which has been broadly utilized in various biomedical problems. The Gene Expression Omnibus (GEO) database (Barrett, Troup et al. 2006) of NCBI holds more than 20 thousands microarray experiments.

A problem of microarray data is that they are usually full of noises. Soong et al. used principal component analysis (PCA) to reduce such noises (Soong, Wrzeszczynski et al. 2008). PCA is a statistical technique used to find hidden factors from observed factors, expression values in this case. Lee and Batzoglou have shown that proteins with extreme principal components are prone to participate in relevant biological processes (Lee and Batzoglou 2003). The transformation of expression values to principal components can be represented as follows:

$$PX = Y \text{,}$$

where P is a $l \times m$ transformation matrix obtained by PCA, X is a $m \times n$ matrix of the raw expression values from m microarrays and n samples while Y is a $l \times n$ matrix containing every sample's l principal components. The final feature vector of two proteins a and b was the concatenation of a's principal components, b's principal components and the Pearson correlation of both.

This section ends with a method based on literature data, which has been discussed in subsection 2.6. Demonstrating literature data in a ML-based method is to reinforce the impression that in principle any data can be used as features with appropriate encoding schemes. Thus, one can consider combining any of the features discussed in section 2 with ML-based tools. Donaldson et al. proposed an extraction procedure for identifying protein interactions in literature (Donaldson, Martin et al. 2003). They first used a parser to collect synonyms for proteins and their encoding loci. The collected protein names were then used to search the title and abstract of articles in the PubMed literature database. An article was encoded by terms it contained. The weight of each term was the *tf-idf* score (term frequency-inverse document frequency), where term frequency is the number of occurrences of the term in the document and inverse document frequency is the inverse of the number of documents having the term. Here a term was a word or two adjacent words (usually called 2-gram) that appear in at least three documents.

4. Machine learning techniques

After encoding proteins into feature vectors, the next step is to choose a ML tool to generate a model describing these feature vectors. The generated model can be used to predict novel protein interactions. Most ML tools provide a user-friendly interface, where all that researchers need to do is encode their data. The remaining task is very trivial: i) a command to train the model and ii) a command to predict with the trained model. In this regard, researchers who want to adopt ML-based methods can focus on features without caring about the ML algorithms. This section briefly lists three ML algorithms that have been used in recent studies of protein interaction, which can be considered as a basic introduction for researchers who have no idea how to choose an appropriate ML tool.

4.1 Decision tree

Decision trees are usually constructed recursively (Witten, Frank et al. 2011). The first step is to select a feature to split samples (branch the decision tree) based on the selected feature. This step divides the original dataset into several disjointed subsets, each of them can be considered as another dataset. Thus, the same procedure can be applied recursively to each subset and the further sub-subsets. Such a recursive fashion stops at several conditions of, for example, all samples in a branch belonging to the same class or all features have been examined. The above descriptions, however, missed an important detail in decision trees: how to select a feature to branch. A trivial strategy is to select the feature that can result in the purest subsets, namely most samples in the same subset belong to the same class. Thus, a measure of set purity is required.

So far, there have been many purity measurements proposed. This subsection introduces the most fundamental one, entropy, as follows. Indeed, larger entropy indicates the less purity. Thus, negative entropy, in definition, is a measurement of purity. Many mature decision tree algorithms use variants of entropy.

$$\text{entropy}(p_1, p_2, \ldots, p_n) = -p_1 \log p_1 - p_2 \log p_2 \ldots - p_n \log p_n,$$

where p_i is the fraction of class-i samples in the subset and n is the number of total classes. For example, suppose that a dataset has nine positive samples and five negative samples. Before any branching, the entropy of the original dataset is $-(9/14)\log(9/14)-(5/14)\log(5/14)$ $= 0.940$, where p_1 is 9/14, p_2 is 5/14 and n is 2 (positive and negative). If after a branch, the 14 samples are split into three nodes that contain 2-3, 4-0, 3-2 positive-negative samples, respectively. The entropies of the three nodes are $-(2/5)\log(2/5)-(3/5)\log(3/5) = 0.971$, $-(4/4)\log(4/4)-0 = 0$ and $-(3/5)\log(3/5)-(2/5)\log(2/5) = 0.971$, respectively. The total entropy of the branched tree became $(5/14)\times0.971+(4/14)\times0+(5/14)\times0.971 = 0.693$, a weighted sum of the three entropies corresponding to the subset size. It is observed that the entropy decreased from 0.940 to 0.693, revealing that this operation of branch did increase the purity of the dataset. For a purity measurement, the following three conditions must be satisfied:

1. when a subset is pure (all samples belong to the same class), the measurement is zero;
2. when all possible classes appear equally, the measurement is maximized;
3. the measurements must be the same without depending on the order of branches.

The third condition requires that if a dataset is first split into two nodes of a-b and c-d, positive-negative samples and then the second node is further split into two more sub-nodes of e-f and g-h, the entropy should be the same as split into three nodes of a-b, e-f and g-h in a single branch while using another feature. Entropy is the only one function that fits all these conditions (Witten, Frank et al. 2011). This explains the high popularity of entropy and its variants in decision trees.

4.2 Support Vector Machine (SVM)

Currently, SVM is the state-of-the-art ML tool. It prevails in biomedical data because of its high accuracy. SVM first transforms the original data to a higher dimensional space with a non-linear transformation and then finds the maximum margin hyperplane to separate samples of different classes in the transformed space (Witten, Frank et al. 2011). This strategy has two advantages: i) it can generate non-linear model and ii) it prevent overfitting as the decision boundary is still linear in the transformed space. Overfitting is a critical issue in ML. It indicates that the constructed model overfit the training dataset, so that which cannot be used to predict novel data. This problem becomes more serious when using more complicated model. However, some complex data do need complicated models to describe. Thus most advanced ML algorithms still favor complicated models and then try to solve the overfitting issue. In this regard, SVM finds an excellent balance, which can generate very complicated models depending on the adopted transformation while choosing a very simple decision, a hyperplane, which equals to a one stage decision tree of two branches.

Mathematically, SVM uses support vectors to model the transformation and hyperplane. That is the reason for the name. Transforming the original data from the sample space with a non-linear function to a new space means that a linear model (a straight line in a two dimensional space, a plane in a three dimensional space and a hyperplane in a higher dimensional space) in the new space becomes non-linear in the original sample space. For example, for a two dimensional sample $x = (a, b)$, a non-linear transformation to a three dimensional space could be $x' = (a^2, ab, b^2)$. If any ML tool finds a decision boundary in the new space, it does not look like a straight line in the original space. Notice that, in principle, any tool, such as a decision tree, could be used to make the decision in the transformed space. SVM advances in already developing a robust mathematical system with efficient optimization algorithms to find good hyperplanes.

4.3 Relaxed Variable Kernel Density Estimation (RVKDE)

The biggest drawback of SVM is the computational cost. Yu et al. reported that using SVM to perform a complete interaction analysis on human genome may take years (Yu, Chou et al. 2010). In this regard, efficient ML algorithms with acceptable accuracy are reasonable alternatives to SVM. The relaxed variable kernel density estimation (RVKDE) algorithm (Oyang, Hwang et al. 2005) has been practically used in recent interaction studies (Chang, Syu et al. 2010; Yu, Chou et al. 2010). The time complexity of RVKDE is an order faster than SVM. Furthermore, unlike other fast ML algorithms, such as decision trees, the descriptive capability of the constructed model of RVKDE is comparable to SVM.

The kernel of RVKDE is an approximate probability density function. Let $\{s_1, s_2, ..., s_n\}$ be a set of samples randomly and independently taken from the distribution governed by f_x in a m-dimensional vector space. RVKDE estimates the value of f_x at point $\mathbf{v}$ as follows:

$$\hat{f}(\mathbf{v}) = \frac{1}{n} \sum_{s_i} \left(\frac{1}{\sqrt{2\pi} \cdot \sigma_i} \right)^m \exp\left(-\frac{\|\mathbf{v} - \mathbf{s}_i\|^2}{2\sigma_i^2} \right), \text{ where}$$

1. $\sigma_i = \beta \dfrac{R(\mathbf{s}_i)\sqrt{\pi}}{\sqrt[m]{(k+1)\Gamma(\frac{m}{2}+1)}}$;

2. $R(\mathbf{s}_i)$ is the maximum distance between $\mathbf{s}_i$ and its ks-th nearest training sample;

3. $\Gamma(\cdot)$ is the Gamma function (Artin 1964);

4. β and ks are parameters to be set either through cross-validation or by the user.

For prediction, a kernel density estimators is constructed to approximate the distribution of each class. Then, a query sample located at $\mathbf{v}$ is predicted to the class that gives the maximum value among the likelihood functions defined as follows:

$$L_j(\mathbf{v}) = \frac{|S_j| \cdot \hat{f}_j(\mathbf{v})}{\sum_h |S_h| \cdot \hat{f}_h(\mathbf{v})},$$

where $|S_j|$ is the number of class-j training samples and $\hat{f}_j(\cdot)$ is the kernel density estimator corresponding to class-j training samples.

RVKDE belongs to the radial basis function network (RBFN), a special type of neural networks with several distinctive features (Mitchell 1997; Kecman 2001). The decision function of two-class RVKDE can be simplified as follows:

$$f_{RVKDE}(\mathbf{v}) = \sum_{\mathbf{s}_i} y_i \cdot \frac{1}{\sigma_i} \cdot \exp\left(-\frac{\|\mathbf{v}-\mathbf{s}_i\|^2}{2\sigma_i^2}\right),$$

where $\mathbf{v}$ is a testing sample, y_i is the class value as either +1 (positive) or -1 (negative) of a training sample $\mathbf{s}_i$, and σ_i is the local density of the proximity of $\mathbf{s}_i$, estimated by the kernel density estimation algorithm. The testing sample $\mathbf{v}$ is classified as positive if $f_{RVKDE}(\mathbf{v}) \geq 0$, and as negative otherwise. Interestingly, the decision function of RVKDE is very similar to that of SVM using the radial basis function (RBF) kernel:

$$f_{SVM}(\mathbf{v}) = \sum_{\mathbf{s}_i} y_i \cdot \alpha_i \cdot \exp\left(-\gamma\|\mathbf{v}-\mathbf{s}_i\|^2\right),$$

where α_i (corresponds to the inverse of σ_i in f_{RVKDE}) and γ (corresponds to $1/2\sigma_i^2$ in f_{RVKDE}) are user-specified parameters. Thus, the mathematical models of RVKDE and SVM are analogous. The main difference between RVKDE and SVM is the criteria used to determine σ_i and α_i.

5. Evaluation

A paradoxical situation is that a benchmark requires negative samples - proteins known not to interact. A benchmark that contains only interacting protein pairs is useless, since a trivial predictor predicting any protein pairs as interacting can achieve a perfect accuracy. However, there are very limited techniques developed to confirm that two proteins do not interact. Recently, several studies have addressed this problem in evaluating computational methods of identifying protein interactions (Yu, Chou et al. 2010; Yu, Guo et al. 2010). This issue is still in a chaos stage and there is no perfect solution that fit everyone's requirements. Instead, this chapter demonstrates this issue via three major contradictions in this area.

1. **Sampled vs. entire data (also efficiency issue)**—most ML-based methods adopted SVM and have to reduce the data size because of its high time complexity. However, sampled data must lose some information and may bias the evaluation. This contradiction is especially important when comparing co-occurrence- and ML-based methods, where the former usually can be applied on entire data. Using more computing power or switching more efficient ML tool is a compromising solution.

2. **Balanced vs. unbalanced** — once sampled data is adopted, (most studies of ML-based methods adopted using sampled data even without carefully considering the previous contradiction), how to sample is another serious problem. Random sampling can preserve the data distribution (ratio of positive and negative samples) but loss too many positive samples. However, balanced sampling, which forces the inclusion of all positive samples and thus change the data distribution, has also been shown bias the evaluation accuracy (Yu, Chou et al. 2010).

3. **Distinct vs. similar** — one philosophy of creating negative data is to choose the samples which can never be positive. For example, proteins appear in different cellular compartments are possible negative samples. An opposite philosophy is that if a method can discriminate between the negative samples that are very similar to the positive ones, then this method can discriminate those dissimilar ones. The first philosophy prevents collecting negative samples that are actually positive but somehow makes the problem easier while the second philosophy has opposite advantage and disadvantage.

6. Conclusions

In this chapter, various computational methods of protein interaction are reviewed. These methods used various data sources, including localization data, structural data, expression data and/or interactions from orthologs. As a result, all of them are limited to the experimental technologies that generate such data and the incompleteness of verified data. Based on current understanding, the size of protein interaction network (PIN) of human comprises ~650,000 interactions (Stumpf, Thorne et al. 2008). However, the Human Protein Reference Database (HPRD) deposits less than 3% of them (Peri, Navarro et al. 2003; Mishra, Suresh et al. 2006). Even under such a challenging circumstance, computational methods have shown to achieve satisfying performance. This encourages more effort in developing computational methods of protein interaction to complement experimental technologies.

7. References

Albert, S., S. Gaudan, et al. (2003). "Computer-assisted generation of a protein-interaction database for nuclear receptors." *Molecular Endocrinology* 17(8): 1555-1567.

Altschul, S. F., T. L. Madden, et al. (1997). "Gapped BLAST and PSI-BLAST: a new generation of protein database search programs." *Nucleic Acids Research* 25(17): 3389.

Artin, E. (1964). *The Gamma Function*. New York, Holt, Rinehart and Winston.

Ashburner, M., C. A. Ball, et al. (2000). "Gene Ontology: tool for the unification of biology." *Nature genetics* 25(1): 25.

Barrett, T., D. B. Troup, et al. (2006). "NCBI GEO: mining tens of millions of expression profiles — database and tools update." *Nucleic Acids Research* 35(suppl 1): D760.

Chang, D., Y. T. Syu, et al. (2010). "Predicting the protein-protein interactions using primary structures with predicted protein surface." *BMC Bioinformatics* 11(Suppl 1): S3.

Chang, D. T. H., Y. T. Syu, et al. "Predicting the protein-protein interactions using primary structures with predicted protein surface." *BMC bioinformatics* 11.

Cohen, A. M. and W. R. Hersh (2005). "A survey of current work in biomedical text mining." *Briefings in bioinformatics* 6(1): 57.

Davey, N. E., K. Van Roey, et al. (2011). "Attributes of short linear motifs." *Mol. BioSyst.*

Doerr, A. (2010). "The importance of being negative." *Nature Methods* 7(1): 10-11.

Donaldson, I., J. Martin, et al. (2003). "PreBIND and Textomy–mining the biomedical literature for protein-protein interactions using a support vector machine." *BMC bioinformatics* 4(1): 11.

Enault, F., K. Suhre, et al. (2003). "Annotation of bacterial genomes using improved phylogenomic profiles." *Bioinformatics* 19(Suppl 1): i105.

Enault, F., K. Suhre, et al. (2003). "Annotation of bacterial genomes using improved phylogenomic profiles." *Bioinformatics* 19(suppl 1): i105.

Enright, A. J., I. Iliopoulos, et al. (1999). "Protein interaction maps for complete genomes based on gene fusion events." *Nature* 402(6757): 86-90.

Espadaler, J., O. Romero-Isart, et al. (2005). "Prediction of protein–protein interactions using distant conservation of sequence patterns and structure relationships." *Bioinformatics* 21(16): 3360.

Guo, Y., L. Yu, et al. (2008). "Using support vector machine combined with auto covariance to predict protein–protein interactions from protein sequences." *Nucleic Acids Research* 36(9): 3025.

Huynen, M. A. and B. Snel (2000). "Gene and context: integrative approaches to genome analysis." *Advances in Protein Chemistry* 54: 345-379.

Jothi, R., P. F. Cherukuri, et al. (2006). "Co-evolutionary analysis of domains in interacting proteins reveals insights into domain-domain interactions mediating protein-protein interactions." *Journal of molecular biology* 362(4): 861-875.

Jothi, R., M. G. Kann, et al. (2005). "Predicting protein–protein interaction by searching evolutionary tree automorphism space." *Bioinformatics* 21(suppl 1): i241.

Kecman, V. (2001). *Learning and soft computing : support vector machines, neural networks, and fuzzy logic models.* Cambridge, Mass., MIT Press.

Krogan, N. J., G. Cagney, et al. (2006). "Global landscape of protein complexes in the yeast Saccharomyces cerevisiae." *Nature* 440(7084): 637-643.

Lee, S. I. and S. Batzoglou (2003). "Application of independent component analysis to microarrays." *Genome Biology* 4(11): R76.

Li, H., J. Li, et al. (2006). "Discovering motif pairs at interaction sites from protein sequences on a proteome-wide scale." *Bioinformatics* 22(8): 989.

Marcotte, E. M., M. Pellegrini, et al. (1999). "Detecting protein function and protein-protein interactions from genome sequences." *Science* 285(5428): 751.

Mishra, G. R., M. Suresh, et al. (2006). "Human protein reference database - 2006 update." *Nucleic Acids Research* 34: D411-D414.

Mitchell, T. M. (1997). *Machine learning.* New York, McGraw-Hill.

Oyang, Y. J., S. C. Hwang, et al. (2005). "Data classification with radial basis function networks based on a novel kernel density estimation algorithm." *IEEE Transactions on Neural Networks* 16(1): 225-236.

Peri, S., J. D. Navarro, et al. (2003). "Development of human protein reference database as an initial platform for approaching systems biology in humans." *Genome research* 13(10): 2363.

Peri, S., J. D. Navarro, et al. (2003). "Development of human protein reference database as an initial platform for approaching systems biology in humans." *Genome Research* 13(10): 2363-2371.

Pevzner, P. A. and S. H. Sze (2000). *Combinatorial approaches to finding subtle signals in DNA sequences*, Citeseer.

Rogozin, I. B., K. S. Makarova, et al. (2002). "Connected gene neighborhoods in prokaryotic genomes." *Nucleic Acids Research* 30(10): 2212.

Salgado, H., G. Moreno-Hagelsieb, et al. (2000). "Operons in Escherichia coli: genomic analyses and predictions." *Proceedings of the National Academy of Sciences* 97(12): 6652.

Salwinski, L., C. S. Miller, et al. (2004). "The database of interacting proteins: 2004 update." *Nucleic Acids Research* 32(suppl 1): D449.

Shen, J., J. Zhang, et al. (2007). "Predicting protein–protein interactions based only on sequences information." *Proceedings of the National Academy of Sciences* 104(11): 4337.

Sheu, S. H., D. R. Lancia, et al. (2005). "PRECISE: a database of predicted and consensus interaction sites in enzymes." *Nucleic Acids Research* 33(suppl 1): D206.

Shoemaker, B. A. and A. R. Panchenko (2007). "Deciphering protein–protein interactions. Part II. Computational methods to predict protein and domain interaction partners." *PLoS computational biology* 3(4): e43.

Smialowski, P., P. Pagel, et al. (2010). "The Negatome database: a reference set of non-interacting protein pairs." *Nucleic acids research* 38(suppl 1): D540.

Snitkin, E., A. Gustafson, et al. (2006). "Comparative assessment of performance and genome dependence among phylogenetic profiling methods." *BMC bioinformatics* 7(1): 420.

Soong, T., K. O. Wrzeszczynski, et al. (2008). "Physical protein–protein interactions predicted from microarrays." *Bioinformatics* 24(22): 2608-2614.

Stumpf, M. P. H., T. Thorne, et al. (2008). "Estimating the size of the human interactome." *Proceedings of the National Academy of Sciences* 105(19): 6959.Sun, J., J. Xu, et al. (2005). "Refined phylogenetic profiles method for predicting protein–protein interactions." *Bioinformatics* 21(16): 3409.

Szklarczyk, D., A. Franceschini, et al. (2011). "The STRING database in 2011: functional interaction networks of proteins, globally integrated and scored." *Nucleic Acids Research* 39(suppl 1): D561.

Tan, S. H., W. Hugo, et al. (2006). "A correlated motif approach for finding short linear motifs from protein interaction networks." *BMC bioinformatics* 7(1): 502.

Van Dijk, A., C. Ter Braak, et al. (2008). "Predicting and understanding transcription factor interactions based on sequence level determinants of combinatorial control." *Bioinformatics* 24(1): 26.

Vapnik, V. and V. Vapnik (1998). *Statistical learning theory*, Wiley New York.

Walhout, A. J. M., R. Sordella, et al. (2000). "Protein interaction mapping in C. elegans using proteins involved in vulval development." *Science* 287(5450): 116.

Witten, I. H., E. Frank, et al. (2011). *Data mining : practical machine learning tools and techniques*. Burlington, MA, Morgan Kaufmann.

Yu, C. Y., L. C. Chou, et al. (2010). "Predicting protein-protein interactions in unbalanced data using the primary structure of proteins." *BMC Bioinformatics* 11(1): 167.

Yu, J., M. Guo, et al. (2010). "Simple sequence-based kernels do not predict protein-protein interactions." *Bioinformatics* 26(20): 2610.

12

Computational Approaches to Elucidating Transient Protein-Protein Interactions, Predicting Receptor-Ligand Pairings

Ernesto Iacucci[1], Samuel Xavier de Souza[2] and Yves Moreau[1]

[1]K.U.Leuven
[2]Universidade Federal do Rio Grande do Norte
[1]Belgium
[2]Brazil

1. Introduction

Protein-protein interactions (PPI) are one of the most important biological events which occur in the cell. As PPIs regulate almost all biological processes in the cell, aberrations in PPI may cause severe health problems. One specific area of PPI is receptor-ligand interactions. These interactions are transient yet account for a large part of cell-to-cell communication. As PPI is an important area of research, many groups have proposed methods to make computational predictions of PPI.

The basis of the majority of these methods rely largely on the phylogenetic profile analysis of candidate interactors. These methods determine the similarity of the phylogenetic history of a protein A and its putative protein partner B, examining the most accurate measure of similarity between the phylogenetic histories of A and B in order to predict interaction. As interacting proteins should co-adapt as they are under the same evolutionary pressures, it is self-evident that interacting receptors and ligands should be identifiable by application of the same methodology.

While several methods, described below, make use of phylogenetic information to predict protein-protein interaction (PPI), more contemporary work has been conducted in the area of data fusion and kernel learning. We describe one method [Iacucci et al. 2011] in detail which does both. In this work, the existing line of phylogenetic research is extended by using phylogenetic data to construct a kernel to train a least square support vector machines (LS-SVM) in order to classify candidate receptors and ligands as *interacting* or *non-interacting*.

In this chapter, we discuss the plethora of various methods for determining protein-protein interactions. In addition, we evaluate the application of LS-SVMs to the sub-problem of receptor-ligand interaction prediction.

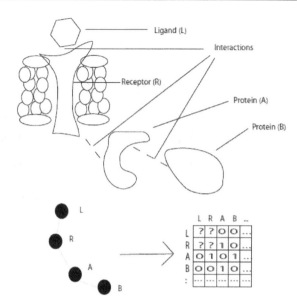

Fig. 1. The Receptor Ligand Schematic. Schematic of receptor-ligand and protein-protein interaction model. Top image is a representation of in-vivo interaction of proteins, receptors, and ligands while bottom image is the graph representation from which a PPI adjacency matrix may be derived. (Figure published in Iacucci *et al.* 2010)

2. Current computational approaches for predicting protein-protein interaction

During the past decade, many methods for prediction of interaction between proteins have been studied due to the crucial role that these interactions have in the understanding of the diverse cellular mechanisms of life forms. Many of these methods involve experimental analysis of specific protein pairs in a smaller scale or, in current high throughput methods [Uetz et al. 2000, Giot et al. 2003], a large amount of protein interactions. The later can be used to detect many interactions with reasonable sensitivity but rather low specificity. Another, relatively inexpensive, way to predict protein-protein interactions does not include wet lab analysis, using instead a variety of computational approaches. These approaches can complement experimental wet lab techniques and are often supported by either the hypothesis of protein co-evolution [Tan et al. 2004, Tillier et al. 2006, Izarzugaza et al. 2006], structural similarities [Gong et al. 2005, Ogmen et al. 2005] or amino-acids sequence conservation [Pitre et al. 2006].

While the entire genomes of many species are already completely sequenced, the interactone of these life forms is often many orders of magnitude larger and yet far from being fully mapped [Claverie et al. 2001, Rubin et al. 2001]. High throughput experimental techniques will certainly help to create this mapping and computational approaches can complement their results identifying false positive interactions, and therefore improving the specificity of these experimental techniques. Apart from the experimental techniques, computational methods are themselves a powerful and affordable alternative to contribute to interactome mapping.

Several computational approaches have been developed in recent years. Many of them are freely available as web tools offering a variety of services to biologists and bioinformatics that range from prediction of interactions between of proteins in pairs or in batch mode, through browsing of consolidated large scale analysis, up to visualization of binding sites and physical interactions in 3-dimensional images.

The methodologies of these many different approaches vary, but they all seem to be supported by the following findings: (a) evidences in favor of the hypothesis of protein co-evolution and the similarities observed in the phylogenetic trees of these proteins; and (b) datasets of already known protein-protein interactions verified by experimental techniques. Co-evolutionary methods find protein pairs with the highest co-evolutionary signal. This information is powerful to predict which members of interacting protein families are associated structurally or functionally although it is not specific enough to predict whether or not two protein families interact. On the other hand, methods supported by verified protein-protein interactions make use of the structural or amino-acid sequence similarities of interacting proteins partners to predict interaction between query protein pairs. This makes such methods more suitable to predict physical interactions rather than functional relationships.

We have reviewed 6 methods and their web tools for predicting protein-protein interactions. Three of them, supported by the protein co-evolution hypothesis, are: TSEMA [Izarzugaza et al. 2006], ADVICE [Tan et al. 2004], Codep [Tillier et al. 2006]. The other three, supported by datasets of verified interactions, are: PIPE [Pitre et al. 2006], PSIbase [Gong et al. 2005], and PRISM [Ogmen et al. 2005]. In the next Sections, we describe each one of these two types of methods.

2.1 Current co-evolutionary methods

Many studies of the problem of predicting protein-protein interactions investigate the similarity of the phylogenetic history of the interaction partners. Many examples of interaction between proteins have presented signs of co-evolution in such a way that members of different interacting protein families present similarity between their phylogenetic trees [Fryxell 1996, Goh et al. 2000, van Kesteren et al. 1996, Moyle et al. 1994, Pazos and Valencia 2001]. The core of co-evolutionary methods is based on measures of similarity for the phylogenetic trees of interacting protein partners.

There are several measures for similarity between phylogenetic trees. The trees can be compared directly [Goh et al. 2000], via distance matrices [Moyle et al. 1994, Goh and Cohen 2002, Ramani and Marcotte 2003, Gertz et al. 2003], or using multiple sequence alignments [Tillier et al. 2006]. In the following Sections, we present three co-evolutionary methods: TSEMA and ADVICE, which uses distance to compare the phylogeneitc trees, and Codep, which computes the correlation between co-evolving partners from their multiple sequence alignments.

2.1.1 Interactive prediction of protein pairing between interacting families TSEMA

TSEMA is a method and web tool to predict mappings between two families of homologous proteins. The probed protein families can either be inputted using the Newick format or in a format comparable with ClustalW, which is used to build the trees. The distances for all pairs of proteins within both families are extracted from their phylogenetic trees by summing the length of the branches separating each pair of proteins in the trees. The algorithm of TSEMA finds the mapping between the two sets proteins which maximizes the matching between the sets of distances using a modified implementation of the Ramani and Marcotte's Monte Carlo Metropolis method [Ramani and Marcotte 2003].

Availability: http://tsema.bioinfo.cnio.es/

2.1.2 Automated Detection and Validation of Interaction by Co-Evolution – ADVICE

ADVICE predicts and validate protein-protein interactions using observed co-evolution between proteins. The web tool retrieves orthologous sequences of a list of input protein sequences and compute the similarities among the proteins evolutionary histories. The tool also provides visualization for the resulting network of co-evolved proteins.

The ADVICE algorithm infers interaction based on the correlation between distance matrices constructed from the evolutionary history using orthologous sequences of top 10 species. The tool uses BLAST [Altschul et al. 1990] to search the orthologous sequences from Swiss-Prot and TrEMBL databases [Boeckmann 2003]. The distance matrices are constructed using only pairs of orthologous sequences occurring together in the same species. By default, only the orthologous sequences of the top 10 species, based on the BLAST E-value, are used to construct the matrices, excluding those species where more than one orthologous sequence of the input sequence is found. The actual distance matrices are build from the respective multiple sequence alignments using ClustalW [Thompson et al. 1994]. The algorithm then calculates the correlation between pairs of matrices measuring the Pearson's correlation coefficients, which has values between -1, implying 100% anti-correlation, and 1,which representing 100% evolutionary history similarity, being velues above 0.8 good indicators of interaction and values below 0.3 a good cut-off value to detect potential spurious interaction.

Availability: http://advice.i2r.a-star.edu.sg

2.1.3 Maximizing co-evolutionary interdependencies to discover interacting proteins – Codep

Codep and the other co-evolutionary methods find proteins with the highest co-evolutionary signals, independent of physical or functional interaction. The main difference of Codep is that it uses multiple sequence alignments directly rather than distances obtained from the sequences. The user inputs two phylogenetic trees with orthologous sequences. The algorithm maximizes interdependency based on the maximal mutual information. It does this by fixing one of the multiple sequence alignments and varying the order of the other via exhaustive search or via simulated annealing.

The rationale to use directly multiple sequence alignments instead of the distance matrices, which provides a faster way to calculate correlation, is that character-state methods in the field of phylogenetic analysis are more powerful than distance method and some information can be lost in transforming character-state data into distance matrices.

Availability: http://www.uhnresearch.ca/labs/tillier/

2.2 Methods based on verified interactions

Another promising computational approach to predict new protein-protein interactions is to look at the physical structure and the conservation of amino-acid sequences in partners of interactions that are already reliably known to exist. Then, use the gathered information to find correlation with query protein partners of a probed interaction. Many methods apply this approach, which have delivered powerful tools for finding new interactions [Pitre et al. 2006] and even to corroborate with the protein co-evolution hypothesis [Kim et al. 2004]. In the next three Sections we describe three of these methods: PIPE, which compares amino-acid subsequences between probed protein partners and partners of verified protein interactions from a database; and PSIbase and PRISM, both which compare structural characteristics of probed and verified interactions.

2.2.1 Protein-Protein Interaction Prediction Engine – PIPE

PIPE is a computational tool that can effectively identify protein-protein interactions among *S. cerevisiae* protein pairs. It relies on previously determined *S. cerevisiae* protein interactions compiled from the DIP [Salwinski et al. 2004] and MIPS [Mewes et al. 2002] databases to construct a graph where the nodes are proteins and the edges represent the relationship of interacting proteins.

The working principle of the PIPE algorithm to probe interaction between the pair of proteins A-B is to compare sliding subsequences of amino-acids of size w from A to subsequences of the same size of all proteins in the graph of known interactions; then compare sliding subsequences of B to the neighbors of all matches of A. If protein pair C-D are connected in the graph, representing a verified interaction, and if A has subsequence matches with C and B has matches with D, then the pair A-B is more likely to present interaction. The accumulation of all matches of subsequence comparisons presented in form of a matrix indicates a predicted interaction when the higher values in this matrix is above a given threshold of M matches.

The algorithm has three tuning parameters: w, M, and S_{PAM}, which is the threshold value that indicates a match between two subsequences of amino-acids. The author of PIPE chose to fix w in 20, and tune the other two parameter either by trial and error or by statistical evaluation.

PIPE is reported to have success rate comparable to biochemical techniques, with a sensitivity of 61% , specificity of 89%, and overall accuracy of 75%. The main disadvantages of PIPE is its heavy computational burden and its limitation to yeast proteins.

Availability: http://pipe.cgmlab.org

2.2.2 Protein Structural Interactome Map – PSIMAP

PSIMAP is a map that describes the information about domain-domain and protein-protein interactions known to exist in the Protein Data Bank of structures. It is based on the principle that interaction between protein structures is conserved as closely as protein structures themselves [Park et al. 2001, Aloy and Rossell, 2002; Aloy et al. 2003]. It that predicts if domains or proteins structures interact calculating if every possible pair of structures has an Euclidean distance below a certain threshold. There are three different methods to do this: Full Atom Contact (FAC); Sample Atom Contact (SAC); and Bounding Box Contact (BBC). FAC is the most accurate, whereas SAC and BBC [Dafas et al. 2004] are faster methods.

PSIMAP extract the molecular interaction information of proteins from the PDB. It associates this information to domains using the Structural Classification of Proteins (SCOP) to assign the domains to the structures.

Availability: http://psimap.org and http://psibase.kaist.ac.kr/

2.2.3 Protein Interactions by Structural Matching – PRISM

The PRISM tool allows the user to explore protein interfaces and predict protein-protein interactions by comparing the structure of query proteins to those of a structurally and evolutionarily subset of biological and crystal interactions present in the Protein Data Bank (PDB) [Berman 2000]. Interfaces are defined as the set of residues forming the region of the structure through which two different protein chains bind to each other. This set consists the contacting residues between the chains and the neighboring residues up to a certain distance threshold.

The interfaces in PRISM were obtained from all higher complexes of proteins available in the PDB [Keskin et al. 2004]. From the 49512 interfaces extracted form the PDB, 8205 clusters were obtained using a sequence order-independent computer vision-based algorithm to structurally compare the interfaces. From these 8205 clusters, PRISM considers only 158 template interfaces (Oct/2011) that were found to have evolutionary hotspots [Keskin et al. 2005].

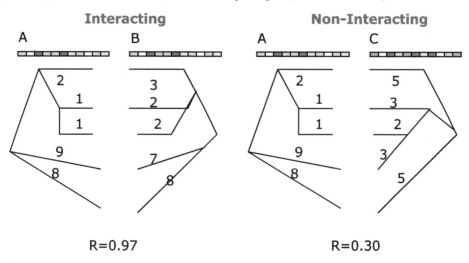

Fig. 2. Phylogenetic Analysis of Proteins

As proteins A and B are interacting proteins, they share a similar phylogenetic history and thus their phylogenetic profiles are highly correlated (R=0.97). Proteins A and C are non-interacting and are thus not strongly correlated (R=0.30).

PRISM algorithm compares the 158 template interfaces to a target dataset of 18698 structures obtained from passing the structures extracted form the PDB through a 50% sequence identity filter, splitting multimeric proteins into constituent chains, and counting homologous chains only once. The user can also probe a protein structure that is originally not present in the target dataset. To compare target proteins to template interfaces PRISM algorithm do as follow: (a) extract target protein surfaces; (b) compare the target surface with all interface complementary partners from the template dataset using MULTIPROT [Shatsky et al. 2004] in order to detect common geometrical cores in a sequence-order-independent way; (c) check for the presence of hotspots in the target structure. The final prediction score is calculated weighting the structural match ratio and the hotspot match ratio.

Availability: http://prism.ccbb.ku.edu.tr/prism/

3. Phylogenetics and beyond, how multiple kernel learning can improve predictions of receptor-ligand pairings

As seen in the sections above, there are several groups which have used phylogenetic analysis to predict PPI. Here we examine the use of multiple kernel learning in the task of PPI prediction. Kernel learning provides the ability to utilize directly and indirectly related data (such as expression measures, domain content, etc.) and perform classification in high dimensional space. When different data sources are used, separate kernel classifiers can be built and the combined output used to provide a final result.

One of the first groups to look at predicting PPI using multiple data sources was Bhardwaj et al. (2003). They use both phylogenetic information as well as expression data to make their predictions. The use of both data sources were proved, in their work, to provide results with greater accuracy than with using phylogenetic analysis alone. Co-expression is a logical source of information for use in this setting as proteins which interact for the purpose of performing a common function are likely to be co-expressed as they will need to be present at the same time in the cell [Bhardwaj et al. 2003, Grigoriev et al. 2001].

The idea of combining expression and phylogenetic information to predict PPI is clearly a step on a path which leads one to consider a wider variety of data integration. Other data sources include domain information as domains are known to interact and it is clear that this data would provide additional insight into the task of protein-protein interaction. Combining the above mentioned data sources can be carried out by using multiple kernel learning.

To examine the utility of multiple kernel learning with respect to this task, it is necessary to cite an example in which it performs better than other settings. One such example exists when one looks at the work of Gertz et al. (2003) and compare it with the work presented in Iacucci et al. (2011). Both groups look at the receptor-ligand prediction task and apply computational methods to the same dataset. The datasets consist of members of the chemokine and tgfβ ligand families with their respective receptor families. In the case of Gertz et al (2003), distances matrices are created for the families and are matched according to their similarity. Using a Metropolis Monte Carlo optimization algorithm, the Gertz et al. (2003) group explored and scored possible matches between the two matrices, until they reached optimal solutions. A limitation of this approach is that it relied on phylogenetic distance information alone.

Contrary to the work of Gertz et al (2003), the work presented by Iacucci et al 2011 proposes that the integration of multiple data sources results in more accurate matches. This work involved the creating of a combined kernel classifier to carry out the learning task. While other kernel-based works have been applied to the PPI task [Kim et al. 2010, Miwa et al. 2009], the work of Iacucci et al (2011) is unique as they apply multiple kernel learning to the receptor-ligand problem. More specifically, they apply the least-squares support vector machines (LS-SVM) method based on the conclusions by Suykens et al. (2001) which shows this implementation to be robust.

The ability of Iacucci et al. (2011) to predict candidate receptor-ligand pairs has been show to outpace that of Gertz et al. (2003) on the same dataset. This work involves using multiple data sources (expression, phylogenetic, and protein-domain content information), computing separate kernels for each data type, creating LS-SVM classifiers and combining the results to predict receptor-ligand pairs. The specifics of these steps will be discussed below.

3.1 Data sources

Several choices for data sources can be considered when addressing the PPI prediction task. While the studies, mentioned above, which use phylogenetic information rely on sequence data, other sources are available. Such sources include domain content data and expression data.

The phylogenetic data used in the Iacucci et al. (2011) study was derived through several steps. First, candidate receptor and ligand sequences were retrieved for seven species (*Rattus norvegicus, Mus* musculus, *Homo sapiens, Pan troglodytes, Canis familiaris, Cavia porcellus,* and *Bos taurus*) from ensemble build 51 [Hubbard et al. 2009]. Following this, the sequences were aligned using ClustalW [Thompson et al. 1994] Once aligned, the sequences were edited so as to eliminate the positions which were not conserved across the seven orthologous sequences. Finally, the pair-wise alignment score was then taken for each possible species to species comparison between the edited orthologous sequences (as seven species are used, a total of 21 pair-wise comparisons for each candidate are created). The distance scores form a phylogenetic vector which was then used to create the phylogenetic kernel.

The expression data used in the Iacucci et al. (2011) work was taken from the well-known GNF human expression atlas (79 tissues) [Su et al. 2004], the data was normalized (values were mean-zeroed and the standard deviation was set to one) and was further transformed into the expression kernel.

For the Iacucci et al. (2011) work, the domain content of each candidate protein (receptor or ligand) was taken from the Interpro Database [Hunter et al. 2009]. A vector for each candidate protein was created where the presence of a protein domain was indicated with a '1' and the absence of a domain was indicated by a '0'. This data was then transformed to create the domain content kernel.

The "Golden Standard" for the verification of the Gertz et al (2003) and the Iacucci (2011) et al. work is based on the Database of Ligand-Receptor Partners (DLRP) [Graeber et al. 2001]. This dataset is an experimentally derived dataset where known receptor-ligand pairs are stored. The information found here was used to train the LS-SVM described below. In addition, it was also used as the "Golden Standard" to determine which predictions, by both groups, were true positives and false positive as well as false negatives and true negatives. These values were then used to calculate specificity and sensitivity of each groups' predictions to ultimately determine which approach provided better results.

3.2 Kernel creation and the LS-SVM

The creation of the kernels and the training of the least-squares support vector machine (LS-SVM) in the work presented by Iacucci et al. (2011) required multiples steps. First, the data sources, discussed above, were used to create data matrices (phylogenetic, expression, and domain content) which were then used to create three kernels for each receptor-ligand family. Following this, the LS-SVMs were trained using the three kernels to predict outcomes for receptor-ligand pairs known from the DLRP "Golden Standard".

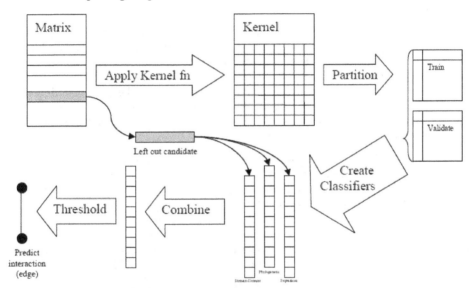

Fig. 3. Work flow of the combined kernel classifier

Data was partitioned into training and validation sets and parameters were tuned using a five fold validation strategy. The final output of the classifiers was achieved by a leave one out strategy. The classifier values were combined for a final result and a threshold was applied to determine which values are predicted edges (Figure published in Iacucci *et al.* 2011).

The kernel function used by Iacucci et al. (2011) measures the similarity between two proteins A and B (K(A,B)), one a candidate receptor A and the other a candidate ligand B. The LS-SVM classifier produced by Iacucci et al. (2011) is a binary predictor which assigns new examples in "interacting" or "non-interacting" classes. Creating the kernels from the various data matrices involved trials with different kernel functions, with linear functions ultimately being found to give the best performance in all cases. Data was partitioned into training and validation sets and parameters were tuned using a five fold validation strategy The final output of the classifiers was achieved by a leave-one-out strategy. The classifier values were scaled (minimum set to zero, maximum set to one). The values were then combined, as defined in (1), for a final result. Figure 3 provides an overview of the workflow as described above.

$$g_{comb}(x) = \frac{g_{phylo}(x) + g_{exp}(x) + g_{dom}(x)}{3} \tag{1}$$

3.3 Results and discussion

The comparison of the phylgenetic based method of Gertz et al. (2003) and the combined kernel classifer method of Iacucci et al. (2011) provides a clear perspective on the advantages of multiple kernel learning in the PPI prediction task. As both groups use the same dataset and have results which can be summarized and contrasted using recall, precision, and the F-measures.

The Iacucci et al. (2011) predictions for the tgfβ family accurately reconstructed over 76% of the supported edges (0.76 recall and 0.67 precision) of the know DLRP receptor-ligand pairs. In this case, the combined kernel classifier was able to relatively improved upon the Gertz et al. (2003) work by a factor of approximately two as the Gertz et al. (2003) work reconstructs 44% of the supported edges (0.44 recall and 0.53 precision) of the know DLRP receptor-ligand pairs. Comparing F-measures, we see that the combined kernel classifer method improved upon that of Gertz et al. (2003) significantly as the Iacucci et al. (2011) method has an F-measure of 0.71 while that of Gertz et al. (2003) has a value of 0.48.

The Iacucci et al. (20011) predictions for the chemokine family accurately reconstructed over 65% of the supported edges (0.65 recall and 0.23 precision) of the know DLRP receptor-ligand pairs. In this case, the combined kernel classifier was able to relatively improved upon the Gertz et al. (2003) work by a factor of approximately three as the Gertz et al. (2003) work reconstructs 22% of the supported edges (0.22 recall and 0.37 precision) of the know DLRP receptor-ligand pairs. Comparing F-measures, we see that the combined kernel classifer method improved upon that of Gertz et al. (2003) significantly as the Iacucci et al (2011) method has an F-measure of 0.33 while that of Gertz et al. (2003) has a value of 0.27.

Qualitatively, the performance of the Iacucci et al (2011) method also seems to be matching the performance of Gertz et al. (2003), as the novel interaction of CCR1 with SCY11 [Gao et al. 1996] reported in their work is also discovered using Iacucci et al (2011) method.

The comparison of the results of the two methods discussed here support the notion that kernel learning presents a useful methodology for elucidating receptor-ligand pairings. The benefits of the combined kernel classifier method over the Gertz et al. (2003) method are clear. Foremost in the advantages are the ability to predict multiple ligands for one receptor, which represents an necessary feature for receptor-ligand research. Also, as the classifier output is continuous, the results can be considered to be prioritized, this presents a major convenience to researchers as often the set of candidate ligands are large and financial and time resources to validate few.

4. Conclusion

The task of PPI prediction is a difficult and important area of bioinformatics research. As the number of possible interacting protein pairs in the cell is huge, wet-lab experimentation validation of all of them is essentially impossible. In addition to being time consuming, in-vivo validation costs are also a consideration. Having a computational method for predicting PPI is therefore a necessary tool for researchers.

Several groups have addressed the PPI prediction task. While several have used phylogenetics to solve the problem, others have used physical protein structures and amino-acid sequence information to assist in making the predictions. We have reviewed these methods and discussed the key differences among them.

Methods, which rely on the physical structure and the conservation of amino-acid sequences in partners of interactions that are already reliably known to exist, also give researchers additional insight to function prediction as the methods are based on known examples. The drawback of these methods is that one has to have a known example for a comparison, which is not always the case when researching candidate receptor-ligand pairs.

Methods which rely on phylogenetic histories to determine PPI are based on a well-establish rational which holds that as interacting proteins co-evolve, there phylogenetic histories should be similar. This explains why the methods which rely on phylogenetic information are largely based on measures of similarity for the phylogenetic trees of interacting protein partners.

The advantage of using multiple kernel learning to predict PPI is apparent when using multiple sources of data. Many of the methods, mentioned above, rely on an ever growing amount of publicly available data. The ever expanding amount of high throughput data which continues to become available to the bioinformatics community represents an excellent opportunity to enhance the kernel classifier method presented in Iacucci et al. (2011).

A practical advantage of using multiple data sources allows one to extend the method as new and higher quality sources become available. For example, if better micro-array dataset becomes available in the future, it is an advantage to be able remove the existing expression-based kernel with one derived from the new dataset without having to the retrain a global classifier. Likewise, if additional data sources become available, adding an additional sub-classifier based on the new data source would take less time to train than adding the data source and retraining the global classifier.

Looking forward many exciting challenges remain to be addressed in this field. While the task of PPI is daunting and complex, the work reviewed above demonstrates that it is also rich with opportunities for improvement and further development.

5. Acknowledgments

Funding: The authors would like to acknowledge support from:

- Research Council KUL:
    ProMeta, GOA Ambiorics, GOA MaNet, CoE EF/05/007 SymBioSys en KUL PFV/10/016 SymBioSys , START 1, several PhD/postdoc & fellow grants
- Flemish Government:
    FWO: PhD/postdoc grants, projects , G.0318.05 (subfunctionalization), G.0553.06 (VitamineD), G.0302.07 (SVM/Kernel), research communities (ICCoS, ANMMM, MLDM); G.0733.09 (3UTR); G.082409 (EGFR)
    IWT: PhD Grants, Silicos; SBO-BioFrame, SBO-MoKa, TBM-IOTA3

FOD:Cancer plans
IBBT
- Belgian Federal Science Policy Office: IUAP P6/25 (BioMaGNet, Bioinformatics
  and Modeling: from Genomes to Networks, 2007-2011) ;
  -EU-RTD: ERNSI: European Research Network on System Identification; FP7-HEALTH
  CHeartED

6. References

Aloy, P., & Russell, R. B. (2002). Interrogating protein interaction networks through
    structural biology. Proceedings of the National Academy of Sciences of the United
    States of America, 99(9), 5896-901. doi:10.1073/pnas.092147999

Aloy, P., Ceulemans, H., Stark, A., & Russell, R. B. (2003). The relationship between
    sequence and interaction divergence in proteins. Journal of molecular biology,
    332(5), 989-98. Retrieved from http://www.ncbi.nlm.nih.gov/pubmed/14499603

Berman, H. M. (2000). The Protein Data Bank. Nucleic Acids Research, 28(1), 235-242.
    doi:10.1093/nar/28.1.235

Bhardwaj N, Lu H: Correlation between gene expression profiles and protein-protein
    interactions within and across genomes. Bioinformatics 2005, 21:2730-2738.

Bleakley K, Yamanishi Y: Supervised prediction of drug-target interactions using bipartite
    local models. Bioinformatics 2009, 25:2397-2403.

Boeckmann, B. (2003). The SWISS-PROT protein knowledgebase and its supplement
    TrEMBL in 2003. Nucleic Acids Research, 31(1), 365-370. doi:10.1093/nar/gkg095

Claverie, J. M. (2001). Gene number. What if there are only 30,000 human genes? Science
    (New York, N.Y.), 291(5507), 1255-7. Retrieved from
    http://www.ncbi.nlm.nih.gov/pubmed/11233450

Dafas, P., Bolser, D., Gomoluch, J., Park, J., & Schroeder, M. (2004). Using convex hulls to
    extract interaction interfaces from known structures. Bioinformatics (Oxford,
    England), 20(10), 1486-90. doi:10.1093/bioinformatics/bth106

Fryxell, K. J. (1996). The coevolution of gene family trees. Trends in Genetics, 12(9), 364-369.
    doi:10.1016/S0168-9525(96)80020-5

Gao JL, Sen AI, Kitaura M, Yoshie O, Rothenberg ME, Murphy PM, Luster AD:
    Identification of a mouse eosinophil receptor for the CC chemokine eotaxin.
    Biochem Biophys Res Commun 1996, 223:679-684.

Ge H, Liu Z, Church GM, Vidal M: Correlation between transcriptome and interactome
    mapping data from Saccharomyces cerevisiae. Nat Genet 2001, 29:482-486.

Gertz J, Elfond G, Shustrova A, Weisinger M, Pellegrini M, Cokus S, Rothschild B: Inferring
    protein interactions from phylogenetic distance matrices. Bioinformatics 2003,
    19:2039-2045.

Giot, L., Bader, J. S., Brouwer, C., Chaudhuri, A., Kuang, B., Li, Y., Hao, Y. L., et al. (2003). A
    protein interaction map of Drosophila melanogaster. Science (New York, N.Y.),
    302(5651), 1727-36. doi:10.1126/science.1090289

Goh, C. S., Bogan, A. A., Joachimiak, M., Walther, D., & Cohen, F. E. (2000). Co-evolution of
    proteins with their interaction partners. Journal of molecular biology, 299(2), 283-
    93. doi:10.1006/jmbi.2000.3732

Gong, S., Yoon, G., Jang, I., Bolser, D., Dafas, P., Schroeder, M., Choi, H. H., et al. (2005).
    PSIbase: a database of Protein Structural Interactome map (PSIMAP).
    Bioinformatics (Oxford, England), 21(10), 2541-3.

doi:10.1093/bioinformatics/bti366

Graeber TG, Eisenberg D: Bioinformatic identification of potential autocrine signaling loops in cancers from gene expression profiles. Nat Genet 2001, 29:295-300.

Grigoriev A: A relationship between gene expression and protein interactions on the proteome scale: analysis of the bacteriophage T7 and the yeast Saccharomyces cerevisiae. Nucleic Acids Res 2001, 29:3513-3519.

Hubbard TJ, Aken BL, Ayling S, Ballester B, Beal K, Bragin E, Brent S, Chen Y, Clapham P, Clarke L et al.: Ensembl 2009. Nucleic Acids Res 2009, 37:D690-D697.

Hunter S, Apweiler R, Attwood TK, Bairoch A, Bateman A, Binns D, Bork P, Das U, Daugherty L, Duquenne L et al.: InterPro: the integrative protein signature database. Nucleic Acids Res 2009, 37:D211-D215.

Izarzugaza, J. M. G., Juan, D., Pons, C., Ranea, J. a G., Valencia, A., & Pazos, F. (2006). TSEMA: interactive prediction of protein pairings between interacting families. Nucleic acids research, 34(Web Server issue), W315-9. doi:10.1093/nar/gkl112

Jacob L, Vert JP: Protein-ligand interaction prediction: an improved chemogenomics approach. Bioinformatics 2008, 24:2149-2156.

Keskin, O., Ma, B., & Nussinov, R. (2005). Hot regions in protein--protein interactions: the organization and contribution of structurally conserved hot spot residues. Journal of molecular biology, 345(5), 1281-94. doi:10.1016/j.jmb.2004.10.077

Keskin, O., Tsai, C.-J., Wolfson, H., & Nussinov, R. (2004). A new, structurally nonredundant, diverse data set of protein-protein interfaces and its implications. Protein science : a publication of the Protein Society, 13(4), 1043-55. doi:10.1110/ps.03484604

Kim S, Yoon J, Yang J, Park S: Walk-weighted subsequence kernels for protein-protein interaction extraction. BMC Bioinformatics 2010, 11:107.

Kim, W. K., Bolser, D. M., & Park, J. H. (2004). Large-scale co-evolution analysis of protein structural interlogues using the global protein structural interactome map (PSIMAP). Bioinformatics (Oxford, England), 20(7), 1138-50. doi:10.1093/bioinformatics/bth053

Mewes, H. W., Frishman, D., Güldener, U., Mannhaupt, G., Mayer, K., Mokrejs, M., Morgenstern, B., et al. (2002). MIPS: a database for genomes and protein sequences. Nucleic acids research, 30(1), 31-4. Retrieved from http://www.pubmedcentral.nih.gov/articlerender.fcgi?artid=99165&tool=pmcentrez&rendertype=abstract

Miwa M, Saetre R, Miyao Y, Tsujii J: Protein-protein interaction extraction by leveraging multiple kernels and parsers. Int J Med Inform 2009, 78:e39-e46.

Moyle, W. R., Campbell, R. K., Myers, R. V., Bernard, M. P., Han, Y., & Wang, X. (1994). Co-evolution of ligand-receptor pairs. Nature, 368(6468), 251-5. doi:10.1038/368251a0

Nagamine N, Sakakibara Y: Statistical prediction of protein chemical interactions based on chemical structure and mass spectrometry data. Bioinformatics 2007, 23:2004-2012.

Ogmen, U., Keskin, O., Aytuna, a S., Nussinov, R., & Gursoy, a. (2005). PRISM: protein interactions by structural matching. Nucleic Acids Research, 33(Web Server), W331-W336. doi:10.1093/nar/gki585

Park, J., Lappe, M., & Teichmann, S. A. (2001). Mapping protein family interactions: intramolecular and intermolecular protein family interaction repertoires in the PDB and yeast. Journal of molecular biology, 307(3), 929-38. doi:10.1006/jmbi.2001.4526

Pazos, F., & Valencia, A. (2001). Similarity of phylogenetic trees as indicator of protein-protein interaction. Protein engineering, 14(9), 609-14. Retrieved from http://www.ncbi.nlm.nih.gov/pubmed/11707606

Pitre, S., Dehne, F., Chan, A., Cheetham, J., Duong, A., Emili, A., Gebbia, M., et al. (2006). PIPE: a protein-protein interaction prediction engine based on the re-occurring short polypeptide sequences between known interacting protein pairs. BMC bioinformatics, 7, 365. doi:10.1186/1471-2105-7-365

Rubin, G. M. (2001). The draft sequences. Comparing species. Nature, 409(6822), 820-1. doi:10.1038/35057277

Salwinski, L., Miller, C. S., Smith, A. J., Pettit, F. K., Bowie, J. U., & Eisenberg, D. (2004). The Database of Interacting Proteins: 2004 update. Nucleic acids research, 32(Database issue), D449-51. doi:10.1093/nar/gkh086

Sato T, Yamanishi Y, Kanehisa M, Toh H: The inference of protein-protein interactions by co-evolutionary analysis is improved by excluding the information about the phylogenetic relationships. Bioinformatics 2005, 21:3482-3489.

Shatsky, M., Nussinov, R., Wolfson, H., Guigó, R., & Gusfield, D. (2002). Algorithms in Bioinformatics. (R. Guigó & D. Gusfield, Eds.) (Vol. 2452, pp. 235-250). Berlin, Heidelberg: Springer Berlin Heidelberg. doi:10.1007/3-540-45784-4

Su AI, Wiltshire T, Batalov S, Lapp H, Ching KA, Block D, Zhang J, Soden R, Hayakawa M, Kreiman G, Cooke,M.P., Walker,J.R., Hogenesch,J.B: A gene atlas of the mouse and human protein-encoding transcriptomes. Proc Natl Acad Sci U S A 2004, 101:6062-6067.

Suykens JA, Vandewalle J, De MB: Optimal control by least squares support vector machines. Neural Netw 2001, 14:23-35.

Tan, S.-H., Zhang, Z., & Ng, S.-K. (2004). ADVICE: Automated Detection and Validation of Interaction by Co-Evolution. Nucleic acids research, 32(Web Server issue), W69-72. doi:10.1093/nar/gkh471

Thompson JD, Higgins DG, Gibson TJ: CLUSTAL W: improving the sensitivity of progressive multiple sequence alignment through sequence weighting, position-specific gap penalties and weight matrix choice. Nucleic Acids Res 1994, 22:4673-4680.

Thompson, J. D., Higgins, D. G., & Gibson, T. J. (1994). CLUSTAL W: improving the sensitivity of progressive multiple sequence alignment through sequence weighting, position-specific gap penalties and weight matrix choice. Nucleic acids research, 22(22), 4673-80. Retrieved from http://www.pubmedcentral.nih.gov/articlerender.fcgi?artid=308517&tool=pmcentrez&rendertype=abstract

Tillier, E. R. M., Biro, L., Li, G., & Tillo, D. (2006). Codep : Maximizing Co-Evolutionary Interdependencies to Discover Interacting Proteins, 831(December 2005), 822- 831. doi:10.1002/prot

Uetz, P., Giot, L., Cagney, G., Mansfield, T. A., Judson, R. S., Knight, J. R., Lockshon, D., et al. (2000). A comprehensive analysis of protein-protein interactions in Saccharomyces cerevisiae. Nature, 403(6770), 623-7. Macmillian Magazines Ltd. doi:10.1038/35001009

van Kesteren, R. E., Tensen, C. P., Smit, A. B., van Minnen, J., Kolakowski, L. F., Meyerhof, W., Richter, D., et al. (1996). Co-evolution of ligand-receptor pairs in the vasopressin/oxytocin superfamily of bioactive peptides. The Journal of biological chemistry, 271(7), 3619-26. Retrieved from http://www.ncbi.nlm.nih.gov/pubmed/8631971

13

G-Protein Coupled Receptors:
Experimental and Computational Approaches

Amirhossein Sakhteman, Hamid Nadri and Alireza Moradi
Faculty of Pharmacy, Shahid Sadoughi University of Medical Sciences, Yazd
Iran

1. Introduction

Guanine Nucleotide binding protein coupled receptors (GPCRs) are among the most important targets in the treatment of cancer, endocrine, neural and many other types of disorders. (Katritch, V. & Abagyan, R. 2011) It is believed that activation of some GPCRs is involved in conditions such as immunosupression and response to ischemia of the brain and heart. Therefore, antagonists and agonists of GPCRs are potential therapeutic agents in treatment of inflammatory and ischemic diseases. (Moro, S., Spalluto, G., & Jacobson, K. A. 2005)

The superfamily of GPCRs consists of about 800 receptors which can be divided into different families regarding the similarities in the protein sequence. (Marshall, F. H. & Foord, S. M. 2010)

1. Family A, including rhodopsin and adrenoreceptor
2. Family B, Secretin vasointestinal peptide (VIP), the members of this family bind to hormones and neuropeptides
3. Family C, which include at least eight subtypes of glutamate receptors, the major excitatory receptor in the CNS.
4. Family D, the fungal pheromone p family
5. Family E, the fungal pheromone A family
6. Family F, CAMP receptors of *Dictyostelium discoideum*

The family A receptors is the best studied family of GPCRs in terms of functional and structural viewpoints and is therefore the most important target of GPCRs in drug discovery. (Moro, S. et al. 2005) It was reported that about 30% of the market prescription drugs act on these targets. (Marshall, F. H. et al. 2010)

2. Features and functions of GPCRs

A common feature in class A GPCRs is a core consisting of seven transmembrane domains (TM) connected by three intracellular loops (IL1, IL2 and IL3) and three extracellular loops (EL1, EL2, EL3).(Fig 1) Another feature observed in this class is the two cysteine residues, one in TM3 and the other in EL2. These two cysteines form a disulfide bridge which is responsible for the packing and stabilization of a restricted number of conformations for the seven TM domains (Fig 1).

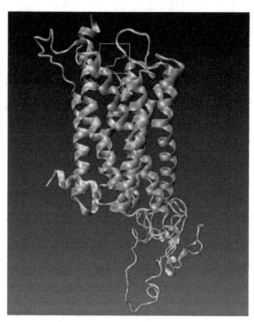

Fig. 1. The seven transmembrane structure of GPCRs. A disulfide bridge between TM3 and EL2 is conserved in most class A GPCRs.

GPCRs differ in the length and function of their N-terminal extracellular domains, C-terminal domain and intracellular loops. (Moro, S. et al. 2005) As an instance, glycoprotein hormone receptor (GPHR) family tether large amino terminal extracellular extensions which are responsible for the recognition and binding of dimeric agonists. Some studies in the area of GPCRs focused on interaction of ligands with GPCRs. There is strong evidence that in case of small molecules like the biogenic amines, the agonists of GPCRs interact directly with specific residues of TM helices of the receptor. On the other hand, for neuropeptides and small protein agonists like neurokinin the interactions involve both exoloops and amino portion of the receptor in association with the residues in TM helices. (Gilbert Vassart & Sabine Costagliola 2003)

2.1 Receptor activation in GPCRs

Many physiological procedures in the body are controlled by the GTPase including signal transduction, control of cellular growth, vesicle and protein transport and cytoskeletal assembly.(Smith, B., Hill, C., Godfrey, E. L., Rand, D., van den Berg, H., Thornton, S. et al. 2009). (Kobilka, B. K. 2007) Activation of a GPCR leads to nucleotide exchange on the $G\alpha$ subunit and cause dissociation of the heterodimer and effector activation. GPCRs are mostly activated by diverse set of signals including small molecules, peptides and light. (Schneider, M., Wolf, S., Schlitter, J., & Gerwert, K. 2011) Members of GPCRs transduce signals by activation of at least one member of homologous hetrotrimeric G proteins. For example in FSH (Follicle Stimulating Hormone), receptor is activated by adrenaline which binds to the TM regions.

It should be noticed that in GPCRs, the ligand binds from the extracellular side and blocks the receptor. The activation or reduction in the basal activity of the heterotrimeric G-protein complex is dependent to the nature of the ligand such as agonists, antagonists and reverse agonists. The activation of G protein is in such a way that an exchange of guanosine diphosphate takes place in a subunit of G-protein. This exchange causes a conformational change in α subunit and leads to dissociation of α subunit from βγ subunits. (Fig 2) The two subunits introduce transduction systems in different ways. (Jaakola, V. P. & Ijzerman, A. P. 2010)

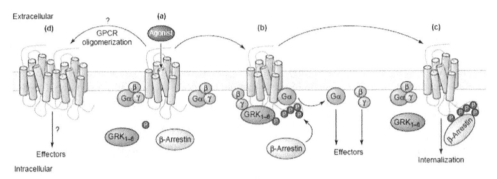

Fig. 2. a) Activation of GPCRs upon agonist binding b) conformational changes in hetrotrimeric G c) dissociation in G protein subunits proteins. d) A common feature observed in most GPCRs is the formation of dimers.

Some evidence verified that GPCRs exist as ensembles of conformations and two factors including binding of agonist and intracellular signaling proteins stabilize the active site of the receptor and accounts for the basal activity of GPCRs in the absence of agonists. The crystal structures of β2AR bound to an agonist and G-protein shows a large conformational change in the intracellular region of the receptor. It is likely that upon agonist binding reshuffling of short range intracellular contact cause large scale domain motions in the receptor. (Vaidehi, N. & Bhattacharya, S. 2011)

With the exception of rhodopsin, most family A GPCRs have considerable basal activity and both modeling and crystallographic data suggest that agonist dependent activation can vary between GPCRs. (Ahuja, S. & Smith, S. O. 2009;Deupi, X. & Kobilka, B. 2007;Katritch, V. et al. 2011) In contrast to rhodopsin where more information is present for the activation state of the molecule, in case of the other GPCRs much less is known about agonist induced conformational changes that occur during the activation of the receptor. (Wess, J., Han, S. J., Kim, S. K., Jacobson, K. A., & Li, J. H. 2008) The most similar parts for the GPCRs are the cytoplasmic ends of the TM segments adjacent to the second and third cytoplasmic domains which interact with G protein.

2.2 Dimerization of GPCRs

An important feature in GPCRs is formation of dimer which affects the receptors in terms of signal trafficking and pharmacology.(Marshall, F. H. et al. 2010). In many cases, the GPCR dimer can alter or regulate coupling or potency of other receptors. As an instance, dimerization of κ-opioid receptor was shown to be in close relation with δ-opioid receptor dimerization. Another consequence of such dimerization is the augmented selectivity of some agonists such as 6-guanidinonal for the dimer with respect to any of the monomers. In addition, it was postulated that the binding of some drugs to more than one type of receptor is the result of dimerization. (Panetta, R. & Greenwood, M. T. 2008)

2.3 GPCRs and drug discovery

Most researches in the area of GPCRs focused on development of more selective or potent compounds of the orthostatic sites, which are apart from the binding sites of endogenous ligands. The allosteric modulators are also considered as promising therapeutic Agents. (Moro, S. et al. 2005)

While, the binding site for most small organic agonists is within TM segments, in case of peptide hormones and proteins the binding site is laid in the extracellular domain.(Kobilka, B. K. 2007) The intrinsic plasticity of GPCRs is a major problem in using their inactive state for agonist design in drug discovery. (Katritch, V. et al. 2011)

3. The role of experimental techniques in structural elucidation of GPCRs

The 3D structures of GPCRs have been identified using different techniques such as electron paramagnetic resonance spectroscopy (EPR), site directed mutagenesis, Fluorescence spectroscopy, cysteine cross-linking studies, Atomic Force Microscopy (AFM) and X-ray crystallography.

The first structure for GPCRs originated from cryoelectron microscopy of 2 dimensional crystals of bovine rhodopsin. Meanwhile, EPR has provided complementary evidence about photoactivation of rhodopsin including rotation and tilting of TM6 with respect to TM3.

It was clarified through electron paramagnetic resonance spectroscopy that photo activation of rhodopsin includes rotation and tilting of TM6 with respect to TM3.

By using some experimental techniques such as site directed mutagenesis data and cysteine scanning mutagenesis, it was possible to detect conformational changes in GPCRs.(Kobilka, B. K. 2007) As an instance, through site directed mutagenesis studies, it was proposed that the rotamer positions of the three residues including Cys 282, Trp 286 and Phe 290 of β2AR modulate the binding of TM6 around the highly conserved proline kink and lead to movement of cytoplasmic end of TM6 (Ahuja, S. et al. 2009;Deupi, X. et al. 2007) (Fig3).

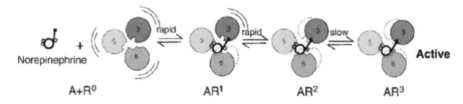

Fig. 3. Movement of TM6 (blue) in β2AR upon binding of an endogenous agonist (Norepinephrine) (Kobilka, B. K. 2007) with permission.

By using site directed mutagenesis data, binding sites of many receptors such as melatonin have also been discovered. (Panetta, R. et al. 2008)

Fluorescence spectroscopy and cross linking studies of some receptors such as muscarine (M3) has also revealed this fact that upon agonist binding, rotation or tilting of cytoplasmic end of transmembrane domains can take place in GPCRs. In cysteine cross-linking studies of M3 muscarinic receptor, it was suggested that the movements of TM5 and TM6 occurs upon agonist binding. (Kobilka, B. K. 2007) Through fluorescence spectroscopy and by tagging fluorophore to the intracellular end of TM6 in β2AR, the receptor revealed a single population of fluoroscence life time. Although, in the presence of antagonists the peak was reinforced, agonists showed an additional peak in the fluorescence life time. (Vaidehi, N. et al. 2011). Documents for the movements of TM6 in rhodopsin have been also provided by chemical reactivity measurements and fluorescence spectroscopy and Zinc cross linking studies of histidines. (Kobilka, B. K. 2007)

The most direct evidence for the structure of a GPCR oligomer comes from AFM (Atomic Force Microscopy). An advantage of AFM is that it provides a 3D profile of the protein. (Simpson, L. M., Taddese, B., Wall, I. D., & Reynolds, C. A. 2010)

Despite the great applicability of the described methods, X-ray crystal structures are normally the starting points in the studies of GPCRs. Since 2000 the high resolution X-ray structures of GPCRs began to emerge among which bovine rhodopsin was the first case study. (Topiol, S. & Sabio, M. 2009)

Crystallography of GPCRs has encountered some limitations due to low level of expression and instability of the proteins outside the membrane. Other problems are also attributed to the folding of the protein and homogeneity during purification. Another limitation in crystallography of the proteins is using detergents for dissociation of the protein from the membrane which might lead to some modifications in 3D structure of the protein. During the procedure for the preparation of crystal structures, a cost effective and straight forward to use system of protein expression is bacteria. The most important problem in this issue is the post translational limitations in the bacteria such as glycosylation which is required for the correct folding of GPCRs. To solve this problem many researchers have focused on using yeasts for the expression of GPCRs. The main problem with using yeasts is the differences observed in yeasts membranes in comparison with those observed in human. The most efficient system for the expression of the GPCRs includes the baculovirus expression system in the insects' cells. This method has been successfully adopted for the receptors including the β-adrenoceptors, the adenosine A2A receptor and the chemokine receptor CXCR4. (Congreve, M., Langmead, C., & Marshall, F. H. 2011)

In crystallography of GPCRs a solution for obtaining the crystal structure in single conformation is addition of the ligand with the prefential affinity to one conformation of the receptor. (Congreve, M. et al. 2011)

Although by X-ray crystallographic data, it is possible to obtain static of the protein complex at near atomistic resolutions biophysical and structural studies are still needed to complete the data of crystallography. (Jaakola, V. P. et al. 2010)

4. Homology modeling studies of GPCRs

Regarding the limitations for using X-ray crystallography in case of GPCRs, molecular modeling techniques such as homology modeling and docking studies are needed to fill the gaps between the primary sequence and secondary structures for drug design studies. There are two major types of drug design including ligand based drug design and structure based drug design. In structure based drug design, the 3D structure of the target molecules is necessary for the study. (Ostopovici-Halip, L., Curpan, R., Mracec, M., & Bologa, C. G. 2011) An important goal of molecular modeling is to provide a microscopic details of the membrane proteins where there is no way to obtain enough information through experimental approaches. (Henin, J., Maigret, B., Tarek, M., Escrieut, C., Fourmy, D., & Chipot, C. 2006)

The crystal structures for seven GPCRs have been represented so far. Most GPCRs are representing the inactive state of the receptor and are therefore suitable for the discovery of antagonists and reverse agonists. (Schneider, M. et al. 2011)

Homology modeling is a knowledge based approach relying upon the crystallographic structure of related receptor and experimental information. This method has limitations when targeting active receptors. (Henin, J. et al. 2006)

Homology modeling made it possible to align the protein of interest with the homologous structure and to subsequently evaluate the model with different scoring functions. Since the most important part of a homology modeling study is the alignment procedure, errors in predicting the structure protein with low homology are normally common. The alignment procedure is usually done by different methods such as cluatalW or T-cofee servers. In order to refine homology based models energy minimization or limited conformational sampling using molecular dynamic simulations are used. (Harterich, S., Koschatzky, S., Einsiedel, J., & Gmeiner, P. 2008)

Due to diversity in loop area of the proteins, a loop refinement is usually necessary in most homology modeling studies. The available loop modeling algorithms are limited to up to 13 residues long. Therefore in loop refinement step considerations must be taken with loops of the long size. (Ostopovici-Halip, L. et al. 2011)

By the emergence of 3D structure for bovine rhodopsin, this receptor was widely used as template structure homology modeling studies of GPCRs. For example, 3D structure of neurotensin, a neuropeptide distributed in the CNS was modeled based on Rhodopsin as template.(Harterich, S. et al. 2008) Two other GPCRs namely, the principal canabinoid receptors CB1 and CB2 are the components of endogenous endocanabinoid systems. The 3D structures of CB1 and CB2 have also been modeled using Rhodopsin as template. (Pei, Y., Mercier, R. W., Anday, J. K., Thakur, G. A., Zvonok, A. M., Hurst, D. et al. 2008)

Another GPCR which has been characterized by homology modeling techniques was melatonin receptor. This receptor is responsible for the effects of melatonin, a compound taking part in resynchronization of biological rhythms such as sleep. In case of MT1 and MT2 receptors, the helices of the receptor were supposed to be superposable with the experimentally known helices of bovine rhodopsin. It was also reported that the identity of MT1 towards rhodopsin is more in respect to MT2 (23% vs 19%).

In other studies the active state of opsin has been used as template to model active structures of β2-adrenergic receptor. Interaction fingerprint studies have been used for dynamic ligand binding study of the interaction of ligands in the active and inactive states. It was concluded that the active structure of opsin is suitable for modeling GPCR agonists. (Schneider, M. et al. 2011)

In spite of the many reports for the usefulness of rhodopsin in homology modeling studies, rhodopsin is merely suitable for antagonist design. The reason can be explained by the fact that rhodopsin is merely crystallized in its inactive state. Therefore, in order to design agonists of GPCRs, it is important to obtain information about the active states of the receptor.(Topiol, S. et al. 2009). There are some other limitations for using rhodopsin as template in homology modeling of GPCRs. One is that it has low homology (less than 25%) with family A GPCRs and no homology with other families of GPCRs such as secretin, adhesion and metabotropic receptors. The other limitation is the very complicated mechanism of activation in rhodopsin in comparison with the other GPCRs. The binding of the ligand to rhodopsin is covalent and signaling is conducted through activation of the ligand by photoisomerism. (Congreve, M. et al. 2011)

Another problem with using rhodopsin is that the binding domains are arranged clockwise in this receptor while sequentially oriented anticlockwise in case of others.(Claude Nofre 2001) Therefore, it is very difficult to obtain a reasonable overview for the activation of the GPCRs from rhodopsin antagonist binding.(Congreve, M. et al. 2011) Based on the findings it was claimed that rhodopsin might not be a suitable template for some GPCRs such as the cholecystokinin CCK1 receptor. (Kobilka, B. K. 2007) A revolution in the field of GPCRs has occurred after publication of the crystal structure of β1 and β2- adrenoceptor.(Congreve, M. et al. 2011) Afterwards, the crystal structures of adenosine A_2A (A_2AR), Chemokine CXCR4, dopamine D3 and histamine H1 in complex with antagonists have been reported which made a reasonable framework for the studies of GPCR functions and drug discovery. (Katritch, V. et al. 2011)

The β2AR is almost a good model for the studies of agonist binding since much information is obtained about the site of interaction between the receptor and catechol amine ligands. (Kobilka, B. K. 2007)

The crystal structures of ß1 and ß2 have been used for homology modeling of 5-HT_2C receptor. They showed similar homology rates of 41% and 62% with the regions of 5-HT_2C. (Renault, N., Gohier, A., Chavatte, P., & Farce, A. 2010)

Another successful example of homology modeling studies using β2-AR as template was in Alpha 2 adrenoreceptor (α2ARs) receptors. These receptors with wide distributions are responsible for many activities such as the control of nervous system and cardiovascular systems. In this study, the resulted models have been minimized using the OPLS2005 force field implemented in schrodinger package. (Ostopovici-Halip, L. et al. 2011)

In many cases of homology modeling, the validity of the structures was verified by ramachandran plot. A common method for docking the homology based models is WHATIF algorithm which generates ramachandran plots to identify outliers in terms of torsion angles and also compares the quality of the model with reliable structures presented in the form of Z-scores. It is also possible to get consensus votes through WHATIF to select between homologous structures. (Abu-Hammad, A., Zalloum, W. A., Zalloum, H., Abu-Sheikha, G., & Taha, M. O. 2009)

Another tool to assess the structural validity of the models is to use hydrophobic moments of the helices. By this method, it is possible to obtain the orientation of hydrophobic moment in transmembrane domains. (Panetta, R. et al. 2008)

5. Simulation studies of GPCRs

In modeling studies for keeping the receptors electrochemically sealed the interaction of the lipids and proteins are needed. (Escriba, P. V., Wedegaertner, P. B., Goni, F. M., & Vogler, O. 2007)

Different molecular dynamic (MD) simulation methods have been used to study GPCRs. While all atom MD simulations in lipid bilayer and water is used to study the dynamics of the membrane proteins, By using targeted MD simulation such as metadynamics, it was possible to study the process of activation in the receptors. In metadynamics a Gaussian term is added to the free energy which disallows the system from returning to previous state. A pitfall of this method is the bias used for forcing the system change its state. The requirement of this method is the primary knowledge needed about the active and inactive states of the receptor. By using this method it is possible to obtain the intermediates in the activation process of the receptor. (Vaidehi, N. et al. 2011)

Since the ligand induced conformational change in GPCRs happen in the range of microseconds, all atom MD simulations are not able to predict large scale simulations such as conformational change in GPCRs. (Vaidehi, N. et al. 2011)

Another simulation method is elastic network model (ENM) in which the protein is represented as a collection of beads connected by springs, where beads refer to protein residues and springs refer to connections. By using this method it is possible to study the micro second simulations. (Vaidehi, N. et al. 2011)

LITicon is a method in which the receptor conformations are permitted to have coarse grain degree of freedom to avoid the built in bias observed in targeted MD simulations. In this method the TMs are considered as rigid bodies connected to each other by flexible loops. The TM helices are rotated in a desired range of rotation angles and the side chain conformations are optimized for each backbone conformation using a rotamer library. Subsequently, the potential energy is minimized using all atom force field function. By this method it is possible to obtain an energy landscape for the GPCRs in the rotational span of the TM helices. After identifying the local minima in the landscape, the global minima state of energy landscape is chosen on the most stable state of the protein. In LITicon, the coarse grain simulation is used to forecast the ensemble of active and inactive states from the inactive crystal structure of the protein. A problem with coarse grain method is some significant barriers which might be missing during the activation pathway.

Monte Carlo (MC) simulations have been also used to calculate the pathway for the activation of some receptors. By MC simulation, it was possible to search the minimum energy from the inactive state towards the ligand stabilized states. An important note to be considered in the computational studies of GPCRs, is the role of water molecules in the activation procedure which has been proposed by many researches to take role in conformational changes of GPCRs. It must be denoted that some water molecules in the crystal structure of the GPCRs might be either absent or not well resolved. It is known that multiscale methods with a combination of coarse grain and fine grain all atom methods are required for understanding the conformational changes of GPCRs. (Vaidehi, N. et al. 2011)

Although some studies have reported the usefulness of MD simulations in studying the changes during dimerization, it was normally difficult to study the GPCRs in dimer form by molecular dynamic simulation methods. (Simpson, L. M. et al. 2010) Recent developments in protein-protein docking made it possible to perform studies on dimer formation.

As an instance, the 5HT4 receptor was subjected to docking approach using GRAMM wherein the interface for dimerization was TM2.4- TM2.4. (Simpson, L. M. et al. 2010)

Docking simulation studies have also been taken to predict the binding mode in GPCRs and estimate the ligand- receptor affinities in case of many receptors such as cholecystokinin (CCK). (Henin, J. et al. 2006)

A further step in modeling based discovery of drugs for GPCRs is to identify potential binding sites in the receptors. The binding site for some GPCRs such as human sweet receptor (HSR) was modeled using ligand based approach. (Claude Nofre 2001)

A successful example for using computational methods was in melanin concentrating hormone (MCH1R). The 3D structure of melanin concentrating hormone which belongs to rhodopsin superfamily was predicted using homology based modeling studies. During the procedure the models were built by the web based model suit SWISS MODEL and scanned for the ligands binding site. The model was then subjected to docking studies of ligands with known activities. The result of the docking step was used for making comparative molecular force field analysis. The combinations of docking/scoring/COMFA were previously reported to be successful in predicting docked conformer/pose closed to that of cocrystalized ligand. In these types of studies, the validity of COMFA models can be verified using ligand based approaches. (Abu-Hammad, A., Zalloum, W. A., Zalloum, H., Abu-Sheikha, G., & Taha, M. O. 2009)

6. Conclusion

Different experimental and computational approaches proposed the role of molecular switches on structural and conformational changes of GPCRs.

By using experimental techniques such as site directed spin labeling, it was observed that in case of receptors such as rhodopsin, a well conserved salt bridge between TM3 and TM6 known as ionic lock is broken during activation. This cleavage leads to flexibility of TM 6 and its movement towards TM3.

Based on molecular modeling studies, it was suggested that in case of some receptors such as MCH, the binding site is a cleft inside the helical domain of the receptor including three hydrophobic regions and a hydrogen bonding polar region.(Abu-Hammad, A. et al. 2009) Other studies revealed that polar interactions of serines with agonists and the movement of TM5 in B_2AR pocket is resulted by shift of TM7 towards TM3 upon agonist binding. An optimal confrontation for this was made based on virtual ligand screening of known ligands. (Katritch, V. et al. 2011)

Biochemical and mutagenesis of B2AR established two major interactions for full agonists in which the amine group forms a salt bridge in Aspll3 while the hydrogen groups of catechol interact with serine in TM5. Analysis of B2AR demonstrated that an inward shift (~2 A) of TM5 is needed for binding of full agonists. The same results have been observed in induced fit docking studies with flexible TM helices. The TM5 shift was caused by conformational freedom in this domain and strong H-binding between catechol OH and Ser 207. The modeling studies based on Adenosine A2A receptor was another example for modeling of agonists. In this case some interactions were revealed to be common for agonists and antagonists such as aromatic ring and amine core contacts. It was seen that adjustment of ligand in an optimal position and engagement of all polar interactions is needed for the shift of the conserved Trp 6.48. (Katritch, V. et al. 2011)

In case of M_3 receptor it was predicted that the binding of Ach to M_3 triggers conformational changes within the TM receptor core. Agonist binding causes the disruption of the existing interhelical interactions and promotes a set of interactions that leads to a new favorable conformational state for the receptor. (Wess, J. et al. 2008)

An important molecular switch is the ionic lock bottom highly conserved D/E motif found in all class A GPCRs. This ionic interaction holds together the cytoplasmic ends of TM3 and TM6 in many amine receptors. Another example for the role of ionic lock is in Angiotensin 1 receptors. The evidence shows that Asn111 interacts with Asn295 in TM7 to stabilize the inactive state of the receptor. (Ahuja, S. et al. 2009;Deupi, X. et al. 2007) In another study it was postulated that reduction of conserved disulfide bridge might be a molecular switch for the activation of the receptor. This study was based on molecular dynamic simulation and virtual screening of dopamine D_2 receptor. It was observed that a predictive model for the catechol binding cavity of D_2 had reduced disulfide bridge. The movement of TM6 towards TM5 was supposed to be the result of cleavage in the conserved disulfide bridge (Fig 4) (Sakhteman, A., Lahtela-Kakkonen, M., & Poso, A. 2011)

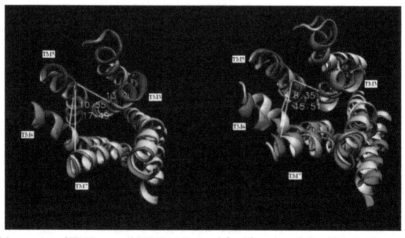

Fig. 4. Movement of TM6 towards TM5 in D_2 model with reduced disulfide bridge.

7. References

Abu-Hammad, A., Zalloum, W. A., Zalloum, H., Abu-Sheikha, G., & Taha, M. O. (2009). Homology modeling of MCH1 receptor and validation by docking/scoring and protein-aligned CoMFA. *Eur.J.Med.Chem.*, Vol.44, No.6, pp. (2583-2596)

Abu-Hammad, A., Zalloum, W. A., Zalloum, H., Abu-Sheikha, G., & Taha, M. O. (2009). Homology modeling of MCH1 receptor and validation by docking/scoring and protein-aligned CoMFA. *Eur.J.Med.Chem.*, Vol.44, No.6, pp. (2583-2596)

Ahuja, S. & Smith, S. O. (2009). Multiple switches in G protein-coupled receptor activation. *Trends Pharmacol.Sci.*, Vol.30, No.9, pp. (494-502)

Claude Nofre. (2001). New hypotheses for the GPCR 3D arrangement based on a molecular model of the human sweet-taste receptor. *Eur.J.Med.Chem.*, Vol.36, pp. (101-108)

Congreve, M., Langmead, C., & Marshall, F. H. (2011). The use of GPCR structures in drug design. *Adv.Pharmacol.*, Vol.62, pp. (1-36)

Deupi, X. & Kobilka, B. (2007). Activation of G protein-coupled receptors. *Adv.Protein Chem.*, Vol.74, pp. (137-166)

Escriba, P. V., Wedegaertner, P. B., Goni, F. M., & Vogler, O. (2007). Lipid-protein interactions in GPCR-associated signaling. *Biochim.Biophys.Acta*, Vol.1768, No.4, pp. (836-852)

Gilbert Vassart & Sabine Costagliola. (2003). The thyrotropin receptor, a GPCR with a built-in inverse agonist. *International Congress Series,* Vol.1249,

Harterich, S., Koschatzky, S., Einsiedel, J., & Gmeiner, P. (2008). Novel insights into GPCR-peptide interactions: mutations in extracellular loop 1, ligand backbone methylations and molecular modeling of neurotensin receptor 1. *Bioorg.Med.Chem.*, Vol.16, No.20, pp. (9359-9368)

Henin, J., Maigret, B., Tarek, M., Escrieut, C., Fourmy, D., & Chipot, C. (2006). Probing a model of a GPCR/ligand complex in an explicit membrane environment: the human cholecystokinin-1 receptor. *Biophys.J.*, Vol.90, No.4, pp. (1232-1240)

Jaakola, V. P. & Ijzerman, A. P. (2010). The crystallographic structure of the human adenosine A2A receptor in a high-affinity antagonist-bound state: implications for GPCR drug screening and design. *Curr.Opin.Struct.Biol.*, Vol.20, No.4, pp. (401-414)

Katritch, V. & Abagyan, R. (2011). GPCR agonist binding revealed by modeling and crystallography. *Trends Pharmacol.Sci.*, Vol.32, No.11, pp. (637-643)

Kobilka, B. K. (2007). G protein coupled receptor structure and activation. *Biochim.Biophys.Acta*, Vol.1768, No.4, pp. (794-807)

Kobilka, B. K. (2007). G protein coupled receptor structure and activation. *Biochim.Biophys.Acta*, Vol.1768, No.4, pp. (794-807)

Marshall, F. H. & Foord, S. M. (2010). Heterodimerization of the GABAB receptor-implications for GPCR signaling and drug discovery. *Adv.Pharmacol.*, Vol.58, pp. (63-91)

Moro, S., Spalluto, G., & Jacobson, K. A. (2005). Techniques: Recent developments in computer-aided engineering of GPCR ligands using the human adenosine A3 receptor as an example. *Trends Pharmacol.Sci.*, Vol.26, No.1, pp. (44-51)

Ostopovici-Halip, L., Curpan, R., Mracec, M., & Bologa, C. G. (2011). Structural determinants of the alpha2 adrenoceptor subtype selectivity. *J.Mol.Graph.Model.*, Vol.29, No.8, pp. (1030-1038)

Panetta, R. & Greenwood, M. T. (2008). Physiological relevance of GPCR oligomerization and its impact on drug discovery. *Drug Discov.Today*, Vol.13, No.23-24, pp. (1059-1066)

Pei, Y., Mercier, R. W., Anday, J. K., Thakur, G. A., Zvonok, A. M., Hurst, D., Reggio, P. H., Janero, D. R., & Makriyannis, A. (2008). Ligand-binding architecture of human CB2 cannabinoid receptor: evidence for receptor subtype-specific binding motif and modeling GPCR activation. *Chem.Biol.*, Vol.15, No.11, pp. (1207-1219)

Renault, N., Gohier, A., Chavatte, P., & Farce, A. (2010). Novel structural insights for drug design of selective 5-HT(2C) inverse agonists from a ligand-biased receptor model. *Eur.J.Med.Chem.*, Vol.45, No.11, pp. (5086-5099)

Sakhteman, A., Lahtela-Kakkonen, M., & Poso, A. (2011). Studying the catechol binding cavity in comparative models of human dopamine D2 receptor. *J Mol Graph.Model.*, Vol.29, No.5, pp. (685-692)

Schneider, M., Wolf, S., Schlitter, J., & Gerwert, K. (2011). The structure of active opsin as a basis for identification of GPCR agonists by dynamic homology modelling and virtual screening assays. *FEBS Lett.*, Vol.585, No.22, pp. (3587-3592)

Simpson, L. M., Taddese, B., Wall, I. D., & Reynolds, C. A. (2010). Bioinformatics and molecular modelling approaches to GPCR oligomerization. *Curr.Opin.Pharmacol.*, Vol.10, No.1, pp. (30-37)

Smith, B., Hill, C., Godfrey, E. L., Rand, D., van den Berg, H., Thornton, S., Hodgkin, M., Davey, J., & Ladds, G. (2009). Dual positive and negative regulation of GPCR signaling by GTP hydrolysis. *Cell Signal.*, Vol.21, No.7, pp. (1151-1160)

Topiol, S. & Sabio, M. (2009). X-ray structure breakthroughs in the GPCR transmembrane region. *Biochem.Pharmacol.*, Vol.78, No.1, pp. (11-20)

Vaidehi, N. & Bhattacharya, S. (2011). Multiscale computational methods for mapping conformational ensembles of G-protein-coupled receptors. *Adv.Protein Chem.Struct.Biol.*, Vol.85, pp. (253-280)

Wess, J., Han, S. J., Kim, S. K., Jacobson, K. A., & Li, J. H. (2008). Conformational changes involved in G-protein-coupled-receptor activation. *Trends Pharmacol.Sci.*, Vol.29, No.12, pp. (616-625)

14

Finding Protein Complexes via Fuzzy Learning Vector Quantization Algorithm

Hamid Ravaee[1,2,*], Ali Masoudi-Nejad[1] and Ali Moeini[2]

[1]Laboratory of Systems Biology and Bioinformatics (LBB), Institute of Biochemistry and Biophysics, University of Tehran, Tehran,
[2]Department of Algorithm and Computations, College of Engineering,
University of Tehran, Tehran,
Iran

1. Introduction

Protein–protein interactions (PPI) make up fundamentals of biological processes inside a cell. PPI has most important roles in cells such as post-translational regulation of protein activity, which is occurred by transient protein-protein interactions and participating in enzymatic complexes ensures substrate channelling which drastically increases fluxes through metabolic pathway (Lin et al., 2006). Metabolic pathways, for instance, consist of several proteins, called enzymes, organize a series of chemical reactions with the intent of altering a variety of chemical substance into the other forms, namely products. Proteins interactions happen in signalling pathways where a set of proteins, by an ordered sequence of reactions, try to convert a type of chemical signal to other form, enabling a cell to obtain environmental information quickly. Proteins interactions can be found in any sort of biological processes within cells. Indeed, existence of these interactions makes a cell function, to grow and more importantly survive (Bader & Hogue, a2003).

The objective in PPI network analysis is the discovering dense highly-connected subgraphs that represent functional modules and protein complexes. For understanding the cell function, it is essential first to find all functional modules in protein interaction networks (Bader & Hogue, b2003). Protein complexes are a group of proteins which have more interactions with each other at the same time and place (Chua et al. , 2008). On the other hand, the functional module consists of proteins that participate in a particular cellular process while interacting with each other at different time and place (Mirny & Spirin, 2003) . In order to simplify the terms, we used protein complex and functional modules as same. Since each protein could be involved in several protein complexes, the partitioning of PPI network to some disjoint groups of subgraphs could not explain the true nature of protein complexes occuring in PPI network. Hence, the finding of vertices group with overlapped boundary can be more useful in analyzing PPI network.

---

* Corresponding Author

In recently years, advances in the high-throughput PPI detection have produced a high volume of PPI datasets freely available to researchers. Therefore many methods and approaches have emerged to analyze experimental PPI data in various organisms. The experimental approaches for discovering protein complexes are more time consuming and expensive. Instead, computational methods which use PPI data are faster and cheaper (Ito et al., 2001).

The most common method of modelling PPI network is using graph theory, which in such a graph G=(V,E) where the nodes correspond to proteins and the edges correspond to interactions. Since the number of proteins and interactions between them in some organism such as yeast or human is remarkably high, the graph modelling PPI is called a complex graph. Partitioning of a complex graph to some disjoint subgraphs is called the graph clustering.

Clustering is the process of grouping data into sets (clusters) which shows more similarity between the objects in the same clusters than they are in different clusters (Schaeffer, 2007). Clustering analysis seeks a set of clusters based on similarity between pairs of elements. Graph clustering is the practice of distribution the vertices of the graph into the clusters taking into consideration the edge connectivity in the graph in such a way that many edges exist within each cluster and relatively few between the clusters. The result of this clustering can define the PPI network's structure and imply functions of proteins in the cluster which were previously uncharacterized (Lin et al., 2006).

Each complex graph modelling a system such as biological systems or social networks has specific properties and characteristics. The properties of graph could be fall into broad categories as the local properties and global properties (Przulj, 2005). The scale-free for distribution of degree and small world properties could be more affective on the result of graph clustering. A scale-free network has a vertex connectivity distribution that follows a power law, with relatively few highly connected vertices and many vertices having a low degree. Most biological networks such as PPI networks have the scale-free property (Pizzuti & Rombo, 2007). In this paper, we convert the normal scale-free PPI network to a non-scale free network by using line graph transformation. In the graph theory, line graph is produced by substituting edges and nodes in the graph. Each interaction is condensed into a node that includes the two interacting proteins. These nodes are then linked by shared protein content.

Important of results of the clustering in PPI network is illustration of structure of the PPI network which can be used to predict the functionality of uncharacterized protein based on other known proteins functions in the same cluster's elements. These clusters correspond to meaningful biological units such as protein complexes and functional modules.

Many clustering approaches (Gao, 2009; Bader & Hogue, b2003; Adamcsek, 2006; Wu et al., 2008 ;Vlasblom, 2009) could not place elements in multiple clusters, which can be unrealistic for biological systems, where proteins may participate in multiple cellular processes and pathways. Since each protein could participate in more than one protein complexes, in the clustering PPI graph, each protein probably have membership to more than one cluster. So in this paper, we present a clustering method that allows to having overlapping founded clusters. Disjoint clusters and overlapping clusters are illustrated in figure 1.

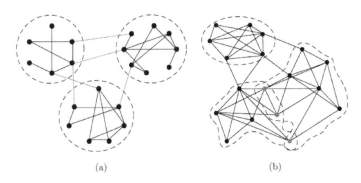

(a)   (b)

Fig. 1. Illustration of the concept of modules. (a) Disjoint modules; (b) Overlapping modules.

K-means (c-means) clustering (Hartigan, 1975) is applied on unlabeled data by partitioning them on predefined number of groups (k) based on the specifying the centers of groups. After each iteration in the k-means algorithm, the distances between each center of group and other data points are calculated and the center points are updated. Learning Vector Quantization uses k-means idea by defining some codebook vectors each of which represents a cluster for n-dimensional input data. The fuzzy clustering based on fuzzy set theory (Zadeh, 1965) is used to deal with indistinct boundaries between clusters. The most widely used fuzzy clustering method is the fuzzy c-means (FCM) algorithm (Bezdek, 1973) which is generalized from hard c-means algorithm. In this paper, extended FLVQ (Bezdek, 1995) as an intelligent computational method has been used for clustering PPIs. The results of this algorithm can be verified by biological and non-biological criteria and we showed that FLVQ technique is more effective and accurate for finding protein complexes in PPI network.

2. Primary definitions

The problem of clustering of PPIs starts with a mathematical representation of PPI networks. A conventional way for representing PPI network is using graph theory concepts. PPI network could be illustrated by a graph G=(V,E) with a set of vertices V and a set of edges E in which each vertex is corresponded by a protein in PPI network and each edge connects to two vertices whose corresponding proteins have physical interaction with each other.

Clusters in the graph could be interpreted as dense subgraphs the number of edges within each subgraph is the maximum number and the number of edges between clusters is the minimum one. Therefore, the PPI clustering is an optimization problem and like other optimization problems, there is a need to an objective function to get optimum point.

PPI networks have scale-free property and finding the dense subgraphs is most difficult task in these networks. So using line graph we eliminate the scale-free property. In each node in the line graph is an edge in original network and every two nodes with common proteins are connected to each other. Figure 2 shows a scale free network and the generated line graph based on original graph.

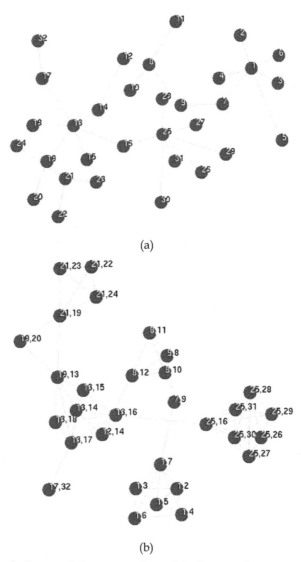

Fig. 2. **a.** Orginal scale-free graph **b.** converted graph by line graph.

2.1 Learning Vector Quantization

Learning Vector Quantization (LVQ) is placed in the competitive learning category and it is closely related to Self-Organizing Map (SOM) (Kohonen, 1990). SOM is a well-developed neural network technique for data clustering and visualization. It can be used for projecting a large data set of a high dimension into a low dimension (usually one or two dimensions) while retaining the initial pattern of data samples. Indeed, SOM has two main principles:

vector quantization and vector projection. Vector quantization makes up a delegate set of vectors called output vectors (codebook vectors) from the input vectors. Let's denote the set of output vectors (codebook vectors) as $Y=\{y_1,y_2,..,y_c\}$ with the same dimension as input vectors. In general, vector quantization reduces the number of vectors, and this can be considered as a clustering process. The maximum number of clusters in a network is defined by user specified value, c. After learning process, it may be possible for some codebook vectors to correspond to empty clusters.

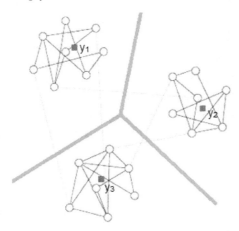

Fig. 3. The red points (y_1,y_2,y_3) corresponded to output vectors indicating a dense subgraph in the sample network.

The LVQ algorithm represents a set of input vectors $x_i \in X \subset \Re^n$ by a set of c prototypes $Y = \{y_1,y_2,..,y_c\} \subset \Re^n$ .. The LVQ is associated with a competitive network which consists of an input layer and an output layer. Each node in the input layer is connected directly to the cells, or units, in the output layer. A weight vector, also referred to as prototype, is assigned to each cell in the output layer (Ravuri & Karayiannis, 1995). The codebook vector having minimum distance with input vector x_i is called winner vector, k, and is defined as:

$$k = \arg \min_l \|y_l - x_i\| \tag{1}$$

Update equation of LVQ algorithm is:

$$y_j(t+1) = y_j(t) + \alpha_t h_{ij,k} \|x_i - y_j(t)\| \tag{2}$$

Here a_t is the scalar-valued learning rate, $0<a_t<1$, and decreases monotonically with time t. The neighborhood function $h_{ij,k}$ denotes the interaction between codebook vector i and j and winner vector k. The simple definition of $h_{ij,k}$ is:

$$h_{ij,k} = \begin{cases} 1 \ if \ j = k \\ 0 \ if \ j \neq k \end{cases} \tag{3}$$

In the LVQ algorithm, neighborhood radius is one and only the winner vector could be updated.

2.2 Fuzzy Learning Vector Quantization

While most typical clustering algorithms assigns each data point to exactly one cluster, fuzzy clustering allows for the extent of membership, to which a data point belongs to different clusters. The FLVQ may be seen as a learning fuzzy c-means using a fuzzification index m. Karayiannis et al (Ravuri & Karayiannis, 1995) presented a broad family of FLVQ algorithms, which were initially introduced on the basis of perceptive arguments. This derivation was based on the minimization of the average generalized distance between the input vectors and the prototype vectors. The fuzzy partitioning algorithm, FCM is run into by minimization problem that is solved by reformation of FCM algorithm to FLVQ algorithm (Bezdek, 1995).

The updated equation for the FLVQ involves the membership functions which are used to determine the strength adjacency between each prototype and input vectors.

$$\alpha_{ij,t} = (u_{ij,t})^{m_t} \tag{4}$$

$$u_{ij} = \left[\sum_{l=1}^{c} (\frac{D_{ij}}{D_{lj}})^{\frac{2}{m-1}} \right]^{-1} \tag{5}$$

$$D_{ij} = \|x_i - y_j\| \tag{6}$$

Where $m = m_t = m_0 - \Delta mt$ and $\Delta m = (m_0 - m_f) / MaxItr$ and D_{ij} is the distance and m_0 is some constant value greater than the final value (m_f) of the fuzzification parameter m. $MaxItr$ is the constant parameter for limitation of iterations.

3. The FLVQ algorithm

The calculation of distances between network vertices and prototype vectors in the FLVQ is critically challenging. In the following algorithm, we used a new definition of vertices based on n-dimensional vectors and; we representing new scalar distance between input vectors and codebooks (output) vectors. Each vertex in PPI graph is modeled by a vector called input vector. Given $G=(V,E)$ represents a PPI network including $|V|$ vertices and $|E|$ edges. An input vector is defined as :

$$x_m = \{x_{m1}, x_{m2}, ..., x_{mn}\} \tag{7}$$

$$x_{ij} = \begin{cases} 0 + \varepsilon & if \quad i \neq j \quad and \quad e_{ij} = 0 \\ 1 - \varepsilon & if \quad e_{ij} = 1 \\ 1 - \varepsilon & if \quad i = j \end{cases}$$

Where $n=|V|$, e_{ij} is element (i,j) in adjacency matrix corresponding the graph G and ε is a real small value between $(0,1)$.

This definition makes possible to use scalar distance measure such as the dot product is possible. There are some distance criteria in vector space to measure similarity (distance) between two vectors. Correlation is a simple way for measuring distance between two vectors in the same dimension. If x_i and x_j are two vectors with the dimension of n, the equation (8) is the inner product of two vectors:

$$S_{ij} = X_i . X_j = \sum_{k=1}^{n} x_{ik} x_{jk} \qquad (8)$$

$$D_{ij} = S_{ij}^{-1} \qquad (9)$$

Where D_{ij} is the distance and S_{ij} is the inner product between X_i and X_j.

The FLVQ algorithm performs clustering of the input graph by training process. Training process consists of some iterations. The number of iteration depends on convergence criteria and can be limited by a user specified constant. Each iteration consists some epochs. The number of epochs is equal by c (number of prototype vectors and the maximum number of clusters). In each epoch, an input vector x_i is selected randomly. A selected input vector is not being selected in a same epoch again. The selected input vector x_i is compared with all the prototype vectors with a similarity measure (ex. dot product) and the prototype vector y_j with most similarity with x_i known as winner vector.

The implementation of the FLVQ algorithm is described as follows:

- **Step 1. Initialization**
  Initialize the c codebook's vectors y={y₁,y₂,..,yₑ} by randomly assigning each element of codebook vectors by a real number between (ε,1-ε). Set iteration counter t=1. Give 0≤ε<1. tₘₐₓ is the iteration limit.
- **Step 2. Learning**
  Repeat until stopping criterion is satisfied:
- **Step 2.1** While there is a unselected input vector
  - *Randomly pick an input x_i*
  - *Compute winner vector based on distance measure of xi and every codebook vectors y_j : j=1..k*
  - *update winner vector y_j based on input vector x_i and learning ratio a*
- **Step 2.2** update learning ratio α

4. Data set

The PPI network is derived from the yeast subset in the Database of Interacting Proteins (DIP) (Xenarios et al., 2002). The dataset of yeast is composed of 4963 proteins and 17570 interactions. Most of these interactions have been derived by yeast two-hybrid screen. For evaluation of finding clusters, we use protein complex data from the MIPS database (Mewes et al., 2004). In the currated complex dataset, there are 404 protein complexes. The protein complex having most proteins is "cytoplasmic ribosomal large subunit" with 88 proteins and there are 169 protein complexes with just two proteins.

5. Experimental result

The FLVQ algorithm is applied on the PPI network of the *Saccharomyces Cerevisiae* (yeast) dataset downloaded from the DIP (Guldener, 2005). After using FLVQ on DIP protein-protein interaction, over than 300 clusters obtained frequency of each based on the number of vertices in is shown in figure (4). As the figure (4) shows most obtained clusters approximately include 9 and 12 vertices. In addition, the number of clusters with size of over 20 are also considerable. This means that the FLVQ algorithm could find larger dense subgraphs in the PPI network. When the cluster size became larger, few graph clustering methods could find these clusters with proper efficiency.

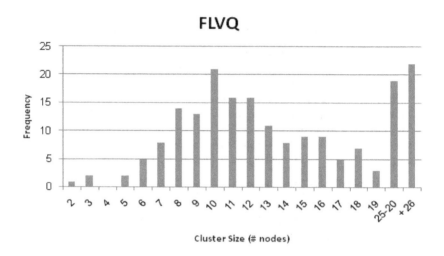

Fig. 4. Number of obtained clusters by FLVQ algorithm based on the cluster size.

The results of the FLVQ algorithm are evaluated by the clustering score used by (Bader & Hogue, a2003; Newman, M. & Girvan. M., 2004). The clustering score for each cluster is defined by the product of size and density of the cluster. The density of cluster is the ratio between number of edges in cluster $|E|$ and maximum number of possible edges in it $|E_{max}|$. The following equation (10) shows clustering score definition.

$$\sigma(\Gamma)=\delta(\Gamma).|V| \qquad (10)$$

Where Γ is a cluster in the clustering result and $\delta(\Gamma)$ is the density of given subgraph Γ and is declared by equation (11) and $|V|$ shows the number of vertices in Γ subgraph.

$$\delta(\Gamma)=2|E|/(|V|(|V|-1)) \qquad (11)$$

Where E is the set of edges that connects the existing vertices in V in given subgraph of Γ. The clustering score for each clusters is shown in figure (5). The cluster score for bigger clusters is more elevated than smaller clusters proving that FLVQ is rather successful to find subgraphs with more higher number of vertices and with most density. Highest clustering score shows that the obtained clusters are more compact and larger.

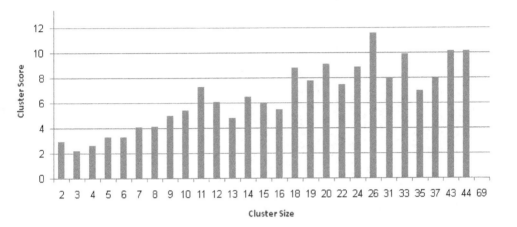

Fig. 5. Amount of clustering score for each obtained cluster in FLVQ algorithm.

The clustering results can be validated by ground truth with Precision and Recall. Assume a module (cluster) X is mapped to a functional module F_i. Recall, also termed the true positive rate or sensitivity, is the proportion of proteins common in both X and F_i to the size of F_i. Precision, which is also termed the positive predictive value, is the proportion of proteins common in both X and F_i to the size of X.

$$precision = \frac{|X \cap F_i|}{|X|} \tag{12}$$

$$recall = \frac{|X \cap F_i|}{|F_i|} \tag{13}$$

The accuracy of clusters is assessed by f-measure. The f-measure is defined as the harmonic mean of recall and precision:

$$f - measure = \frac{2(precision.recall)}{precision + recall} \tag{14}$$

Figure (6) shows the average of f-measure based of protein complex size for the FLVQ algorithm. In figure (6), the f-measure of each obtained cluster is measured based on experimental protein complexes MIPS. The value of f-measure could be between 0 and 1.

The highest f-measure value indicates the most conformity between experimental protein complex and obtained complex by the algorithm.

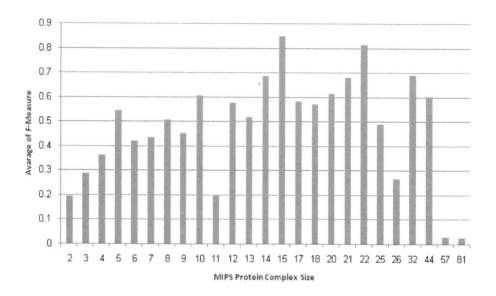

Fig. 6. f-measure between finding subgraphs and experimental protein complexes based on its size.

6. Conclusion

In this paper, we presented a FLVQ algorithm as a robust tolerable method to find dense subgraphs in PPI networks as protein complexes. The algorithm identifies more than 200 dense subgraphs having more overlap among experimentally known protein complexes. By clarifying the structure of protein interactions network, uncharacterized proteins could be predicted by the functions of other known proteins which belong to same clusters. By using line graph transformation, we eliminated the scale-free degree distribution in PPI network which caused larger number of dense highly connected subgraph revealed. There is overlapping between found subgraphs that express the results are more conforming with the reality nature of protein complexes.

7. References

Adamcsek, B. (2006). Cfinder: Locating Cliques And Overlapping Modules In Biological Networks, *Bioinformatics*, Vol. 22, pp. 1021-1023.

Bader, G. & Hogue, C. (2003). An Automated Method For Finding Molecular Complexes In Large Protein Interaction Networks, *BMC Bioinformatics*, Vol. 4

Bader, G. & Hogue, C. (2003). Analyzing Yeast Protein-Protein Interaction Data Obtained From Different Sources, *Nat. Biotechnol*, Vol. 20, pp. 991-997

Bezdek, C. &Hathaway,J. (1995). Optimization Of Clustering Criteria By Reformulation, *IEEE Transactions on Fuzzy Systems*, Vol. 2, pp. 241-246.

Bezdek, C.; (1973). Fuzzy Mathematics In Pattern Classification, *Ph.D. dissertation, Dept. Appl. Math.,Cornell Univ., Ithaca, NY*

Chua, H.; Ning, K.; Sung, W.; Leong. H., (2008). Using Indirect Protein–Protein Interactions For Protein Complex Prediction, *Journal of Bioinformatics and Computational Biology*, Vol. 6., pp. 435-466.

Gao, L.; Sun, p.; Song, j. (2009). Clustering Algorithms For Detecting Functional Modules In Protein Interaction Networks, *Journal of Bioinformatics and Computational Biology*

Guldener, U. (2005). CYGD: The Comprehensive Yeast Genome Database, *Nucleic Acids Res*, Vol. 33, pp. 364–368.

Hartigan, J.A. (1975). Clustering Algorithms. *New York : Wiley*

Ito, T.; Chiba, T.; Ozawa, R.; Yoshida, M. ; Hattori, M.; Sakaki, Y. (2001). A Comprehensive Two-Hybrid Analysis To Explore The Yeast Protein Interactome, *PNAS*, Vol. 98, pp. 4277-4278.

Kohonen, T. (1990). The Self Organizing Map, *IEEE Proc*, Vol. 78

Lin, C.; Cho, Y.; Hwang, W.; Pei, P.; Zhang, A. (2006). Clustering Methods In Protein-protein Interaction Network, In: *Knowledge Discovery in Bioinformatics: Techniques, Methods and Application*. John Wiley & Sons

Mewes, H.W. et al., (2004). MIPS: Analysis And Annotation Of Proteins From Whole Genomes, *Nucleic Acids Res*, Vol. 32, pp. D41-D44

Mirny, V. & Spirin, L. (2003). Protein Complexes And Functional Modules In Molecular Networks, *Proc. Natl Acad. Sci*, Vol. 100(21), pp. 12123–12126

Newman, M. & Girvan. M., (2004). Finding And Evaluating Community Structure In Networks, *Physical Review*, 2004.

Palla, G.; Der´enyi, I.; Farkas, I.; Vicsek, T., (2005). Uncovering The Overlapping Community Structure Of Complex Networks In Nature And Society. , *Nature*, Vol. 435, pp. 814–818

Pizzuti, C.; Rombo, S. (2007). Multi-functional Protein Clustering in PPI Networks, *International Conference on Intelligent Data Engineering and Automated Learning*

Przulj, N. (2005). *Knowledge Discovery in Proteomics Graph Theory Analysis of Protein-protein Interactions*, Department of Computer Science, University of Toronto

Ravuri, N. & Karayiannis, M. (1995). An Integrated Approach To Fuzzy Learning Vector Quantization And Fuzzy C-Means Clustering, *New York : Intelligent Engineering Syslems Through Artificial Neural Networks*, Vol. 4, pp. 247-252.

Schaeffer, S. (2007). Survey Graph Clustering, *Elsevier*

Vlasblom, J. & Wodak., S. (2009). Markov Clustering Versus Affinity Propagation For The Partitioning Of Protein Interaction Graphs, *BMC BIOINFORMATICS*, Vol. 10.

Wu, M.; Li, X.; Kwoh. C. (2008). Algorithms For Detecting Protein Complexes In PPI Networks: An Evaluation Study, *PRIB08*, pp. 135-146.

Xenarios, I.; Salwinski, L.; Duan, X.; Higney, P.; Kim, S.; Eisenberg, D. (2002) DIP, The Database Of Interacting Proteins: A Research Tool For Studying Cellular Networks Of Protein Interaction, *Nucleic Acids Res*, Vol. 30, pp. 303-305.

Zadeh L., A. (1965) Fuzzy Sets, *Inf. Control*, Vol. 8, pp. 338-353.

Zhang, A., (2009). Modularity Analysis of Protein Interaction Networks, *Protein Interaction Networks Computational Analysis* , pp. 66-77

Permissions

The contributors of this book come from diverse backgrounds, making this book a truly international effort. This book will bring forth new frontiers with its revolutionizing research information and detailed analysis of the nascent developments around the world.

We would like to thank Weibo Cai, PhD andHao Hong, PhD, for lending their expertise to make the book truly unique. They have played a crucial role in the development of this book. Without their invaluable contribution this book wouldn't have been possible. They have made vital efforts to compile up to date information on the varied aspects of this subject to make this book a valuable addition to the collection of many professionals and students.

This book was conceptualized with the vision of imparting up-to-date information and advanced data in this field. To ensure the same, a matchless editorial board was set up. Every individual on the board went through rigorous rounds of assessment to prove their worth. After which they invested a large part of their time researching and compiling the most relevant data for our readers. Conferences and sessions were held from time to time between the editorial board and the contributing authors to present the data in the most comprehensible form. The editorial team has worked tirelessly to provide valuable and valid information to help people across the globe.

Every chapter published in this book has been scrutinized by our experts. Their significance has been extensively debated. The topics covered herein carry significant findings which will fuel the growth of the discipline. They may even be implemented as practical applications or may be referred to as a beginning point for another development. Chapters in this book were first published by InTech; hereby published with permission under the Creative Commons Attribution License or equivalent.

The editorial board has been involved in producing this book since its inception. They have spent rigorous hours researching and exploring the diverse topics which have resulted in the successful publishing of this book. They have passed on their knowledge of decades through this book. To expedite this challenging task, the publisher supported the team at every step. A small team of assistant editors was also appointed to further simplify the editing procedure and attain best results for the readers.

Our editorial team has been hand-picked from every corner of the world. Their multi-ethnicity adds dynamic inputs to the discussions which result in innovative outcomes. These outcomes are then further discussed with the researchers and contributors who give their valuable feedback and opinion regarding the same. The feedback is then collaborated with the researches and they are edited in a comprehensive manner to aid the understanding of the subject.

Apart from the editorial board, the designing team has also invested a significant amount of their time in understanding the subject and creating the most relevant covers. They scrutinized every image to scout for the most suitable representation of the subject and create an appropriate cover for the book.

The publishing team has been involved in this book since its early stages. They were actively engaged in every process, be it collecting the data, connecting with the contributors or procuring relevant information. The team has been an ardent support to the editorial, designing and production team. Their endless efforts to recruit the best for this project, has resulted in the accomplishment of this book. They are a veteran in the field of academics and their pool of knowledge is as vast as their experience in printing. Their expertise and guidance has proved useful at every step. Their uncompromising quality standards have made this book an exceptional effort. Their encouragement from time to time has been an inspiration for everyone.

The publisher and the editorial board hope that this book will prove to be a valuable piece of knowledge for researchers, students, practitioners and scholars across the globe.

List of Contributors

Nicolas Férey
CNRS - Laboratoire d'Informatique pour la Mécanique et les Sciences de l'Ingénieur - Université Paris XI Bâtiment 508, 512 et 502 bis, 91403 Orsay Cedex, France

Aleksey Porollo and Jaroslaw Meller
University of Cincinnati, USA

Woojin Jung and KiYoung Lee
Department of Biomedical Informatics, Ajou University School of Medicine, Republic of Korea

Hyun-Hwan Jeong
Department of Biomedical Informatics, Ajou University School of Medicine, Republic of Korea
Department of Computer Engineering, Ajou University, Republic of Korea

Zelmina Lubovac-Pilav
University of Skövde, Systems Biology Research Centre, Sweden

Takatoshi Fujiki, Etsuko Inoue, Takuya Yoshihiro and Masaru Nakagawa
Wakayama University, Japan

Lusheng Wang
Department of Computer Science, City University of Hong Kong, Hong Kong

Shuichi Hirose
Nagase & Co. Ltd. Research & Development Center, 2-2-3 Murotani, Nishi-ku, Kobe, Hyogo, Japan
Computational Biology Research Center, Advanced Industrial Science and Technology, 2-4-7 Aomi, Ko-toku, Tokyo, Japan

Jian Huang, Beibei Ru and Ping Dai
School of Life Science and Technology, University of Electronic Science and Technology of China, China

Fan Bai and Jianshi Jin
Biodynamic Optical Imaging Centre, Peking University, People's Republic of China, Beijing, China

Zhanghan Wu and Jianhua Xing
Department of Biological Sciences, Virginia Tech, USA

Phillip Hochendoner
Department of Physics, Virginia Tech, USA

J. Planas-Iglesias, J. Bonet, M.A. Marín-López, E. Feliu, A. Gursoy and B. Oliva
Structural Bioinformatics Lab. Universitat Pompeu Fabra, Catalunya, Spain

Darby Tien-Hao Chang
National Cheng Kung University, Taiwan

Ernesto Iacucci and Yves Moreau
K.U.Leuven, Belgium

Samuel Xavier de Souza
Universidade Federal do Rio Grande do Norte, Brazil

Amirhossein Sakhteman, Hamid Nadri and Alireza Moradi
Faculty of Pharmacy, Shahid Sadoughi University of Medical Sciences, Yazd, Iran

Hamid Ravaee
Laboratory of Systems Biology and Bioinformatics (LBB), Institute of Biochemistry
and Biophysics, University of Tehran, Tehran,
Department of Algorithm and Computations, College of Engineering,
University of Tehran, Tehran,
Iran

Ali Masoudi-Nejad
Laboratory of Systems Biology and Bioinformatics (LBB), Institute of Biochemistry
and Biophysics, University of Tehran, Tehran,

Ali Moeini
Department of Algorithm and Computations, College of Engineering,
University of Tehran, Tehran,
Iran

Printed in the USA
CPSIA information can be obtained
at www.ICGtesting.com
JSHW011459221024
72173JS00005B/1140

9 781632 395252